DPH PHYSICS SERIES

# TEXT BOOK OF SIMPLE HARMONIC MOTION AND WAVE THEORY

*By*
D.K. Jha

DISCOVERY PUBLISHING HOUSE
NEW DELHI-110002

Reprinted – 2025

ISBN: 978-81-8356-003-0



**Text Book of Simple Harmonic Motion and Wave Theory**

*Published by:*

DISCOVERY PUBLISHING HOUSE
4383/4B, Ansari Road, Darya Ganj
New Delhi-110 002 (India)
*Phone*: +91-11-23279245; 23253475; 43596065
*Mobile*: +91 9811179893 / +91 9871656464
*E-mail*: discoverybooksindia@gmail.com
orderdphbooks@gmail.com
namitwasan9@gmail.com
*web*: www.discoverypublishinggroup.com

*Printed at:*
Infinity Imaging Systems
Delhi (INDIA)

# Preface

There are number of books on Simple Harmonic and Wave Motion in the market for the use of degree students in various universities in India. It is the experience of author that the average students need the treatment of theory in a way that should be easily comprehensible to him. Therefore an effort has been made in this book to put the matter in a very lucid and simple way to that even a beginner has no difficulty is grasping the subject. Each chapter for this book contains complete theory and fairly large number of solved examples sufficient problems have also been selected from various university examination paper. At the end of each chapter an exercise containing objective questions only has been given.

The answer to almost all solved problems have been checked and every care has been taken to avoid printing and other mistakes. It is sincerely hoped that this book will satisfy the needs of the student and if it gives them even part of pleasure that the author had in its preparation he will consider his labour amply rewarded.

The author will feel amply rewarded if the book serve the purpose for which it is means suggestion for the improvement of this book are always welcome.

I am very thankful to Mr. Tilak Wasan (Managing Director), Discovery Publishing House, for their valuable efforts to complete this book.

**D.K. Jha**

# Contents

Superposition of Waves, Interference—Beats, Vibrations of Strings, Propagation of Transverse Vibration (or a Transverse Wave) Along a String, Velocity of a Transverse Wave Along a String, Longitudinal Waves (or Sound Waves) in a Gaseous (or a Fluid) Medium, La' Place's Correction of Newton's Formula—Effect of Pressure and Temperature on the Velocity of Sound in a Gas (or Air), Longitudinal Waves in Rods, Stationary or Standing Waves inn a Linear Bounded Medium, Different Modes of Vibration of Strings, Rods and Air Columns, Fourier's Theorem, Some Illustrative Examples of the Application of Fourier's Theorem, Group Velocity-Its Relationship with Wave (or Phase) Velocity.

# HARMONIC OSCILLATOR

## PERIODIC MOTION

We as human being are caught up in a world full of motion. A part from linear or translatory motion, the other commonly occurring type of motion is vibratory or oscillatory motion. A motion that repeats itself over and over again is equal internal of time is called a *periodic motion.* Some common examples of periodic motion are :

(a) A swinging pendulum.

(b) The motion of the hands of the clock.

(c) Motion of a particle along a horizontal circle with uniform speed.

(d) Motion of planets around the sun.

(e) The piston of an automobile engine.

(f) The balance wheel of a watch.

(g) The motion of a vertical spring with a mass hanging.

(h) Oscillations of electric charge in a L.C.R. circuit.

(i) Vibrations of atoms in a solid about the respective mean positions.

A particular case of periodic motion is that the body moves back and fourth about a mean or equilibrium position. This type of motion is referenced to as an oscillatory or vibratory motion. It must be pointed out here that all periodic motion are not necessarily oscillatory. In the examples of periodic motion listed above, the examples (c) and (d) are not oscillatory motions.

The simplest type of oscillatory motion is referred to as simple harmonic motion usually designated as S.H.M. When a particle moves so that its acceleration is always directed towards a fixed point in its path and varies as its displacement from that fixed point, the particle is said to be executing the simple harmonic motion is of fundamental importance, in the study of Physics for the following reasons :

(1) All physical systems, for small displacements execute motion, which can be approximated to the first degree or approximation very close to simple harmonic motion.

(2) Any complex periodic motion can be shown to be made up of a sum a large number of simple harmonic motions.

(3) The study of simple harmonic motion is of basic importance in understanding the process of wave propagation.

## HARMONIC OSCILLATOR

We have seen already, under how the position of minimum potential energy of a particle is its position of stable equilibrium and how its displacement remains confined to a little distance on either side of this equilibrium or mean position, within what is called a *potential well* and also, how a restorting force $F = -dV/dx$ acts upon it at every point of its displacement except at the mean position (where it is *zero*), tending to bring it back to its mean or equilibrium position. It has also been shown there how the energy function $U = f(x)$ may, in general, be expanded about the position $x_0$ of stable equilibrium by means of *Taylor's theorem*, which then gives

$$U = U_0 + \frac{Cx^2}{2!} + \frac{C_1 x^3}{3!} + ....$$

and, therefore, force acting on the particle is given by

$$F = -\frac{dU}{dx} = -Cx - \frac{C_1}{2}x^2 - \frac{C_2}{6}x^3 + ...$$

So that, if the displacement be small and $C_1$, $C_2$ etc., are negligible, we have $U = U_0 + \frac{1}{2}Cx^2$ and, therefore, force acting on the particle, $F = -Cx$, where C is a *positive constant*, called the *force constant*.

This relation indicates that a restoring force acts on the particle, tending to bring it back to its mean or equilibrium position and that the potential energy curve is parabolic in form.

Such an oscillating particle is called a harmonic oscillator and in case the limits ($x_1$ and $x_2$) are equally spaced about the equilibrium position, it is called a *simple harmonic oscillator* and its motion, a *simple harmonic motion*, (S.H.M., for short).

Thus, in the case of a simple harmonic motion, the P.E. *curve* varies as the square of the displacement and *the force acting on the particle, and hence its acceleration, is proportional to displacement but is directed oppositely to it (towards the mean or equilibrium position), the maximum displacement (or amplitude) of the particle being the same on either side of the mean or equilibrium position.*

In case the displacement is not the same on either side of the equilibrium position, the motion is harmonic all right but not simple harmonic. And, if the values of $C_1$, $C_2$ etc. in the relation for above, cannot be neglected, the P.E. curve would no longer be parabolic but will assume the form shown dotted. Even so, it will be parabolic in form at the bottom of the curve, implying that whatever be the nature of the potential function $U = U_{(x)}$, *small oscillations may always be regarded as simple harmonic.*

Common in nature, the S.H.M. is, in fact, the most fundamental type of periodic motion and all other periodic motions (*harmonic as well as non-harmonic*) can be obtained by a suitable combination of two or more simple harmonic motions. Let us, therefore, study this motion in some detail.

## SIMPLE HARMONIC MOTION

From what we have just seen in above, we may give a clear-cut definition to S.H.M. as follows:

*A particle may be said to execute a simple harmonic motion if its acceleration is proportional to its displacement from its equilibrium position, or any other fixed point in its path, and is always directed towards it.*

Thus, if F be the force acting on the particle and x, its displacement from its mean or equilibrium position, we have $F = -Cx$.

Now, in accordance with Newton's second law of motion, $F = ma$. So that, substituting $-Cx$ for F and $d^2x/dt^2$ for acceleration a, we have

$$-Dx = m\frac{d^2x}{dt^2}.$$

Or,
$$\frac{d^2x}{dt^2} + \frac{C}{m}x = 0.$$

This equation is called the *differential equation of motion* of a simple harmonic oscillator or a simple harmonic motion, because by solving it we can find out how the displacement of the particle depends upon time and thus know the correct nature of the motion of the particle.

To solve the equation, we may put it in the form $d^2x/dt^2 = (C/m)x$, the negative sign, as we know, indicating that the acceleration is directed oppositely to displacement x.

Or, putting $(C/m) = \omega^2$, where $\omega$ is the angular velocity of the particle, the equation takes the form

$$\frac{d^2x}{dt^2} = -\omega^2 x = -\mu x, \qquad ...(i)$$

where $\mu$ is a constant, equal to $\omega^2$. Or, since $d^2x/dt^2 = -\mu$ if $x = 1$, we may define $\mu$ as the *acceleration per unit displacement of the particle.* Multiplying both sides of the equation by 2 dx/dt, we have

$$2\frac{dx}{dt}\frac{d^2x}{dt^2} = -\omega^2.2x\frac{dx}{dt},$$

integrating which with respect to t, we have

$$\left(\frac{dx}{dt}\right)^2 = -\omega^2 x^2 + A, \qquad ...(ii)$$

where A is a constant of integration.

Since at the *maximum displacement* (or *amplitude*) a of the oscillator (or the oscillation), the velocity dx/dt = 0, we have

$$0 = -\omega^2a^2 + A, \text{ whence, } A = \omega^2a^2.$$

Substituting this value of A in relation (ii), therefore, we have

$$\left(\frac{dx}{dt}\right)^2 = -\omega^2x^2 + \omega^2a^2 = \omega^2\left(a^2 - x^2\right),$$

whence, the *velocity of the particle at an instant t*, is given by

$$\frac{dx}{dt} = \omega\sqrt{a^2 - x^2}. \qquad ...(iii)$$

Putting equation (iii) as $dx/\sqrt{a^2 - x^2} = \omega dt$ and integrating again with respect to t, we have

$$\sin^{-1}\frac{x}{a} = \omega t + \phi.$$

or, $$x = a \sin(\omega t + \phi) \qquad ...(iv)$$

This gives the displacement of the particle at an instant (in terms of its amplitude (a) and its total phase ($\omega t + \phi$), made up of the phase angle $\omega t$ and what is called the *initial phase, phase constant* or the *epoch* $\phi$ of the particle, usually denoted by the letter e. This *initial phase* or *epoch* arises because of our starting to count time, not from the instant that the particle is in some standard position, like its mean position or one of its extreme positions, but from the instant when it is anywhere else in between.

Thus, if we start counting time *when the particle is in its mean position, i.e.*, when $x = 0$ at $t = 0$, we have $\phi = 0$ and, therefore,

$$x = a \sin \omega t.$$

And, if we start counting time *when the particle is in one of its extreme positions, i.e.*, when $x = a$ at $t = 0$, we have $a = a \sin (0 + \phi) = a \sin \phi$, *i.e.*, $\sin \phi = a/a = 1$ or $\phi = \pi/2$. So that, $x = a \sin (\omega t + \pi/2) = a \cos \omega t. = a \cos \omega t$.

If, on the other hand, we start counting time from an instant t′, *before the particle has passed through its mean position*, we have $x = 0$ when $t = t'$, so that, $0 = a \sin (\omega t' + \phi)$. Or, $\omega t' + \phi = 0$, whence, $\phi = -\omega t' = -e$, say. And, therefore, $x = a \sin (\omega t - e)$.

Similarly, if we start counting time from an instant t′, *after the particle has passed through its mean position*, we have $x = a \sin (\omega t + e)$.

In all these cases, however, the particle executes an oscillation or vibration such that its displacement varies cyclically with time and it is said to execute a simple harmonic motion.

It will be readily seen that if we put $\phi = \phi' + \pi/2$, we shall have

$$x = a \sin\left(\omega t + \phi' + \frac{\pi}{2}\right). \quad \text{Or, } x = a \cos (\omega t + \phi').$$

Thus, a simple harmonic motion may be expressed either in terms of a sine or a cosine function. Only, the initial phase or the phase constant will have different values in the two cases.

In fact, the general solution of equation (i), *viz.*, $d^2x/dt^2 = -\omega^2 x$, is of the form $x = a \sin \omega t + b \cos \omega t$, which is a combination of both the sine and the cosine terms.

Coming back to our relation $x = a \sin (\omega t + \phi)$, we find that if we increase the time t by 2p/w, we have

$$x = a \sin\left[\omega\left(t + \frac{2\pi}{\omega}\right) + \phi\right].$$

Or, $x = a \sin(\omega t + 2\pi + \phi)$

$= a \sin(\omega t + \phi)$.

The same, as before, thus indicating that the *particle repeats its movements after every* $2\pi/\omega$ sec, or that, in other words, the **time-period of the particle,**

$$T = \frac{2\pi}{\omega} = 2\pi\sqrt{\frac{1}{\omega^2}} = 2\pi\sqrt{\frac{1}{\mu}}.$$

Or, $$T = 2\pi\sqrt{\frac{I}{\text{acceleration per unit displacement}}}.$$

Or, $$T = 2\pi\sqrt{\frac{\text{displacement}}{\text{acceleration}}}.$$

Or, since $\omega^2 = \frac{C}{m}$,

we also have $T = 2\pi\sqrt{\frac{m}{C}}$.

Since T is quite independent of both a and $\phi$ (or e), it is clear that the oscillations of the particle are *isochronous*, (*i.e.*, take the same time irrespective of the values of a and e).

The number of oscillations (or vibrations) made by the particle per second is called its *frequency of oscillation* or, simply, its *frequency*, usually denoted by the letter n. Thus, *frequency is the reciprocal of the time-period, i.e.,*

$$n = \frac{1}{T} = \frac{\omega}{2\pi} = \frac{1}{2\pi}\sqrt{\frac{C}{m}},$$

whence $\omega = 2\pi n = \frac{2\pi}{T}$.

Since $\omega$ is the angle described by the particle per second, it is also referred to as the *angular frequency* of the particle.

All these results are equally true for *angular simple harmonic motion*, if we consider angular velocity, displacement, acceleration etc., instead of the linear ones.

*Alternatively,* we may obtain the same results, perhaps a, trifle more neatly, by solving the differential equation of motion of the particle with the help of *complex numbers*. Thus, taking the solution of the equation to be $x = Ae^{ipt}$, where A and p are arbitrary constants and $i = \sqrt{-1}$, we have

$$\frac{dx}{dt} = ipAe^{ipt}$$

and $$\frac{d^2x}{dt^2} = -p^2Ae^{ipt}.$$

Substituting these values in the differential equation $d^2x/dt^2 = -\omega^2x$, we have

$$-p^2Ae^{ipt} = -\omega^2Ae^{ipt},$$

whence, $$p^2 = \omega^2$$

and, therefore, $$p = \pm\,\omega.$$

There are thus two solutions to the equation, *viz.*,

$$x = A_1e^{iwt}$$

and $$x = A_2e^{-iwt}.$$

The *most general solution*, therefore, is $x = A_1e^{iwt} + A_2e^{-iwt}$, where the constants $A_1$ and $A_2$ can be determined from the initial conditions.

Now, as we know,

$$e^{\theta i} = (\cos\theta + i\sin\theta).$$

We, therefore, have

$$x = A_1(\cos\omega t + i\sin\omega t) + A_2(\cos\omega t - i\sin\omega t)$$

$$= (A_1 + A_2)\cos\omega t + i(A_1 - A_2)\sin\omega t.$$

Or, putting $(A_1 + A_2) = a\sin\phi$ and $i(A_1 - A_2) = a\cos\phi$, we have

$$x = a\sin\phi\cos\omega t + a\cos\phi\sin\omega t,$$

*i.e.*, $x = a\sin(\omega t + \phi)$, *the same relation as obtained above.*

Here, obviously, $a = \sqrt{(a\sin\phi)^2 + (a\cos\phi)^2}$

$$= \sqrt{(A_1 + A_2)^2 + \left[i\sqrt{(A_1 - A_2)}\right]^2} = \sqrt{(A_1 + A_2)^2 - (A_1 - A_2)^2}$$

and $\phi = \tan^{-1}\dfrac{A_1 + A_2}{i(A_1 - A_2)}$.

## PERIODIC AND HARMONIC MOTION

A motion which repeats itself over and over again after regularly recurring intervals of time, called its time-period, is referred to as a periodic motion.

If a particle, undergoing periodic motion, covers the same path back and forth about a mean position, it is said to be executing an *oscillatory* (or vibratory) motion or an *oscillation* (or a *vibration*). Such a motion is not only periodic but also *bounded, i.e.*, the displacement of the particle on either side of its mean position remains confined within a well-defined limit.

And since, of all the trigonometrical ratios, the *sines* and the *cosines* alone are periodic as well as bounded, the displacement of a particle executing an oscillatory motion is usually expressed in terms of sines or cosines or a combination of both. This, coupled with the fact that this type of motion is, in general, associated with musical instruments, is probably the reason why it is also spoken of as *harmonic motion.*

## ENERGY OF A HARMONIC OSCILLATOR

The acceleration of a harmonic oscillator, *i.e.*, of a particle executing a simple harmonic motion, is, as we know, directed towards its mean or equilibrium position, *i.e.*, opposite to the direction in which its displacement x increases. Work is, therefore, done during the displacement of the particle and it thus possesses *potential energy*. Also, the particle has velocity and, therefore, it possesses kinetic energy. Thus, a particle executing simple harmonic motion has, in general, both kinetic and potential energies. And if there be no dissipative forces (like friction etc) at work, the sum total of the two remains constant, *i.e., the mechanical energy of the oscillator {or the particle) remains conserved* though, as the displacement increases, the P.E. increases and the K.E. decreases. Thus at its maximum displacement (a), when it is momentarily at rest, the whole of its energy is present in the potential form, the kinetic energy being *zero*, and at its mean position, where its velocity is the maximum, the whole of the energy is present in the kinetic form, its potential energy now being *zero*. In between these two extreme positions, the energy of the oscillator or the particle is partly kinetic and partly potential, the sum total of the two remaining constant throughout. This may be seen from the following:

Since the acceleration of the particle, $d^2x/dt^2 = -\omega^2 x$, the force F required to maintain the displacement x is $m\omega^2 x$, where m is the mass of the particle.

$\therefore$ work done for a small displacement dx of the particle

$$= Fdx = m\omega^2 xdx.$$

This work is obviously a measure of the P.E. of the particle at this displacement, so that *P.E. of the particle at displacement*

$$dx = m\omega^2 xdx.$$

$\therefore$ work done for the whole displacement x of the particle or its

*P.E. at displacement x,* say, $U = \int_0^x m\omega^2 xdx$

$$= \frac{1}{2}m\omega^2 x^2 = \frac{1}{2}m\left(\frac{C}{m}\right)x^2 = \frac{1}{2}Cx^2, \qquad \left[\because \omega^2 = \frac{C}{m}\right].$$

*i.e.,* *P.E of the particle* $\alpha\ x^2$.

The maximum value of the potential energy is thus at x = a and is, obviously, $\frac{1}{2}m\omega^2 a^2 = \frac{1}{2}Ca^2$.

Now, *velocity of the particle at displacement x is* $v = \frac{dx}{dt} = \frac{d}{dt}$ a sin (wt + f) = $\omega\sqrt{a^2 - x^2}$.

$\therefore$ *K.E. of the particle at displacement* $x = \frac{1}{2}m\omega^2\left(a^2 - x^2\right)$

$$= \frac{1}{2}C\left(a^2 - x^2\right)$$

*The maximum value of K.E. is obviously at x = 0 and is also equal to* $\frac{1}{2}m\omega^2 a^2 = \frac{1}{2}Ca^2$.

*Hence, total energy of the particle at displacement x, i.e.,*

$$E = K.E. + P.E.$$

$$= \frac{1}{2}m\omega^2\left(a^2 - x^2\right) + \frac{1}{2}m\omega^2 x^2 = \frac{1}{2}m\omega^2 a^2 = \frac{1}{2}Ca^2.$$

As will be readily seen, *the total energy of the particle is quite independent of its displacement* x and thus remains the same throughout.

Since $\omega = 2\pi/T$, where T is the *time-period* of the particle, we may also express *total energy of the particle as*

$$E = \frac{1}{2}m\left(\frac{2\pi}{T}\right)^2 a^2 = \frac{2\pi^2 ma^2}{T^2}.$$

Or, since $1/T = n$, the *frequency* of the particle, we also have

$$E = 2\pi^2 n^2 ma^2.$$

Now, since in a conservative system, such as this, the sum total of the kinetic and potential energies of the system remains a constant, it follows that any one of them can increase only at the expense of the other and thus attains its maximum value when that of the other is the minimum or zero.

This means, in other words, that *the maximum value of any one of the two forms of energy is the same as the total energy of the system.*

Thus, *maximum value of K.E.* = *maximum value of P.E.* = *total energy*

$$E = \frac{1}{2}m\omega^2 a^2 = \frac{1}{2}Ca^2.$$

And, since during motion of the particle,

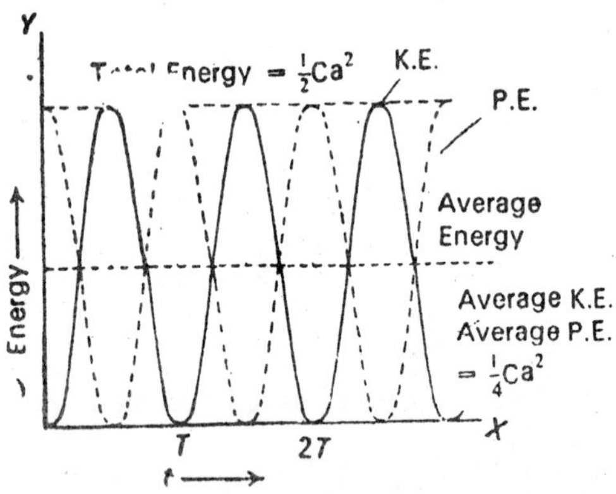


**Fig. 1.1**

the kinetic energy varies from 0 to the maximum, the graph showing the relation between K.E. and time (t) is as shown in full line in Fig. 1.1.

Similarly, during motion of the particle, the P.E. of the particle varies between 0 and the same maximum $\frac{1}{2}m\omega^2a^2$ or $Ca^2$ but reciprocally with K.E., *i.e.*, when K.E. is the maximum, P.E. is zero and vice versa. Hence the graph between P.E. and t, shown dotted in Fig. 1.1, is the reciprocal of that for K.E.

The total energy is represented by the upper horizontal line parallel to the time axis and touching the two curves at points representing the maximum values of kinetic and potential energies respectively.

If, on the other hand, we plot the potential energy (U) of the particle against its displacement (x), the relation between which is given by

$$U = \frac{1}{2}Cx^2,$$

we obtain a *parabola* with its vertex at x = 0, as shown in full line in Fig. 1.2.

The maximum displacement of the particle on either side of its mean position is the *same*, + a and − a respectively. The value of the P.E. is, therefore, the maximum at these points and equal to the total energy of the particle. The upper horizontal line passing through these points of maximum potential energy and parallel to the displacement axis thus represents the total energy curve of the particle.

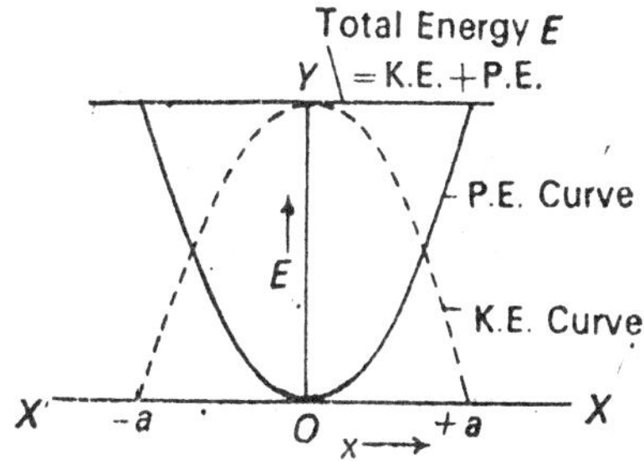


**Fig. 1.2**

A graph between K.E. and displacement (x) of the particle will obviously be the reciprocal of the P.E. curve, as shown dotted in the figure, with the vertex of the parabola touching the, upper horizontal line or the total energy curve at x = 0.

Since the total energy curve is a horizontal line in either case (Fig. 1.1 and 1.2), being parallel to the time-axis in the former and to the displacement axis in the latter case, it follows that *the total energy of the particle remains constant throughout and is independent of both time and displacement*, (being equal to the sum of the K E. and the P.E. of the particle at any given instant).

It is also clear from Fig. 1.2 that at x = a or – a, the total energy is wholly potential in form and at x = 0, it is entirely kinetic in form, being partly kinetic and partly potential at all other points.

## AVERAGE VALUES OF KINETIC AND POTENTIAL ENERGIES OF A HARMONIC OSCILLATOR

It can be seen at a glance from Fig. 1.1 that the *average value of K.E. of the particle is equal to the average value of its potential energy*

$$= \frac{1}{4} m\omega^2 a^2 = \frac{1}{4} Ca^2.$$

The same result may, however, be obtained directly as follows:

We have *P.E. of the particle at displacement* $x = \frac{1}{2} m\omega^2 x^2$

$$= \frac{1}{2} m\omega^2 a^2 \sin^2(\omega t + \phi).$$

∴ *average P.E. of the particle over a complete cycle or a whole time-period*

$$T = \frac{1}{T}\int_0^T \frac{1}{2} m\omega^2 a^2 \sin^2(\omega t + \phi)dt.$$

$$= \frac{m\omega^2 a^2}{4T}\int_0^T 2\sin^2(\omega t + \phi)dt = \frac{m\omega^2 a^2}{4T}\int_0^T 1 - \cos 2(\omega t + \phi)dt.$$

Since the average value of both a sine and a cosine function for a complete cycle or a whole time-period T is 0, we have *average P.E. of the particle*

$$= \frac{1}{4T} m\omega^2 a^2 [t]_0^T - 0 = \frac{1}{4T} m\omega^2 a^2 T$$

$$= \frac{1}{4} m\omega^2 a^2 = \frac{1}{4} Ca^2.$$

And, since *K.E. of the particle at displacement*

$$x = \frac{1}{2}m\left(\frac{dx}{dt}\right)^2$$

$$= \frac{1}{2}m\left[\frac{d}{dt}((a\sin(\omega t + \phi)))\right]^2 = \frac{1}{2}m\omega^2 a^2 \cos^2(\omega t + \phi),$$

we have average *K.E. of the particle over a complete cycle or a whole* time-period

$$T = \frac{1}{T}\int_0^T \frac{1}{2}m\omega^2 a^2 \cos^2(\omega t + \phi)dt$$

$$= \frac{m\omega^2 a^2}{4T}\int_0^T [1 + \cos 2((\omega t + \phi))]dt.$$

Again, the average value of a sine or a cosine function over a complete cycle or a whole time-period T being zero, we have

*average K.E. of the particle*

$$= \frac{m\omega^2 a^2}{4T}[t]_0^T = \frac{m\omega^2 a^2}{4T}T.$$

$$= \frac{1}{4}m\omega^2 a^2 = \frac{1}{4}Ca^2.$$

Thus, *average P.E. of the particle = average K.E. of the particle*

$$= \frac{1}{4}m\omega^2 a^2 = \frac{1}{4}Ca^2$$ *half the total energy.*

## SOME EXAMPLES OF S.H.M

Let us now proceed to examine some important examples of simple harmonic motion..

### The Simple Pendulum

A simple (or a mathematical) pendulum is just a heavy particle (ideally, a *point-mass*) suspended from one end of an *intextensible, weightless* string whose other end in fixed in a rigid support, this point being referred to as the point of suspension of the pendulum.

Obviously, it is simply impossible to obtain such an idealised simple pendulum. In actual practice, therefore, we take a small and heavy spherical bob tied to a long and fine silk thread, the other end of which passes through a split cork securely clamped in a suitable stand, the length ($l$) of the pendulum being measured from the point of suspension to the centre of mass of the bob.

In Fig. 1.3, let S be the point of suspension of the pendulum and O, the mean or equilibrium position of the bob. On taking the bob a little to one side and then gently releasing it, the pendulum starts oscillating about its mean position, as indicated by the dotted lines.

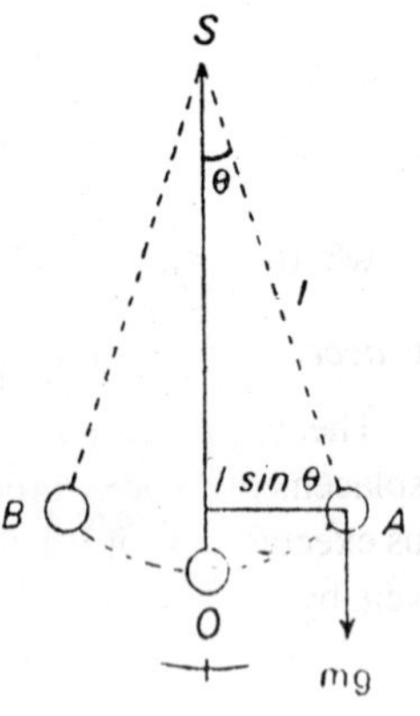

Fig. 1.3

At any given instant, let the displacement of the pendulum from its mean position SO into the position SA be θ. Then, the weight mg of the bob, acting vertically downwards, exerts a torque or a moment $- mg\, l \sin\theta$ about the point of suspension, tending to bring it back to its mean position, the negative sign of the torque indicating that it is oppositely directed to the displacement (θ).

If $d^2\theta/dt^2$ be the acceleration of the bob, towards O, and I its M.I. about the point of suspension (S), the moment of the force or the torque acting on the bob is also equal to $I.d^2\theta/dt^2$. We, therefore, have

$$I\frac{d^2\theta}{dt^2} = -mg\,l\sin\theta$$

Now, expanding sin θ into a power series, in accordance with *Maclaurin's theorem*, we have

$$\sin\theta = \theta - \frac{\theta^3}{3!} + \frac{\theta^5}{5!} - \ldots$$

If, therefore, θ be small, *i.e.*, if the amplitude of oscillation be small, we may neglect all other terms except the first and take sin θ = θ. So that,

$$I\frac{d^2\theta}{dt^2} = -mg\,l\theta,$$

whence,
$$\frac{d^2\theta}{dt^2} = -\frac{mgl}{I}\theta.$$

Or, since M.I. of the bob (or the point mass) about the point of suspension (S) is $ml^2$, we have

$$\frac{d^2\theta}{dt^2} = \frac{mgl}{ml^2}\theta = \frac{g}{l}\theta = \mu\theta,$$

where $\frac{g}{l} = \mu,$

the *acceleration per unit displacement.*

The acceleration of the bob is thus proportional to its angular displacement θ and is directed towards its mean position O. The pendulum thus executes a simple harmonic motion and its time-period is, therefore, given by

$$T = 2\pi\sqrt{\frac{1}{\mu}} = 2\pi\sqrt{\frac{1}{g/l}} = 2\pi\sqrt{\frac{l}{g}},$$

it being clearly understood that the amplitude of the pendulum is small.

The displacement here being angular, instead of linear, it is obviously an example of an angular simple harmonic motion.

### *Drawbacks of a Simple Pendulum*

Although, one of the simplest methods for determining the value of g at a place, a simple pendulum suffers from a number of drawbacks, the more important of which are the following:

(i) It is just an ideal conception, not realisable in actual practice, since it is impossible to have both a point-mass and a weightless string. So that, the string too has a moment of inertia about the axis of suspension.

(ii) The resistance and the buoyancy of the air appreciably affect the motion of the bob.

(iii) The expression for the time-period $\left(T = 2\pi\sqrt{l/g}\right)$ is true only for oscillations of infinitely small amplitude.

(iv) The motion of the bob is not strictly linear. It has also a rotatory motion about the axis of suspension.

(v) The bob also has a relative motion with respect to the string at the extremities of its amplitude on either side.

For all these reasons, a *compound pendulum* is preferred to a simple (pendulum for the determination of the value of g.

## The Compound Pendulum

Also called a *physical pendulum* or a *rigid pendulum*, a compound pendulum is just a *rigid body, of whatever shape, capable of oscillating about a horizontal axis passing through it.*

The point in which the vertical plane passing through the *e.g.*, of the pendulum meets the axis of rotation is called its *point or centre of suspension* and the distance between the point of suspension and the *e.g.*, of the pendulum measures the *length* of the pendulum.

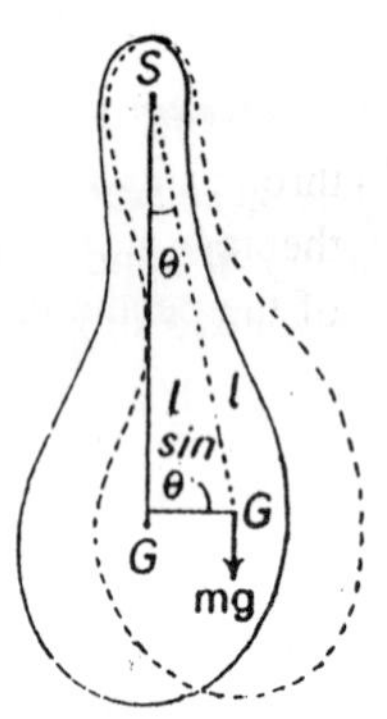


Fig. 1.4

Thus, Fig. 1.4 shows a vertical section of a rigid body or a (compound pendulum, free to rotate about a horizontal axis passing through the point or centre of suspension S In its normal portion of rest, its *e.g.*, G, naturally lies vertically below S, the distance between S and G giving the *length l* of the pendulum.

Let the pendulum be given a small angular displacement θ into the dotted position shown, so that its *e.g.*, takes up the new position G′ where, of course, SG′ = *l*. The weight of the pendulum, mg, acting vertically downwards at G′ and its reaction at the point of suspension S constitute a *couple* (or a *torque*), tending to bring the pendulum back into its original position.

Obviously, *moment of this restoring couple* = – mg *l* sin θ, the negative sign indicating that the couple is oppositely directed to the displacement θ

If I be the *moment of inertia* of the pendulum about the axis of suspension (through S) and $d^2\theta/dt^2$, its angular acceleration, the couple is also equal to $I.d^2\theta/dt^2$. So that, we have

$$\frac{I.d^2\theta}{dt^2} = -mgl\sin\theta.$$

Again, $\sin\theta = \theta - \theta^3/3! + \theta^5/5! \ldots,$

so that, if θ be small, $\sin \approx \theta$

and, therefore, $I.d^2\theta/dt^2 = -mgl\theta.$

whence, $d^2\theta/dt^2 = -(mgl/I)\theta = -\mu\theta$,

where $mgl/I = \mu$, the *acceleration per unit displacement.*

The pendulum thus executes a simple harmonic motion and its *time-period* is given by

$$T = 2\pi\sqrt{\frac{1}{\mu}} = 2\pi\sqrt{\frac{1}{mgl/I}} = 2\pi\sqrt{\frac{I}{mgl}}.$$

Now, if $I_0$ be the moment of inertia of the pendulum about an axis through its *e.g.*, G, parallel to the axis through S, we have, from the theorem of parallel axes, $I = I_0 + ml^2$. And if k be the *radius of gyration* of the pendulum about this axis through G, we have $I_0 = mk^2$. So that,

$$I = mk^2 + ml^2 = m(k^2 + l^2).$$

$$\therefore\ T = 2\pi\sqrt{\frac{m(k^2 + l^2)}{mgl}} = 2\pi\sqrt{\frac{k^2 + l^2}{gl}} = 2\pi\sqrt{\frac{k^2/l + l}{g}}.$$

Thus, *the time-period of the pendulum is the same as that of a simple pendulum of length* $L = (k^2/l + l)$ or $(k^2 + l^2)/l$. This length L is, therefore, called the *length of an equivalent simple pendulum* or the *reduced length* of the compound pendulum. Since $k^2$ is always greater than zero, *the length of the equivalent simple pendulum (L) is always greater than l, the length of the compound pendulum.*

### *Centre of Oscillation*

A point O on the other side of the *e.g.*, (G) of the pendulum in a line with SG and at a distance $k^2/l$ from G is called the *centre of oscillation* of the pendulum (Fig 1.5) arid a horizontal axis passing through it, parallel to the axis of suspension (through S) is called the *axis of oscillation* of the pendulum.

Now, clearly, $GO = k^2/l$ and $SG = l$. So that, $SO = SG + GO = l + k^2/l = L$, the length of the equivalent simple pendulum, *i.e.*, *the distance between the centres of suspension and oscillation is equal to the length of the equivalent simple pendulum or the reduced length (L) of the pendulum* and we, therefore, have $T = 2\pi\sqrt{L/g}$

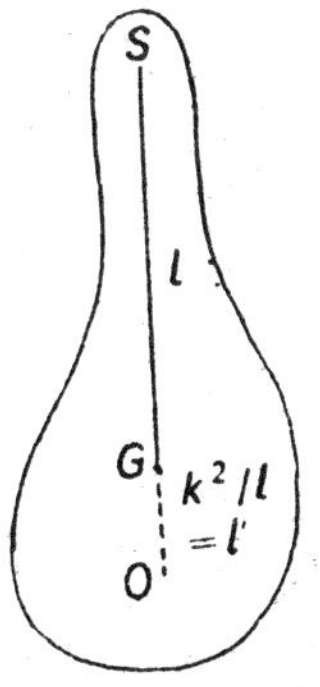


Fig. 1.5

*Interchangeability of centres of suspension and oscillation:* If we put $k^2/l = l'$,

we have $L = l + \frac{k^2}{l} = l + l'$ and, therefore

$$T = 2\pi\sqrt{\frac{(l + l')}{g}}.$$

If now we invert the pendulum, so that it oscillates about the axis of oscillation through O, its, time period, T′, say, is given by

$$T' = 2\pi\sqrt{\frac{(k^2 + l'^2)}{l'g}}.$$

Since $\frac{k^2}{l} = l'$, we have $k^2 = ll'$.

Substituting $ll'$ for $k^2$ in the expression for T′, therefore, we have

$$T' = 2\pi\sqrt{\frac{(ll' + l'^2)}{l'g}} = 2\pi\sqrt{\frac{(l + l')}{g}} = T,$$

*i.e.*, the same as the time-period about the axis of suspension.

Thus, *the centres of suspension and oscillation are interchangeable or reciprocal to each other, i.e., the time-period of the pendulum is the same about either.*

In fact, there are two other points on either side of G, about which the time-period of the pendulum is the same as about S and O. For, if with G as centre and radii equal to $l$ and $k^2/l$ respectively, we draw two circles, so as to cut SG produced in S and 0' above, and at O and S′ below G, as shown in Fig. 1.6, we have

$$SG = GS' = l$$

and $$GO' = GO = k^2/l = l'$$

$$\therefore \quad O'S' = GS' + GO$$

$$= l + k^2/l = l + l' = SO.$$

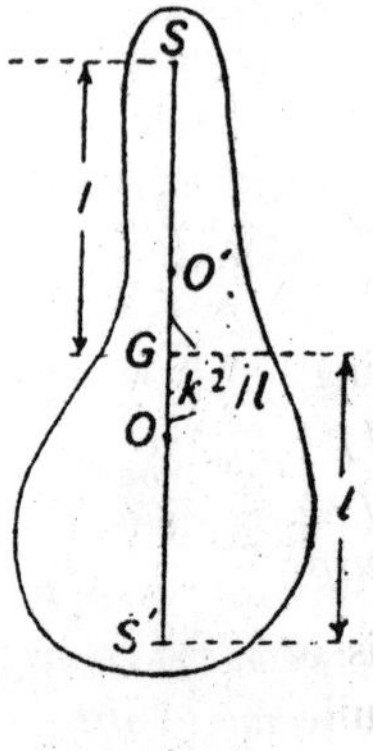

Fig. 1.6

Thus, *There are four points in all, viz., S, O, S′ and O′, collinear with the e.g., of the pendulum (G) about which its time-period is the same.*

*Maximum and minimum time-periods of a compound pendulum:* For the time-period, of a compound pendulum, we have the relation

$$T = 2\pi\sqrt{\frac{\left(k^2 + l^2\right)}{lg}},$$

squaring which, we have

$$T^2 = \frac{4\pi^2\left(k^2 + l^2\right)}{lg} = \frac{4\pi^2}{g}\left(\frac{k^2 + l^2}{l}\right) = \frac{4\pi^2}{g}\left(\frac{k^2}{l} + l\right)$$

Differentiating with respect to $l$, we have

$$2T\frac{dT}{dl} = \frac{4\pi^2}{g}\left(-\frac{k^2}{l^2} + 1\right),$$

a relation showing the variation of T with length ($l$) of the pendulum.

Clearly, T will be a maximum or a minimum when $dT/dl = 0$, *i.e.*, when F = $k^2$ or $l = \pm$ k or when $l$ = k, because the negative value of k is simply meaningless.

Since $d^2T/dl^2$ comes out to be *positive, it is clear that T is a minimum when l = k, i.e., the time-period of a compound pendulum is the* minimum *when its length is equal to its radius of gyration about the axis through its e.g.,* And the value of this minimum time-period will obviously be

$$T_{min} = 2\pi\sqrt{\frac{\left(k^2 + k^2\right)}{kg}} = 2\pi\sqrt{\frac{2k}{g}}.$$

On the other hand, we see from the expression for T above that if $l = 0$ or $\infty$, T= $\infty$ or a maximum.

Ignoring $l = \infty$ as absurd, we thus find that, *the time-period of a compound pendulum is the maximum when its length is zero, i.e., when the axis of suspension passes through its e.g., or the e.g., itself is the point of suspension.*

This is obviously so because the pendulum is then in a state of neutral equilibrium, with no restoring action due to gravity on it.

### *Determination of the Value of g*

From the interchangeability of the points of suspension and oscillation it would appear that the easiest method of determining the value of g at a place would be to locate two points on either side of the *e.g.*, of the pendulum about which the time-period of the pendulum is the same. These points would then correspond to the centres of suspension and oscillation of the pendulum and the distance between them would give L, *the length of the equivalent simple pendulum*. So that, if T be the time-period of the pendulum about either of these, we shall have

$$T = 2\pi\sqrt{L/g},$$

and, therefore, $g = L/T^2$.

This is, however, easier said than done, for it is extremely difficult, if not impossible, to locate two such points in the pendulum. We, therefore, take recourse to one of the following two methods, using special forms of the pendulum.

(i) *By means of a bar pendulum:* A bar pendulum is the simplest form of a compound pendulum and consists of a uniform metal bar AB (Fig. 1.7), having equally spaced holes drilled along its length on either side of its *e.g.*, G, so that any one of them may be slipped on to a horizontal knife-edge K and the bar made to oscillate about it in the vertical plane.

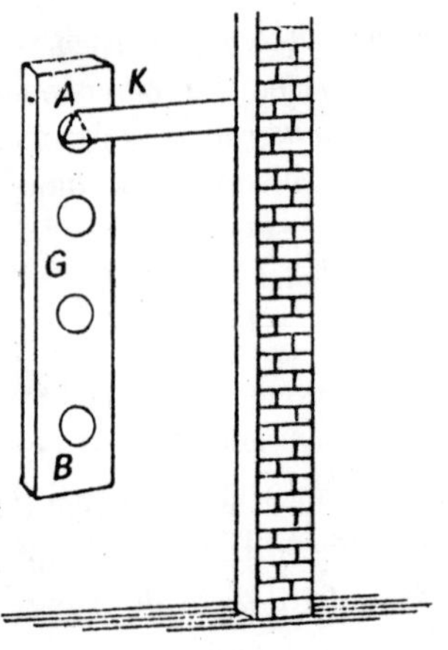


**Fig. 1.7**

First, the time-periods of different lengths of the pendulum are determined by slipping on to the knife-edge one hole after another from A to G and each length (*i.e.*, the distance between the point of suspension and the *e.g.,*) carefully noted.

A graph ABD is then plotted between distances of the holes from the *e.g.*, (or the lengths of the pendulum) along the x-axis and the time-periods (T) along they-axis, as shown in Fig. 1.8.

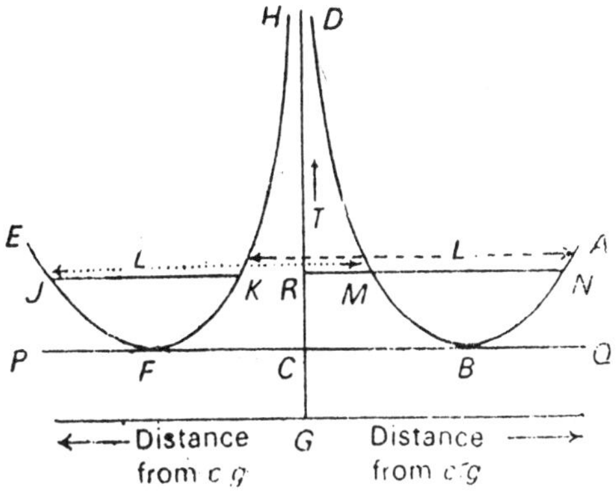


Fig. 1.8

The experiment is next repeated with the holes on the other side of the *e.g.*, of the pendulum and a similar graph EFH drawn alongside the first on the same graph paper, using the same scale, or, this graph may be drawn symmetrically with the first without actually repeating the experiment with the holes on the other side of the *e.g.*, This graph (EFH) will be a mirror image of the first (ABD), as is clear from the figure.

It will be seen at once that as the *e.g.*, of the bar (G) is approached (*i.e.*, as the centre of suspension comes closer to the *e.g.*,), the time-period first decreases, acquires a minimum value and then increases until it becomes infinite at the *e.g.*, itself.

Let a horizontal line JN be drawn, parallel to the x-axis, so as to cut the two curves in points J, K, M and N. Then, clearly, the time-period of the pendulum for lengths corresponding to all these points is the *same*. They thus correspond to the four points S, O', O and S', collinear with the *e.g.*, G, in Fig. 1.6 above, about which the time- period of the pendulum is the same. Clearly, therefore, JM = KN = L, the length of the equivalent simple pendulum, for JR here corresponds to $l$ and RM, to $k^2/l$

so that, $JM = JR + RM = l + k^2/l = L.$

And, similarly, $KN = KR + RN = k^2/l + l = L.$

Thus, knowing the value of T corresponding to these points (J,K, M and N) from the graph, we can easily obtain the value of g at the place from the relation $T = 2\pi\sqrt{L/g}$, which gives $g = 4\pi^2 L/T^2$.

It will also be seen that if we draw a tangential line PQ, touching the two curves at B and F, then B and F represent the points where the centres of suspension and oscillation coincide with each other in the two cases respectively. The time-period of the pendulum corresponding to these points is, therefore, the minimum. In other words, at these points $k^2/l = l$ or $k^2 = l^2$ and, therefore, $k = l$. So that, each of the distances (or lengths of the pendulum) CB and CF is equal to k, or BF = 2k, whence, the value of the radius of gyration (k) of the pendulum about the axis through its *e.g.*, can also be easily obtained.

Now, instead of calculating, the value of g as above, a better method, suggested by *Ferguson* in the year 1928, is to plot $lT^2$ along the axis of x and $l^2$ along the axis of y, which, from the relation $l^2 + k^2 = (lT^2/4\pi^2)g$ must give a straight line graph, as shown in Fig. 1.9. The slope of the curve is $g/4\pi^2$, whence the value of g may be easily obtained.

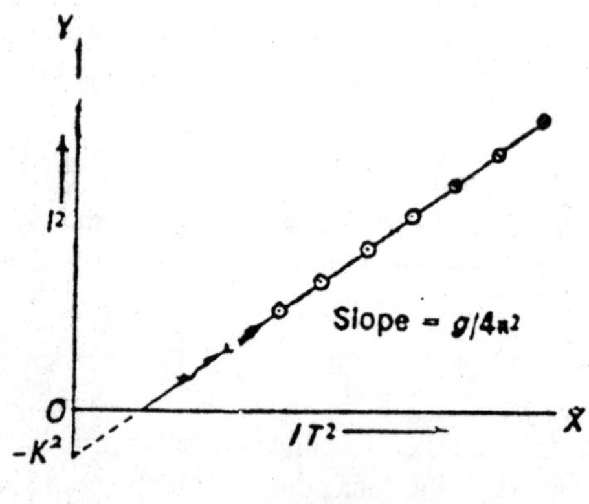


**Fig. 1.9**

Further, the intercept of the curve on the axis of y gives $-k^2$ and thus the values of both g and k can be obtained at once.

A drawback of the method is that it being well nigh impossible to pin-point the position of the *e.g.*, of the bar or the pendulum (as, infact, of any other body), the distances measured from it are not totally accurate. Any error due to this is, however, eliminated automatically as the graph is smoothed out into the form of a straight line.

(ii) *By means of a Kater's pendulum:* A Kater's pendulum is a rigid bar AB of brass or steel, fitted with adjustable and mutually facing knife edges $K_1$, and $K_2$, near its two ends, as shown in Fig. 1.10, so that it may be made to oscillate about any one of them, as desired – hence the name *reversible pendulum* given to it.

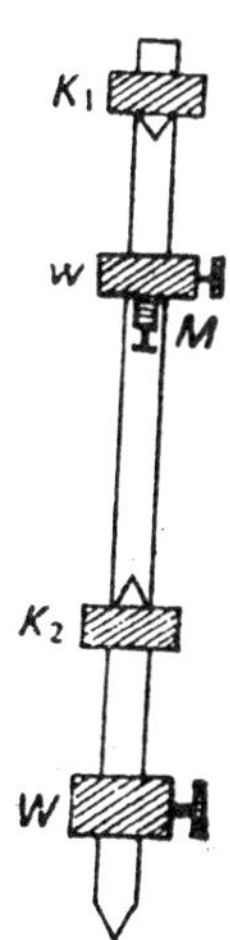

**Fig. 1.10**

Two cylindrical weights w and W, sliding along the bar, can be clamped in any desired position, on it such that *the e.g., of the pendulum lies in between the two knife-edges and nearer to one of them than to the other.* For this purpose, the heavier weight (W) is clamped near one end of the bar and the lighter weight (w) in between the knife-edges, its position being adjusted by means of a micrometer screw M.

In case the weights be so arranged that the time-period of the pendulum is *exactly the same* about either knife-edge, clearly, their positions correspond to those of the centres of suspension and oscillation respectively and the distance between them directly gives the length L of the equivalent simple pendulum. We, therefore have

$$T = 2\pi\sqrt{L/g},$$

whence, $$g = 4\pi^2 L/T^2.$$

As mentioned earlier, however, it is extremely difficult and tiring to thus adjust the positions of the two weights and, as *Bassel* pointed out, not really necessary. It is quite enough to adjust their positions such that the time-periods of the pendulum about the two knife-edges are only very nearly equal. For, then, if $l_1$ and $l_2$ be the lengths of the pendulum (*i.e.*, distances of knife-edges $K_1$ and $K_2$ respectively from the *e.g.,*) and $T_1$, and $T_2$ (very nearly equal), their respective time periods, we have

$$T_1 = 2\pi\sqrt{\frac{k^2 + l_1^2}{l_1 g}}$$

and $$T_2 = 2\pi\sqrt{\frac{k^2 + l_2^2}{l_2 g}}$$

Squaring and rearranging, we have

$$T_1^2 l_1 g = 4p^2 \left(k^2 + l_1^2\right)$$

and $$T_2^2 l_2 g + 4\pi^2 \left(k^2 + l_2^2\right).$$

Subtracting the second relation from the first, we have

$$g\left(T_1^2 l_1 - T_2^2 l_2\right) = 4\pi^2\left(k^2 + l_1^2\right) - 4\pi^2\left(k^2 + l_2^2\right)$$

$$= 4\pi^2\left(l_1^2 - l_2^2\right) = 4\pi^2 (l_1 + l_2)(l_1 - l_2)$$

Or, $$\frac{4\pi^2}{g}(l_1 + l_2) = \frac{T_1^2 l_1 - T_2^2 l_2}{l_1 - l_2} = \frac{2\left(T_1^2 l_1 - T_2^2 l_2\right)}{2(l_1 - l_2)}$$

$$= \frac{\left(T_1^2 + T_2^2\right)(l_1 - l_2) + \left(T_1^2 - T_2^2\right)(l_1 + l_2)}{2(l_1 - l_2)}$$

whence, $$\frac{4\pi^2}{g}(l_2 + l_2) = \left(\frac{T_1^2 + T_2^2}{2}\right) + \left(\frac{T_1^2 - T_2^2}{2}\right)\left(\frac{l_1 + l_2}{l_1 - l_2}\right) \quad ...(i)$$

Here, $(l + l_2)$ is clearly the distance between the two-Knife-edges and can, therefore, be accurately measured, $(l_1 - l_2)$ being the difference between the distances of the two knife-edges from the *e.g.*, of the pendulum can not, of course, be measured accurately on account of the difficulty in locating the exact position of the *e.g.*, However, since $T_1$ and $T_2$ are very nearly equal, $\left(T_1^2 - T_2^2\right)$ is much two small and the second term on the right hand side, involving $(l_1 - l_2)$ becomes negligible compared with the first term. We, therefore, have

$$\frac{4\pi^2}{g}(l_1 + l_2) = \frac{T_1^2 + T_2^2}{2}, \quad ...(ii)$$

whence, $$g = \frac{8\pi^2 (l_1 + l_2)}{T_1^2 + T_2^2}.$$

The accurate value of g at the place can thus be obtained.

Comparing expression I above with that for a simple pendulum, *viz.*, $4\pi^2 g/l = T^2$, we find that if T be the time period for a length $(l_1 + l_2)$ of the pendulum, we have

$$T^2 = \left(\frac{T_1^2 + T_2^2}{2}\right) + \left(\frac{T_1^2 - T_2^2}{2}\right)\left(\frac{l_1 + l_2}{l_1 - l_2}\right)$$

Here T is called the *computed time* of the pendulum.

*Superiority of a compound pendulum over a simple pendulum:* The main points of the superiority of a compound pendulum over a simple pendulum are the following:

(i) Unlike the ideal simple pendulum, a compound pendulum is easily realisable in actual practice.

(ii) It oscillates as a whole and there is no lag like that between the bob and the string in the case of a simple pendulum.

(iii) The length to be measured is clearly defined (*viz.*, the distance between the two knife-edges in a Kater's pendulum). In the case of a simple pendulum, the point of suspension and the *e.g.*, of the bob, the distance between which gives the length of the pendulum, are both more or less indefinite points, so that the distance between them, *i.e.*, $l$, cannot be measured accurately.

(iv) On account of its large mass, and hence a large moment of inertia, it continues to oscillate for a longer time, thus enabling the time for a large number of oscillations to be noted and its time-period calculated more accurately

## Loaded Spring

We have seen earlier how when a particle is displaced from its mean or equilibrium position, a *linear restoring* force acts upon it, tending to bring it back into its original position.

Thus, suppose we have a mass m attached to the free end of a massless flat spiral spring, with its other end fixed to a rigid support, like a wall etc. (Fig. 1.11). If the mass be displaced through a distance x, as shown, a *linear restoring force* F = – Cx at once starts acting on the spring, tending to bring it back into its original condition, where C is the *force constant* of the spring. The –ve sign of the force simply indicates that it is directed oppositely to the displacement of the mass.

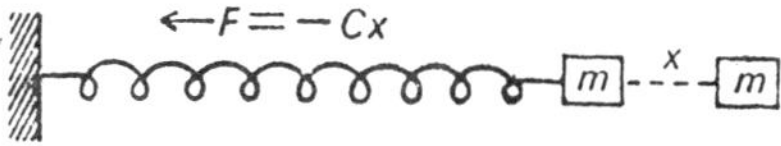


**Fig. 1.11**

Imagining the system to lie on a smooth horizontal surface, if the mass be released, it starts oscillating back and forth due to the spring getting alternately compressed and extended under the action of this force.

If $d^2x/dt^2$ be the acceleration set up in the spring, the force acting on the mass is also equal to $m.d^2x/dt^2$. We, therefore, have

$$m\frac{d^2x}{dt^2} = -Cx.$$

or, $$\frac{d^2x}{dt^2} = -\frac{C}{m}x = -\mu x, \qquad ...(i)$$

where $\frac{C}{m} = \mu$, the acceleration per unit displacement.

Thus, $\frac{d^2x}{dt^2}$ $\alpha$ x and is directed oppositely to it.

The mass m thus executes a *simple harmonic motion* and its *time-period* is given by

$$T = 2\pi\sqrt{\frac{1}{\mu}} = 2\pi\sqrt{\frac{1}{C/m}} = 2\pi\sqrt{\frac{m}{C}}. \qquad ...(II)$$

The solution of equation of motion I above is also of the form

$$x = a\sin(\omega t + \phi), \text{ where } \sqrt{\frac{C}{m}} = \omega.$$

Or, if, as is usually the case, the time is counted from the moment that the mass just crosses its mean or equilibrium position towards the positive direction of x, *i.e.*,

if $x = 0$,

at $t = 0$

we have $\phi = 0$ and

$\therefore$ $x = a \sin \omega t$.

In actual practice, the spiral spring is arranged vertically with either its lower end fixed and the mass m placed on its upper end or with its upper end fixed and the mass suspended from its lower end (particularly if the wire of the spring be thin), as shown in Figs. 1.12 (a) and (b) respectively.

Taking the more usual case (b), if the spring be extended through a distance *l*, say, due to the weight mg of the mass, a linear restoring force $Cl$ at once comes into play in the opposite direction, so that the equilibrium position is attained when the two forces just balance each other, *i.e.*, when mg = $Cl$, whence, C = mg/$l$.

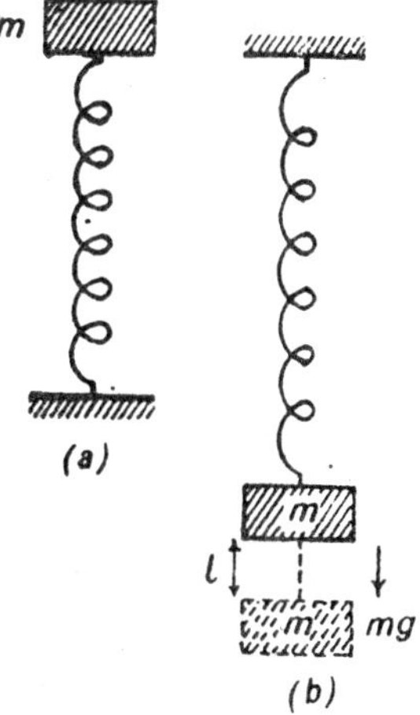


Fig. 1.12

If the mass be now pulled down through a distance x from this equilibrium position, the linear restoring force F = mg − C($l$ + x) = mg − (mg/$l$) ($l$ + x) = mg − mg − (mg/$l$)x = − (mg/l)x = − Cx, *i.e.*, we have $md^2x/dt^2 = -Cx$, as before, clearly showing that *there is no effect of gravity on the force constant C and hence an the period of oscillation of the mass*, given by T = 2pT $= 2\pi/\sqrt{m/C}$.

**Torsion Pendulum**

A heavy body, like a cylinder or a disc, fastened at its mid-point to a fairly, long and thin wire, suspended from a rigid support, constitutes a *torsional pendulum*, (Fig. 1.13). It is so called because, if the cylinder or the disc be turned in its own (*i.e.*, the horizontal) plane to twist the wire a little and then released, it executes torsional vibrations or oscillations about the wire as axis.

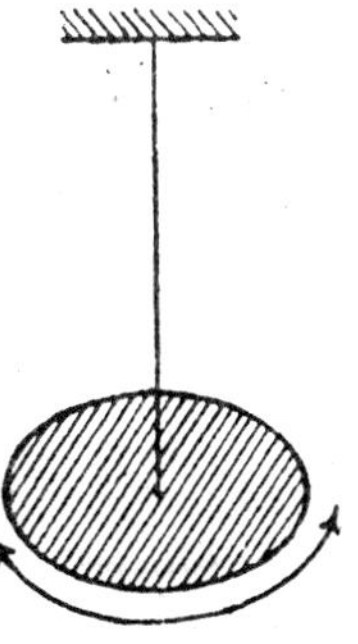

Fig. 1.13

Thus, if the disc (or cylinder) be turned through an angle θ, say, the suspension wire too gets twisted through the same angle $\theta$ and this gives rise to a restoring torsional couple $C\theta$ in it, tending to bring it back into its original condition. Here, C is the *torsional couple per unit twist* of the wire, equal to $\pi nR^4/2L$, where R is the *radius* and L, the length of the wire and n, the modulus of rigidity of its material.

It I be the *moment of inertia* of the disc (or cylinder) about the wire as axis, passing through its centre, and $d^2\theta/dt^2$, its angular acceleration, the couple acting on it is also equal to I $d^2\theta/dt^2$. We therefore, have $Id^2\theta/dt^2 = -C\theta$, the –ve sign indicating that the restoring couple or torque is oppositely directed to the angular displacement. So that,

$$\frac{d^2\theta}{dt^2} = -\left(\frac{C}{I}\right)\theta = \mu\theta, \qquad ...(i)$$

where $C/I = \mu$, the acceleration per unit angular displacement.

Or, $d^2\theta/dt^2 \alpha \theta$ is directed oppositely to it. The disc (or the cylinder) thus executes *an **angular** simple harmonic motion* and its time-period is given by

$$T = 2\pi\sqrt{\frac{1}{\mu}} = 2\mu\sqrt{\frac{1}{C/I}} = 2\pi\sqrt{\frac{I}{C}}. \qquad ...(ii)$$

It may be noted that *no approximations* whatever have been used in arriving at this relation for T, unlike in the case of a simple or a compound, pendulum. *The time-period of a torsional pendulum, therefore, remains unaffected (i.e., the oscillations remain isochronous) even if the amplitude be large, provided, of course, the elastic limit of the suspension wire is not exceeded.* Further, the solution of relation I above is also of the form

$$\theta = \theta_m(\omega t + \phi),$$

where $\theta_m$ is the maximum angular displacement or amplitude of the disc (or the cylinder), $\omega$, its angular velocity equal to $\sqrt{C/I}$ and $\phi$ the initial phase or phase constant.

It will be readily seen that the angular S.H.M. is analogous to linear S.H.M. Only, we take angular in place of linear displacement, moment of inertia instead of mass and torsional constant instead of force constant.

**Note :** It need hardly be pointed out that a torsion pendulum is entirely different from a simple or a compound pendulum. For, here, the *e.g.,* of the suspended body. Instead of moving in an are, remains fixed in its position, the body merely rotating about the axis through it and, unlike there, the time-period of the pendulum is quite independent of the value of g.

*Inertia table:* A simple application of the torsional pendulum is what is called the *Inertia table* which is just a device to determine the moments of inertia of bodies of both regular geometrical shapes and irregular shapes.

It is merely a large disc D of aluminium, about 15 cm in diameter, fitted with two small vertical pillars P, P at the extremity of one diameter and connected at the top by a cross-bar, as shown in Fig. 1.14.

The disc, along with the pillars and the cross bar, is suspended by means of a *long* and *thin* wire from the top of a large framework F, mounted on a heavy circular base, provided with levelling screws. A small mirror m, fixed right at the mid-point of the cross bar enables the oscillations to be observed by a lamp and scale arrangement.

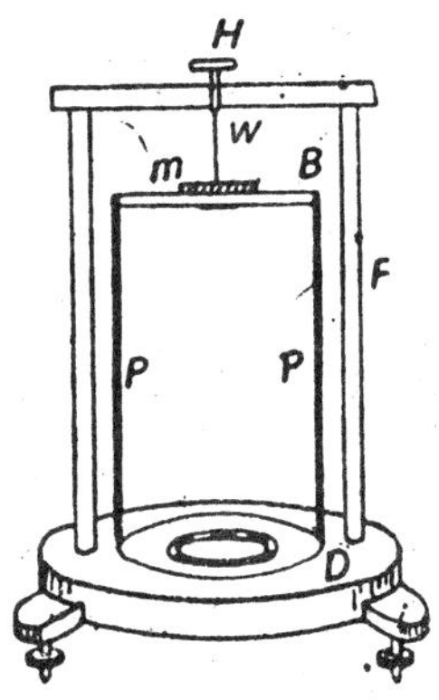


**Fig. 1.14**

To ensure that the disc remains horizontal, a concentric circular groove cut into it, as shown, such that three *balancing weights* can be made to slide into the desired positions inside it. Further, a number of concentric circles are usually drawn on the disc on the table to enable the body, whose moment of inertia is to be determined, to be so placed on it that its centre of mass lies on the suspension wire, which thus becomes the axis of rotation of the body through its centre of mass and perpendicular to its plane. The whole apparatus is enclosed in a glass envelope as a safeguard against disturbance due to air draughts.

*Procedure:* First, the inertia table is made perfectly horizontal by means of the levelling screws and the sliding balancing weights, as indicated. It is then set into torsional vibration all by itself alone and its time-period $T_0$ determined. If $I_0$ be the moment of inertia of the Inertia table about the suspension wire as axis, we have

$$T_0 = 2\pi\sqrt{\frac{I_0}{C}}, \qquad \text{...(i)}$$

where C is the torsional couple per unit twist of the wire.

The body whose moment of inertia I is to be determined is now placed *centrally* on the intertia table in the manner explained (*i.e.*, with its centre of mass lying on the suspension wire), taking care to see that the table remains horizontal, so that its moment of inertia ($I_0$) about the suspension wire does not get altered. If $T_1$ be the time-period of the Inertia table, thus loaded, we have

$$T_1 = 2\pi\sqrt{\left(\frac{I_0 + I}{C}\right)}. \quad ...(ii)$$

Finally, the given body is replaced by another of a known moment of intertia $I_1$ (about the axis passing through its centre of mass and perpendicular to its plane) and, again taking care to keep the table horizontal (by changing the positions of the balancing weights, if necessary), the time-period of the loaded table determined. Let it be $T_2$. Then, we have

$$T_2 = 2\pi\sqrt{\left(\frac{I_0 + I_1}{C}\right)}. \quad ...(iii)$$

Now, squaring and dividing relation (ii) by (i) we have

$$\frac{T_1^2}{T_0^2} = \frac{(I_0 + I)}{I_0}$$

or,
$$\frac{I}{I_0} = \frac{\left(T_1^2 - T_0^2\right)}{T_0^2}. \quad ...(iv)$$

Similarly, squaring and dividing relation (iii) by (i), we have

$$\frac{T_2^2}{T_0^2} = \frac{(I_0 + I_1)}{I_0}$$

or,
$$\frac{I_1}{I_0} = \frac{\left(T_1^2 - T_0^2\right)}{T_0^2}. \quad ...(v)$$

So that, dividing relation (iv) by (v), we have

$$\frac{I}{I_1} = \frac{\left(T_1^2 - T_0^2\right)}{\left(T_2^2 - T_0^2\right)}$$

or,
$$I = \left(\frac{T_1^2 - T_0^2}{T_2^2 - T_0^2}\right) I_1^*,$$

whence, the moment of intertia (I) of the given body about the wire as axis can be easily obtained.

The amplitude ($\theta$) here need not be small since the restoring couple, as pointed out above, is found to be proportional to $\theta$ even if $\theta$ be large. The assumption, however, that even with different loads suspended from the wire and with different longitudinal tensions in it, the value of C remains unaffected is not strictly true.

An obvious disadvantage of the method is that it can be used to determine the moment of inertia of a body only about one particular axis, *viz*,, the axis perpendicular to its plane and passing through its centre of mass.

## Helmholtz Resonator

A, *resonator* is a device to analyse a complex note of sound, *i.e.*, to find out what particular frequencies are present in the given note. For this purpose it is necessary that the resonator should exhibit a sharp resonance, *i.e.*, it should resound with a note *of only one particular frequency*, namely. Its own natural frequency. Thus, resonators of different natural frequencies are used to detect the different frequencies comprising the given note.

*Helmholtz* showed that sharpness of resonance in what is called a *volume resonator* is ensured if the resonator is a large vessel, spherical or cylindrical, of glass or metal, containing air, with a narrow neck N, through which it communicates with the outside air and receives the complex note to be analysed, and a narrow aperture 0 at the opposite end, to be plugged into the ear, [Figs. 1.15 (a) and (b)]. Whereas the spherical type of resonator (a) can be used to detect just *one frequency equal to its own,* the *cylindrical type* (b), being made up of two parts, one sliding over the other, can have its frequency altered at will and can thus be used to detect the different frequencies of the given note.

The *principle* underlying the working of the resonator is that the *air in its neck serves as an air plug, as it were, and performs an oscillatory motion like a piston in a cylinder containing air, or a mass suspended from a spring.*

Thus, if $l$ be the *length of the neck* and $\alpha$, its *area of cross section*, the *mass of the air plug or the air piston in the neck* = $l\alpha\rho$, where $\rho$ is the *density of the air* in the vessel.

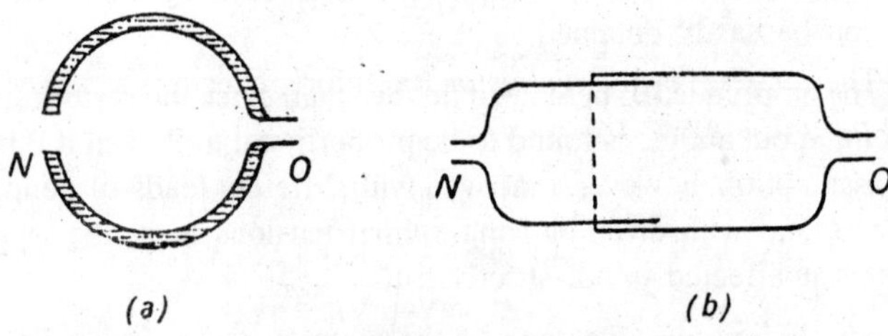


**Fig. 1.15**

If, therefore, this air plug be forced inwards through a small distance x, the decrease in the volume of air inside the vessel is, say, $-\delta V = -\alpha x$ (the – ve sign indicating a decrease), resulting in a slight increase in pressure $\delta p$ on the air inside the vessel. So that, if K be the *volume elasticity of the air and* V, the total volume of the vessel (or the initial volume of the air in it), we have

$$K = -\frac{\delta p}{\delta V/V} = -\delta p \frac{V}{\delta V},$$

whence, $$\delta p = -K\frac{\delta V}{V}.$$

∴ *force acting on the air plug outwards*

$$= \delta p \alpha = -K\frac{\delta V}{V}\alpha = -\frac{K\alpha^2 x}{V}$$

If $\frac{d^2x}{dt^2}$ be the acceleration of the air plug, the force acting on it is also

$$(l\alpha\rho)\frac{d^2x}{dt^2}.$$

We, therefore, have

$$l\alpha\rho\frac{d^2x}{dt^2} = -\frac{K\alpha^2 x}{V},$$

whence, $$\frac{d^2x}{dt^2} = -\frac{K\alpha^2 x}{V} \times \frac{1}{l\alpha\rho} = -\frac{K\alpha}{Vl\rho}x = -\mu x,$$

where $\frac{K\alpha}{Vl\rho} = \mu$ = acceleration per unit displacement.

Thus, $\frac{d^2x}{dt^2} \propto x$ and is directed oppositely to it.

The *air plug* or the *air piston*, therefore, executes a S.H.M. and its *time-period* is given by

$$T = 2\pi\sqrt{\frac{1}{\mu}} = 2\pi\sqrt{\frac{1}{\frac{K\alpha}{V/\rho}}} = 2\pi\sqrt{\frac{V/\rho}{K\alpha}}$$

Now, the velocity of sound in air is given by the relation

$$v = \sqrt{\frac{K}{\rho}}.$$

So that, substituting v for

$$\sqrt{\frac{K}{\rho}}$$

in the expression for T, we have

$$T = \frac{2\pi}{v}\sqrt{\frac{Vl}{\alpha}}$$

and *frequency of oscillation*,

$$n = \frac{1}{T} = \frac{v}{2\pi}\sqrt{\frac{\alpha}{Vl}}.$$

This, *the frequency of the resonator depends upon the total volume of the vessel and the length and area of cross section of its neck.* By suitably adjusting these, therefore, resonators of any desired frequencies may be constructed.

## Inductance-Capacitance or L.C. Circuit

Just as we have harmonic oscillators in mechanical systems performing S.H.M., so also we come across harmonic oscillators in electrical systems where charge, current or voltage executes S.H.M. under suitable conditions. As an example, we shall take up here an inductance-capacitance circuit, written for short as an L.C. circuit.

Let a capacitor of capacitance C, an *inductance coil* of inductance L, (with negligible resistance) be connected up with a battery through a *Morse key*, as shown in Fig. 1.16.

On pressing the knob K of the Morse key to contact stud a, the capacitor gets directly connected to the battery and thus gets charged. On releasing the knob, it gets disconnected from the battery but gets connected to the inductance coil, as shown, through which therefore it discharges itself.

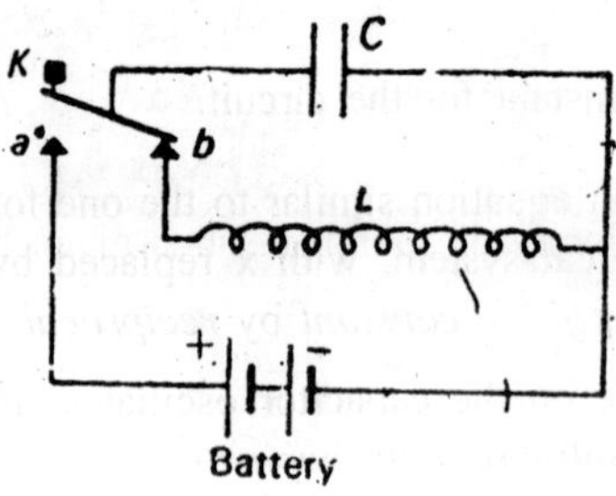


**Fig. 1.16**

Now, as we know, the inductance in an electric circuit plays the same part as mass and moment of inertia do in translatory and rotatory motion respectively and thus *opposes both growth and decay of current in the circuit*. So that, as the capacitor discharges itself through the inductance coil and the current in the latter grows, the increasing magnetic flux due to it gives rise to an induced emf in the circuit opposing the growth of the current in it. Thus, if I be the instantaneous value of the current in the coil (or the circuit) at any given instant, the opposing emf set up across the coil is – LdI/dt, where dI/dt is the rate of change of current through the coil or the circuit.

And, if Q be the charge on the capacitor at the instant considered, the voltage across it tending to drive the current through the coil (or the circuit) is Q/C. Since the external emf is now zero (the battery being cut off), we have net emf in the

$$\textit{net emf in the circuit} = Q/C + LdI/dt = 0 \qquad ...(i)$$

A current being just the rate of flow of charge, we have

$$I = \frac{dQ}{dt}.$$

The above relation thus takes the form

$$\frac{Q}{C} + \frac{Ld^2Q}{dt^2} = 0$$

or, $$\frac{Q}{LC} + \frac{d^2Q}{dt^2} = 0$$

or, $$\frac{d^2Q}{dt^2} = -\left(\frac{1}{LC}\right)Q = -\mu Q, \qquad \text{...(ii)}$$

where $\frac{1}{LC} = \mu$, a constant for the circuit.

This is clearly an equation similar to the one for simple harmonic motion of a mechanical system, with x replaced by Q, mass (m) by *inductance* L and the *force constant* by *reciprocal capacitance* 1/C.

Thus, the charge on the capacitor oscillates simple harmonically with time, *i.e.*, *the discharge of the capacitor is oscillatory in character*, its *time-period* being

$$T = 2\pi\sqrt{\frac{1}{\mu}} = 2\pi\sqrt{LC}.$$

And, therefore, its *frequency*

$$n = \frac{1}{T} = \frac{1}{2\pi}\sqrt{LC}.$$

As will be readily seen, the solution of relation (ii) above is

$$Q = Q_0\sin(\omega t + \phi), \qquad \text{...(iii)}$$

where $Q_0$ is the *maximum value* or the *amplitude* of the charge, $\omega = 1/\sqrt{LC}$, the *angular frequency* (of the variation of charge) and $\phi$, the *phase constant* which depends, as usual, on the initial conditions.

The charge in the circuit thus oscillates between $+Q_0$ and $-Q_0$, with a frequency $n = 1/2\pi\sqrt{LC}$.

Differentiating equation (iii) with respect to t, we have *instantaneous value of the current*,

$$I = \frac{dQ}{dt} = Q_0\omega\cos(\omega t + \phi),$$

where the *maximum value or the amplitude of the current*

$$= Q_0\omega, \text{ } i.e., \text{ when } \cos(\omega t + \phi) = 1.$$

Denoting this by $I_0$, we have

$$I = I_0\cos(\omega t + \phi),$$

showing that the current in The circuit too is oscillatory in character and has the same frequency as the charge, *viz.*, $n = 1/2\pi\sqrt{LC}$.

A system of two bodies connected by a spring so that both are free to oscillate simple harmonically along the length of the spring constitutes a *two-body harmonic oscillator* or a *coupled oscillator.*

A mass m, attached to the free end of a spring whose other end is fixed to a rigid support like a wall, constitutes a harmonic oscillator. This too is in fact a two-body oscillator, only one of the bodies, *viz.*, the wall, is rigidly connected to the earth and has thus effectively an infinite mass and is, therefore, immovable. Since the end of the spring connected to the wall does not move, the change in the length of the spring is given by the displacement of mass in itself. In short, the extension of the spring is determined by the motion of mass m alone, the infinite mass at its other end remaining fixed in its position.

In general, however, we seldom come across such infinite and immovable masses and have, therefore, to consider the motion of both the masses connected to the two ends of the spring. Among examples of such two-body oscillators may be cited the diatomic molecules like those of hydrogen, carbon monoxide, HCl etc. which can oscillate along their respective axes of symmetry. Although, of course, the two atoms in their molecules are coupled by electromagnetic forces, we may, for all practical purposes, imagine them to be connected together by means of tiny, massless springs.

Let us, however, consider the general case of a two-body oscillator consisting of two masses $m_1$ and $m_2$ (Fig. 1.17), connected by a horizontal massless spring of force constant C, so as to be free to oscillate along the length of the spring on a frictionless horizontal surface.

Let the normal length of the spring be $l$ and let, at any given instant, the coordinates of the two ends of the spring be $x_1$ and $x_2$, as shown. Then, clearly,

*extension of the spring*, $x = (x_1 - x_2) - l$,

where x is *positive* if the spring is *stretched, zero,* if the spring has its *normal length* and *negative* if it be *compressed.* Here, we assume it to be positive. The forces (F) exerted by the spring on the two masses are obviously equal in magnitude but opposite in sign, as indicated in the figure, the magnitude of each being Cx.

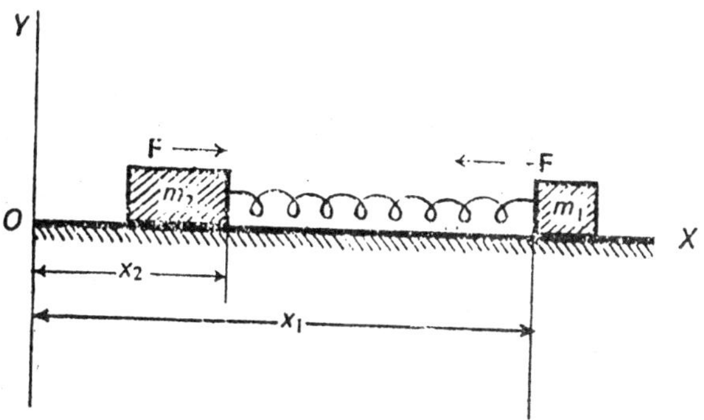


Fig. 1.17

If, therefore, $\frac{d^2x_1}{dt^2}$ be the acceleration of mass $m_1$ and $\frac{d^2x_2}{dt^2}$, that of mass $m_2$, we have

$$\frac{m_1 d^2 x_1}{dt^2} = -Cx \qquad ...(i)$$

and

$$\frac{m_2 d^2 x_2}{dt^2} = Cx. \qquad ...(ii)$$

Multiplying relation (i) by $m_2$ and relation (ii) by $m_1$ and subtracting the latter from the former, we have

$$m_1 m_2 \frac{d^2 x_1}{dt^2} - m_1 m_2 \frac{d^2 x_2}{dt^2} = -m_2 Cx - m_1 Cx.$$

Or,

$$m_1 m_2 \frac{d^2}{dt^2}(x_1 - x_2) = -Cx(m_1 + m_2).$$

Or,

$$\left(\frac{m_1 m_2}{m_1 + m_2}\right)\frac{d^2}{dt^2} = -Cx. \qquad ...(iii)$$

Now, $l$ being a constant, $\frac{d^2(x_1 - x_2)}{dt^2} = \frac{d^2x}{dt^2}$. So that

$$\left(\frac{m_1 m_2}{m_1 + m_2}\right)\frac{d^2x}{dt^2} = -Cx.$$

Or, putting $\frac{m_1 m_2}{(m_1 + m_2)} = \mu$, called the reduced mass of the system (being smaller than either of the two masses $m_1$ and $m_2$), we have

$$\frac{d^2x}{dt^2} = -\frac{C}{\mu}x.$$

This is identical in form with the equation of motion obtained for a single-body oscillator under (3), with the difference that μ here is the *reduced mass of the system* instead of mass m of the single-body there and x here is the *relative displacement of the two masses from their equilibrium positions* instead of the displacement of mass m alone from its equilibrium position there.

The system thus oscillates simple harmonically with a time-period

$$T = 2\pi\sqrt{\frac{1}{C/\mu}} = 2\pi\sqrt{\frac{\mu}{C}}$$

and *frequency* $n = \frac{1}{T} = \frac{1}{2\pi}\sqrt{\frac{C}{\mu}}$

This means that *the two-body system oscillates along the axis of the spring with the same time-period and frequency as a one-body oscillator of mass μ and force constant C.*

Each of the two masses oscillates relatively to the other as though the latter were fixed and its own mass were reduced to μ but, since their directions of motion are opposite to each other (both moving inwards during compression and outwards during extension of the spring), they differ in phase by π.

Also, since no external force acts on the system, its centre of mass remains stationary and the amplitudes of masses $m_1$ and $m_2$, therefore, are respectively $m_2/(m_1 + m_2)$ and $m_1/(m_1 + m_2)$ times the extension x of the spring.

Further, as in the case of a singe-body oscillator, so also here, the P.E. of *the system* is given by $U = \frac{1}{2}Cx^2$ and hence the P.E. curve here too is *parabolic* in form. Here, x, however, depends upon the relative positions of the two masses and is equal to $(r - r_0)$ where $r_0$ is the distance between the two masses when they are at rest and r, the distance between them when the spring has been extended by a length x. So that,

$$U = \frac{1}{2}C(r - r_0)^2.$$

Obviously, therefore, the *curve between* U and r (*i.e.*, the P.E. curve) is parabolic in form as in the case of a single-body system. *The potential energy here, however, is a characteristic of the system as a whole and not of the individual masses constituting it.*

## OSCILLATION OF A DIATOMIC MOLECULE

As mentioned earlier, under above, a diatomic molecule behaves as a two-body oscillator, with its two atoms connected by a small, massless spring, as it were.

The potential energy U of the system or the molecule changes with distance r between the atoms or rather between their nuclei. Thus, putting $r - r_0 = x$, where $r_0$ is the distance between the atoms when they are in their normal positions, or stationary, (and hence a constant), we have

*force acting on the atoms, i.e.,*

$$F = -\frac{dU}{dr} = -\frac{dU}{dx},$$

whence, $dU = -Fdx.$

$$\therefore \quad U = -\int Fdx = \int Cxdx = \frac{1}{2}Cx^2 + A,$$

where A is a constant of integration.

Now, assuming the P.E. to be *zero* at infinite separation of the atoms, it will be negative for ordinary finite distances (as measured on the atomic scale) and will have its *minimum value*, $-U_0$, say, at $x = 0$, *i.e.*, when $r = r_0$ or the atoms are in their normal, stationary state.

We, therefore, have $-U_0 = 0 + A$, Or, $A = -U_0$.

Substituting the value of A in the expression for U above, we have

$$U = -U_0 + \frac{1}{2}Cx^2 = -U_0 + \frac{1}{2}C(r - r_0)^2,$$

indicating, that U r or the *potential energy curve* for the diatomic molecule should also be parabolic in form like that of a two-body or a one-body harmonic, oscillator.

The actual P.E. curve of a diatomic molecule, however, comes out to be of the form shown in Fig. 1.18, which departs somewhat from the true parabolic form (shown dotted) near the top of the potential well, where it more or less spreads out. This departure from the true parabolic form is due to two reasons:

(i) *a rather large amplitude of the oscillations,* and

(ii) *the rotation of the molecule about its centre of mass.* Nevertheless, the bottom of the potential well is almost truly parabolic in form, indicating that for amplitudes about the equilibrium position ($r_0$), the system oscillates like a two-body harmonic oscillator, with its natural frequency

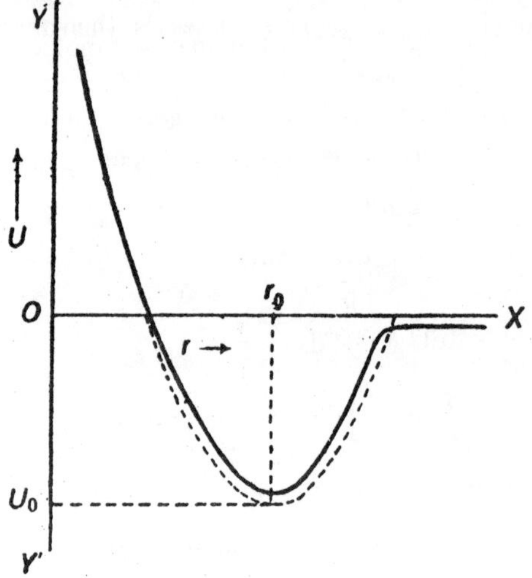


**Fig. 1.18**

$$\nu_0 = \frac{1}{2\pi}\sqrt{\frac{C}{\mu}},$$

where $\mu$ is its *reduced mass.*

## Energy Level Diagram of a Diatomic Molecule

Classical mechanics, as we know, puts no limit to the energy of vibration possessed by a harmonic oscillator which must go on increasing continuously with its amplitude. But quantum mechanics, with its principle of quantisation of energy, asserts—and this assertion is borne out by experiment—that the oscillator can possess only certain discrete values

energy level, it parts with energy which too, naturally, is an integral multiple of $hv_0$. This energy is usually given out in the form of a *photon* or electromagnetic radiations, which are a direct consequence of the oscillations of electric charges in the molecule. Such radiations, however, are given out by what are called *polar molecules* (*i.e.*, molecules in which there are free electric poles or charges that can oscillate) like those of HCl, for example, but not by symmetrical molecules like those of hydrogen (because of the absence of free electrical poles or charges). Here too, there are permissible and forbidden transitions, the permissible ones that occur, most frequently are those in which the value of n changes by 1. So that, if v be the frequency of the radiation emitted, we have *energy given out*, *i.e.*,

$$hv = E_{Vn} - E_{V(n-1)} = \left(n + \frac{1}{2}\right)hv_0 - \left(n - 1 + \frac{1}{2}\right)hv_0 = hv_0,$$

*i.e.*, the frequency v of the radiation emitted is equal to the classical frequency

$$v_0 = \frac{1}{2\pi}\sqrt{\frac{C}{\mu}}.$$

And, therefore, the corresponding *wave number* v (*i.e.*, (number of waves contained in a length of 1 cm)

$$= \frac{1}{\lambda} = \frac{v}{c} = \frac{v_0}{c}$$

$$= \frac{1}{2\mu c}\sqrt{\frac{C}{\mu}},$$

where c is the velocity of light in free space (equal to $3 \times 10^{10}$ cm/sec).

Some of the *forbidden transitions*, *i.e.*, those corresponding to the changes of more than 1 in the value of n, also occur quite often, giving electromagnetic radiations in the form of faint spectral lines.

It may also happen sometimes that the molecule oscillates and rotates simultaneously. In such cases, obviously, many more of the spectral lines are emitted.

## ANHARMONIC OSCILLATOR

We have how that P.E. of a particle at a distance x from its position of stable equilibrium is given by

$$U = U_0 + \frac{Cx^2}{2!} + \frac{C_1x^3}{3!} + \frac{C_2x^4}{4!} + \ldots \qquad \ldots(i)$$

and the force acting on the particle by

$$F = -Cx - \frac{C_1x^2}{2!} - \frac{C_2x^3}{3!} \qquad \ldots(ii)$$

It will also be recalled that in the case of a harmonic oscillator, the terms containing $C_1$, $C_2$ etc, *viz.*, $\frac{C_1x^3}{3!}$, $\frac{C_2x^4}{4!}$ etc., are all equal to *zero,* so that $U = \frac{1}{2}Cx^2$ and hence the P.E. curve is a *parabola.* And the force $F = -Cx$ and hence proportional to *displacement* of the particle or the oscillator.

In case any of the terms containing $C_1$, $C_2$ etc is not equal to zero, the motion of the particle or the system no longer remains harmonic and it is, therefore, called an *anharmonic oscillator.* And since it is the presence of the terms containing $C_1$, $C_2$ etc that really makes it so, they are referred to as *anharmonic terms.*

Let us study the anharmonic oscillator in some detail.

With the force acting on an anharmonic oscillator given by relation II above, its equation of motion is

$$m\frac{d^2x}{dt^2} = -Cx - \frac{C_1x^2}{2!} - \frac{C_2x^3}{3!} - \ldots$$

which may be put as $\frac{d^2x}{dt^2} + \frac{C}{m}x = -\frac{C_1x^2}{(2\times1)m} - \frac{C_2x^3}{(3\times2\times1)m} - \ldots$

Or, putting $\frac{C}{m} = \omega_0^2$, $\frac{C_1}{2m} = \alpha$, $\frac{C_2}{6m} = \beta$ ..., we have

$$\frac{d^2x}{dt^2} + \omega_0^2x = -\alpha x^2 - \beta x^3 - \ldots \qquad \ldots(iii)$$

There is really no general solution to this equation. Approximate solutions may however be obtained in specific cases if, as is usually the case, the anharmonic terms are small and, in particular , if the terms containing the 4th and the higher powers of x are negligible.

Now, as mentioned earlier, any complex periodic motion is really made up of a number of simple harmonic motions (called *harmonics*) whose frequencies are integral multiples of the lowest or the fundamental frequency. This means, in other words, that the *displacement x is a periodic function of* t and can, as such, be expressed as a *Fourier* series as follows:

$$x = A_0 + A_1 \sin \omega t + A_2 \sin 2\omega t + \ldots A_n \sin (n\,\omega t) + B_1 \cos \omega t + B_2 \cos 2\omega t + \ldots B_n \cos (n\,\omega t). \quad \ldots(iv)$$

As will be easily seen, many of the coefficients in this equation will be equal to *zero* in view of the sheer symmetry of the motion. Thus, for example, if we start counting time from an instant such that the motion is an even function of time, *i.e.*, when x has equal values at times t and –t and, therefore, its maximum or minimum value at t = 0, all sine terms will be absent from the equation because the general *sine term*, sin (n ωt), being an odd function of time t, its value changes sign with that of t. We, therefore, take as a possible solution the expression

$$x = A_0 + B_1 \cos \omega t + B_2 \cos 2\omega t + B_3 \cos 3\omega t + \ldots, \quad \ldots(v)$$

where the values of ω and the coefficients $A_0$, $B_1$, $B_2$, $B_3$ etc. are to be determined.

As we know, in the event of α and β etc being each equal to zero, we have (from relation (iii) above) a harmonic oscillator, with the solution $x = B_1 \cos \omega_0 t$.

Here, although α and β etc are not equal to *zero*, they are nevertheless supposed to be small, so that all coefficients other than $B_1$, *viz*., $A_0$, $B_2$, $B_3$ etc., must also be small and hence the products and squares of α, β, $A_0$, $B_2$, $B_3$ etc must be negligibly small. Keeping this in mind and taking the value of x as given by relation (IV) above, we have

$$\frac{d^2x}{dt^2} + \omega_0^2 x = -\omega^2(B_1 \cos \omega t + 4B_2 \cos \cos 2\omega t + 9B_3 \cos 3\omega t + \ldots)$$

$$+ \omega_0^2(A_0 + B_1 \cos \omega t + B_2 \cos 2\omega t + B_3 \cos 3\omega t + \ldots)$$

$$= \omega_0^2 A_0 + \left(\omega_0^2 - \omega^2\right)B_1 \cos\omega t + \left(\omega_0^2 - 4\omega^2\right)B_2 \cos 2\omega t + \left(\omega_0^2 - 9\omega^2\right)B_3 \cos 3\omega t + \ldots$$

and $\alpha x^2 - \beta x^3 = -\alpha(A_0 + B_1 \cos \omega t + B_2 \cos 2\omega t + B_3 \cos 3\omega t + \ldots)^2$

$$-\beta(A_0 + B_1 \cos \omega t + B_2 \cos 2\omega t + B_3 \cos 3\omega t + ...)^2$$

$$= -\alpha B_1^2 \cos^2 \omega t + \beta B_1^3 \cos^3 \omega t - \alpha \frac{B_1^2}{2}(1+\cos 2\omega t)$$

$$-\beta \frac{B_1^3}{4}(3\cos\omega t + \cos 3\omega t)$$

$$= -\alpha \frac{B_1^2}{2} - 3\beta \frac{B_1^3}{4}\cos\omega t - \alpha \frac{B_1^2}{2}\cos 2\omega t - \beta \frac{B_1^2}{4}\cos 3\omega t.$$

Substituting .these values in expression (iii) above, we have

$$\omega_0^2 A_0 - \left(\omega_0^2 - \omega^2\right)B_1 \cos\omega t + \left(\omega_0^2 - 4\omega^2\right)B_2 \cos 2\omega t$$

$$+\left(\omega_0^2 - 9\omega t\right)B_3 \cos 3\omega t + ...$$

$$= \frac{\alpha B_1^2}{2} - \frac{3\beta B_1^2}{4}\cos\omega t - \frac{\alpha B_1^2}{2}\cos 2\omega t - \frac{\alpha B_1^2}{4}\cos 3\omega t \qquad ...(vi)$$

Clearly, in order that relation (iii) may hold good, relation (v) must be satisfied for all values of t. This will obviously happen when the coefficients of the respective cosine functions on both sides of equation (vi) are equal, *i.e.*, when

$$\omega_0^2 A_0 = -\frac{\alpha B_1^2}{2} \qquad ...(a)$$

$$\left(\omega_0^2 - \omega^2\right)B_1 = \frac{3\beta B_1^2}{4} \qquad ...(b)$$

$$\left(\omega_0^2 - 4\omega^2\right)B_2 = -\frac{\alpha B_1^2}{2} \text{ and} \qquad ...(c)$$

$$\left(\omega_0^2 - 9\omega^2\right)B_3 = -\frac{\alpha B_1^2}{4} \qquad ...(d)$$

The terms containing the 4th and higher powers of x being assumed to be negligible, the coefficients $B_4$, $B_5$, $B_6$ etc are all equal to *zero*.

Now, relation (b) above gives

$$\omega^2 = \omega_0^2 + \frac{3}{4}\beta B_1^2.$$

Or,
$$\omega = \left(\omega_0^2 + \frac{3}{4}\beta B_1^2\right)^{1/2}$$

Or, $$\omega = \omega_0\left(1+\frac{3}{4}\frac{\beta B_1^2}{\omega_0^2}\right)^{1/2} = \omega_0\left(1+\frac{1}{2}\cdot\frac{3}{4}\frac{\beta B_1^2}{\omega_0^2}\right).$$

Or, $$\omega = \omega_0 + \frac{3}{8}\frac{\beta B_1^2}{\omega_0}. \qquad \text{...(vii)}$$

Assuming $\omega$ to be very nearly equal to $\omega_0$, we have from relation (c),

$$-3\omega_0^2 B_2 = -\frac{\alpha B_1^2}{2},$$

whence, $$B_2 = \frac{\alpha B_1^2}{6\omega_0^2}.$$

Similarly, from relation (d), we have

$$B_3 = \frac{\beta B_1^2}{32\omega_0^2},$$

and from relation (a), we have

$$A_0 = -\frac{\alpha B_1^2}{2\omega_0^2}.$$

Substituting these values of the coefficients in relation (v) above, we have

$$x = -\frac{\alpha B_1^2}{2\omega_0^2} + B_1\cos\omega t + \frac{\alpha B_1^2}{6\omega_0^2}\cos 2\omega t + \frac{\beta B_1^2}{32\omega_0^2}\cos 3\omega t + \ldots, \qquad \text{(viii)}$$

where the value of $\omega$ is as given by relation (vii) above.

Here, the term $B_1 \cos \omega t$ is referred to as the *fundamental component of oscillation* of the particle or the system and $B_1$ as its *amplitude* and the terms in cos $2\omega t$ and cos $3\omega t$ as the *second* and *third harmonics*, with frequencies twice and thrice respectively of the fundamental.

As mentioned earlier also, if $\alpha = \beta = 0$, we have the case of a *harmonic* oscillator, with $x = B_1 \cos \omega_0 t$ for then $\omega = \omega_0$ [from relation (vii) above.]

The same happens when $B_1$ is small so that all terms containing $B_2$ may be neglected. Relation (viii) then reduces again to $x = B_1 \cos \omega_0 t$. [because $\omega = \omega_0$ from relation (vii)].

This shows that *provided the amplitude be small, the oscillations even in the case of an anharmonic potential well are simple harmonic in character*, as already mentioned.

If, however, the amplitude increases, the following changes occur:

(i) As can be seen from relation (vii) above, if the amplitude ($B_1$) be large, $\omega$ will no longer be equal to $\omega_0$ and hence the *time-period and, therefore, also the frequency, of the oscillation, will change.* Thus, if T be the *time-period* of the oscillations, we shall have

$$T = \frac{2\pi}{\omega} = \frac{2\pi}{\omega_0\left(1+\frac{3\beta B_1^2}{8\omega_0^2}\right)} = \frac{2\pi}{\omega_0}\left(1+\frac{3\beta B_1^2}{8\omega_0^2}\right)^{-1}$$

$$= \frac{2\pi}{\omega_0}\left(1-\frac{3}{8}\frac{\beta B_1^2}{\omega_0^2}\right)$$

Since $\frac{2\pi}{\omega_0} = T_0$, the time-period of the small amplitude (or simple harmonic) oscillations, we have

$$T = T_0\left(1-\frac{3}{8}\frac{\beta B_1^2}{\omega_0^2}\right).$$

(ii) The amplitudes of the second, third and higher harmonics, having twice, thrice etc frequencies of the fundamental, will increase still more rapidly, being proportional to $B_1^2$, $B_1^3$ etc.

(iii) The amplitude of the particle will clearly not be equal to $B_1$, the amplitude of the fundamental component of the oscillation, nor will it move equally on either side of its equilibrium position, (see (iv) below).

(iv) Since the average values of the cosine functions (cos $\omega$t, cos 2$\omega$t etc) over a time-period $T = 2\pi/\omega$ are zero, the average value of the displacement x will become

$$\langle x \rangle = -\frac{\alpha B_1^2}{2\omega_0^2},$$ with $\alpha$, of course, not being zero.

Thus, there will be a shift in the mean position of the particle, proportional to $B_1^2$, indicating a greater restoring force acting on one side of its equilibrium position than on the other. This means, in other words, that *the potential energy curve or the potential well will no longer remain symmetrical and the particle, in consequence, will move more to one side of its equilibrium position than to the other. Its oscillation will thus no longer be simple harmonic.*

An interesting example of the shift in the mean position of a particle is provided by the oscillating atoms in a solid material. The atoms experience a considerable repulsive force when they are closer together, so that they can move apart more easily than they can come nearer to each other. When the solid is heated, therefore, the amplitude of their oscillation increases and their mean position shifts outwards, resulting in the *linear expansion* of the solid, – the greater the shift in the mean positions of the atoms, .the greater the linear expansion of the solid. Thus, *linear expansion of the solid* $\propto$ *shift in the mean position of the atoms.*

Now, as we have just seen above, this shift in the mean position is proportional to the square of the amplitude of the oscillating atom which, in its turn, is proportional to the energy of the atom. And, since the average energy of the atom depends upon the temperature of the solid, it follows that the linear expansion of a solid is proportional to its temperature.

## TIME-PERIOD OF A PENDULUM FOR LARGE AMPLITUDE OSCILLATION

A simple or a compound pendulum oscillating with a large amplitude is a familiar example of an anharmonic oscillator. The oscillations are simple harmonic only if the angular amplitude θ be infinitely small, *i.e.*, when in the expansion of sin 9 into a power series (*viz.*, $\sin\theta = \theta - \theta^3/3! + \theta^5/5! \ldots$) all other terms except the first are negligibly small. So that, the equation of motion of a simple pendulum comes out to be $d^2\theta/dt^2 = -(g/l)\,\theta$ and of a *compound pendulum*,

$$\frac{d^2\theta}{dt^2} = -\left(\frac{mgl}{I}\right)\theta = -\left[\frac{mgl}{\left(I_0 + ml^2\right)}\right]\theta = -\left[g\left(\frac{K^2}{l} + l\right)\right]\theta.$$

Or, putting $\left(\frac{k^2}{l}\right) + l = L$, the *length of the equivalent simple pendulum,* we have

$$\frac{d^2\theta}{dt^2} = -\left(\frac{g}{L}\right)\theta,$$

the two cases being thus identical except that we have L in place of $l$ in the case of a compound pendulum.

If $\theta$ be appreciably large, so that the second term $\left(\frac{\theta^3}{3!}\right)$ in the power series can not be neglected, we have (taking the case of a simple pendulum),

$$\frac{d^2\theta}{dt^2} = -\frac{g}{l}\theta + \frac{g}{(3\times2\times1)l}\theta^3$$

Now, referring back to equation (iii) of above, and taking $\theta$ in place of x we see that

$$\omega_0^2 = \frac{g}{l}, \ \alpha = 0,$$

$$\beta = -\frac{g}{6l} = -\frac{\omega_0^2}{6}.$$

$\therefore$ from relation (viii) of taking amplitude $\theta_1$ in place of $B_1$, we have

$$\theta = \theta_1 \cos\omega t + \frac{\theta_1^3\beta}{32\omega_0^2}\cos 3\omega t = \theta_1 \cos\omega t - \frac{\theta_1^3\omega_0^2}{6\times32\omega_0^2}\cos 3\omega t$$

$$= \theta_1 \cos\omega t - \frac{\theta_1^3}{192}\cos 3\omega t,$$

where (as given by relation (viii)

$$\omega = \omega_0\left(1 - \frac{3}{8}\frac{\omega_0^2}{6}\cdot\frac{\theta_1^2}{\omega_0^2}\right)$$

$$= \omega_0\left(1 - \frac{\theta_1^2}{16}\right) = \sqrt{\frac{g}{l}}\left(1 - \frac{\theta_1^2}{16}\right).$$

Hence, *time-period of the simple pendulum, i.e.,*

$$T = \frac{2\pi}{\omega} = \frac{2\pi}{\sqrt{\frac{g}{l}}\left(1 - \frac{\theta_1^2}{16}\right)} = 2\pi\sqrt{\frac{l}{g}}\left(1 + \frac{\theta_1^2}{16}\right)$$

Putting $2\pi\sqrt{\frac{l}{g}}$, *the time-period when the amplitude is small*, equal to $T_0$, we have

$$T = T_0\left(1 + \frac{\theta_1^2}{16}\right).$$

Similarly, for a compound pendulum, we shall obtain

$$T = 2\pi\sqrt{\frac{L}{g}}\left(1 + \frac{\theta_1^2}{16}\right).$$

Or, $$T = 2\pi\sqrt{\frac{\left(\frac{k^2}{l^2}\right) + l}{g}}\left(1 + \frac{\theta_1^2}{16}\right).$$

Since $2\pi\sqrt{\frac{\left(\frac{k^2}{l^2}\right) + l}{g}}$ is the time-period $T_0$ for oscillations of small amplitude, we have

$$T = T_0\left(1 + \frac{\theta_1^2}{16}\right).$$

indicating that the *time-period increases with amplitude.*

Since the amplitude of the pendulum in both cases does not remain constant but goes on progressively decreasing from $\theta_1$ in the beginning to, say, $\theta_2$, at the end, we may take $\theta_1\theta_2$ in place of $\theta_1^2$. So that, in either case,

$$T_0 = T\left(1 + \frac{\theta_1\theta_2}{16}\right) = T\left(1 - \frac{\theta_1\theta_2}{16}\right).$$

## SOLVED EXAMPLES

***Example 1:***

*Show that if the displacement of a moving point at any time is given by an equation of the form $x = a \cos \omega t + b \sin \omega t$, the motion is simple harmonic. If $a = 3$, $b = 4$ and $\omega = 2$, determine the period, amplitude, maximum velocity and maximum acceleration of the motion.*

***Solution:***

Since the *displacement of the particle or the point* is given by $x = a \cos \omega t + b \sin \omega t$, its *velocity*,

$$v = \frac{dx}{dt} = -a\,\omega \sin \omega t - b\,\omega \cos \omega t$$

and its *acceleration,*

$$a = \frac{d^2x}{dt^2} = -\omega^2(a \cos \omega t + b \sin \omega t) = -\omega^2 x.$$

The particle, therefore, executes a *simple harmonic motion* of *amplitude* $\sqrt{a^2 + b^2} = \sqrt{3^2 + 4^2} = 5\text{cm}.$

The *time-period of the particle,* $T = \frac{2\pi}{\omega} = \frac{2\pi}{2} = \pi = 3.142$ sec.,

*maximum velocity of the particle* $= \omega a = 2 \times 5 = 10$ cm/sec.

and its *maximum acceleration* $= \omega^2 a = (2)^2 \times 5 = 20$ cm/sec$^2$.

***Example 2:***

*The angular vibrational frequency of the carbon monoxide molecule (CO) is $0.6 \times 10^{15}$/sec. Calculate (i) the reduced mass of the molecule in gms, (ii) the force constant for stretching the molecule and (iii) the work done in stretching die molecule by 0.5 Angstrom unit.*

***Solution:***

(i) Clearly, *reduced mass of the CO molecule, i.e.,*

$$\mu = \frac{m_1 m_2}{m_1 + m_2} = \frac{12 \times 16}{12 + 16}$$

$$= \frac{192}{28}\text{amu} = \frac{192}{28} \times \textit{mass of hydrogen atom}$$

$$= \frac{192}{28} \times 1.67 \times 10^{-24}$$

$$= 1.145 \times 10^{-23} \text{ gm.}$$

(ii) As we know, the vibrational frequency of the molecule is given by

$$n = \left(\frac{1}{2\pi}\right)\sqrt{\frac{C}{\mu}}.$$

Or, $$2\pi n \sqrt{\frac{C}{\mu}}.$$

Now, $\sqrt{\frac{C}{\mu}} = \omega$, whence, $C = \mu\omega^2$,

where, $\mu = 1.145 \times 10^{-23}$ gm

and $\omega = 0.6 \times 10^{15}$ (given).

∴ *force constant* $C = (1.145 \times 10^{-23})\ (0.6 \times 10^{15})^2$

$= 1.145 \times 0.36 \times 10^{7}$

$= 4.122 \times 10^{6}$ dynes/cm.

(iii) Work done in stretching the molecule by $(r - r_0)$ is given by

$$\frac{1}{2}C\ (r - r_0)^2.$$

Here, $C = 4.122 \times 10^{6}$ dynes/cm and

$(r - r_0) = 0.5\ A = 0.5 \times 10^{-8}$ cm.

∴ *work done in stretching the molecule through 0.5 A*

$$= \frac{1}{2} \times 4.122 \times 10^{5} \times (0.5 \times 10^{-8})^2$$

$$- 2.061 \times 10^{6} \times 0.25 \times 10^{-26}$$

$$= 5.15 \times 10^{-11} \text{ erg.}$$

***Example 3:***

*HCl gas absorbs light of wave length λ, thereby causing the transition in the vibrational energy level from n = 0 to n = l. Calculate the force constant of the HCl molecule and the wavelength of the light absorbed if the vibrational energy levels be separated by 0.36 eV.*

***Solution:***

Here, the transition in the energy level occurring from n = 0 to n = 1, *i.e.*, with n changing by 1, the *energy absorbed* = hv = $hv_0$, where h is the *Planck's Constant* = $6.6 \times 10^{-27}$ erg-sec and v = frequency of the light absorbed.

Since $hv = 0.36\ eV = 0.36 \times 1.6 \times 10^{-12}$ ergs, we have

$v = v_0 = 0.36 \times 1.6 \times 10^{-12}/6.6 \times 10^{-27}$

$= 8.728 \times 10^{13}$/sec.

Now, $v = v_0 = \left(\frac{1}{2\pi}\right)\sqrt{\frac{C}{\mu}}$, where C is the *force constant* and μ, the *reduced mass of the molecule.*

Clearly, $\mu = \frac{1 \times 35.5}{1 + 35.5} \times 1.67 \times 10^{-24}$

$= 0.97 \times 1.67 \times 10^{-24}$ gm.

We, therefore, have

$$8.728 \times 10^{13} = \frac{1}{2\pi}\sqrt{\frac{C}{0.97 \times 1.67 \times 10^{-24}}},$$

whence, $C = (8.728 \times 10^{13})^2 \times 4\pi^2 \times 0.97 \times 1.67 \times 10^{-24}$.

Or, *force constant* $C = (8.728)^2 \times 0.97 \times 1.67 \times 4\pi^2 \times 10^2$

$= 4.87 \times 10^5$ dynes/cm.

And, *wavelength of the light absorbed,*

$$\lambda = \frac{\text{velocity of light (c)}}{\text{frequency (v)}} = \frac{3 \times 10^{10}}{8.728 \times 10^{13}}$$

$= 3.438 \times 10^{-4}$ cm $= 34380$ A.

***Example 4:***

*(a) If the transition from one energy level to an adjacent one involves an energy change of 0.5 eV in the case of hydrogen, calculate the energy change involved in a similar transition in the case of HCl*

*(b) While comparing the potential energy curves of two diatomic molecules A and B, it is observed that the potential well of the latter is narrower. What conclusions can be drawn from this? Can an estimate of the binding energies of the two molecules-be-made from these curves?*

***Solution:***

(a) The energy change involved in transition between adjacent energy levels is, as we know, hv, where $hv = hv_0 = \left(\frac{h}{2\pi}\right)\sqrt{\frac{C}{\mu}}$, with C as the *force constant of the molecule* and μ, its *reduced mass.*

Now, in the case of hydrogen, hv = 0.5 eV, and, taking the mass of an atom of hydrogen to be m amu, $\mu = m \times m/(m + m) = m/2$ amu. And, therefore,

$$0.5 = \frac{h}{2\pi}\sqrt{\frac{C}{m/2}} = \frac{h}{2\pi}\sqrt{\frac{2C}{m}}. \qquad ...(i)$$

In the case of HCl, $\mu = \frac{m \times 35.5m}{m + 35.5m} = \frac{35.5m^2}{36.5m} = 0.97$ m amu.

If, therefore, the energy change involved in transition between adjacent energy levels be E, we have

$$E = \frac{h}{2\pi}\sqrt{\frac{C}{0.97m}}. \qquad ...(ii)$$

$\therefore$ dividing relation (ii) by (i), we have

$$\frac{E}{0.5} = \sqrt{\frac{C}{0.97m} \times \frac{m}{2C}} = \sqrt{\frac{1}{2 \times 0.97}}.$$

Or, $$E = \sqrt{\frac{1}{2 \times 0.97}} \times 0.5 = 0.359 = 0.36 \text{ eV}.$$

Thus, *the energy change involved in transition between adjacent energy levels in the case of HCl is 0.36 electron volts.*

(b) A narrower potential well in the case of molecule B signifies a higher value of $d^2U/dr^2$ and, in consequence, *a larger value of the force constant* C. So that, for the same value of the reduced mass ($\mu$) in the two cases, the vibrational frequency n [equal to $(1/2\pi)\sqrt{C/\mu}$ ] will be higher in the case of molecule B.

An estimate of the binding energies of the two molecules can be made by observing the minima of the potential energies in the two cases. For, *if we ignore the zero-point energy, the numerical values of the minimum potential energy and the binding energy of a molecule are the same.*

### *Example 5:*

*The vibrational frequency of $H^2Cl^{35}$ molecule is 8990 × $10^{10}$ cycles/sec. Deduce the vibrational frequency of $H^2Cl^{35}$ molecule, assuming the force constant to be the same in the two cases.*

### *Solution:*

Since the vibrational frequency of a diatomic molecule is given by $(1/2\pi)\sqrt{C/\mu}$, it is clear that for the same value of the force constant (C), $n \propto \sqrt{1/\mu}$.

If, therefore, $\mu_1$ and $\mu_2$ be the reduced masses of the HCl molecule in the two cases, and $n_1$ and $n_2$, their vibrational frequencies respectively, we have

$$\frac{n_2}{n_1} = \sqrt{\frac{\mu_1}{\mu_2}}, \text{ whence, } n_2 \sqrt{\frac{\mu_1}{\mu_2}} n_1 \text{ [C being the same for both]}$$

Now, $\mu_1 = 1 \times \frac{35}{(1+35)} = \frac{35}{36}$,

$$\mu_2 = 2 \times \frac{35}{(2+35)} = \frac{70}{37} \text{ and}$$

$$n_1 = 8990 \times 10^{10} \text{ c.p.s. (given).}$$

$$\therefore \quad n_2 = \sqrt{\frac{35}{36} \times \frac{37}{70}} \times 8990 \times 10^{10} = \sqrt{\frac{37}{72}} \times 8990 \times 10^{10}$$

$$= 6445 \times 10^{10} \text{ c.p.s.}$$

Thus, the vibrational frequency of the $H^2Cl^{35}$ molecule

$$= 6445 \times 10^{10} \text{ cycles/sec.}$$

***Example 6:***

*Calculate the percentage change in the time-period of a simple pendulum if the angular amplitude of the pendulum be (i) 30°, (ii) 60°.*

***Solution:***

Let the length of the simple pendulum be $l$. Then, its *true time-period* (for infinitely small amplitudes) is

$$T_0 = 2\pi\sqrt{\frac{l}{g}}.$$

If its amplitude be $\theta_1$, the time-period

$$T = T_0\left(1+\frac{\theta_1^2}{16}\right).$$

∴ *percentage change in the time-period*

$$= \left(\frac{T-T_0}{T}\right) \times 100 = \frac{\theta_1^2}{16} \times 100.$$

Now, in case (i) $\theta_1 = 30° = \frac{\pi}{6}$ radian. So that, we have

*percentage change in the time-period*

$$= \frac{(\pi/6)}{16} \times 100 = \frac{25\pi^2}{144} = 1.7.$$

And, in case (ii), $\theta_1 = 60° = \frac{\pi}{3}$ radian and we, therefore, have *percentage change in the time-period*

$$= \frac{(\pi/3)^2}{16} \times 100 = \frac{25\pi^2}{36} = 6.85.$$

***Example 7:***

*An oscillating particle of mass 1 gm has potential energy equal to $5000x^2 + 200x^4$ ergs what is its frequency for small oscillations? What will be its frequency if the amplitude of its fundamental component be 3 cm? Also calculate the frequency and the amplitude of the first overtone in the latter rase.*

***Solution:***

We have here $U = 5000x^2 + 200x^4$ ergs.

$$\therefore \qquad F = -\frac{dU}{dx} = -10{,}000x - 800x^3 \text{ dynes.}$$

For small oscillations, the second term in $x^3$ may be neglected. So that,

$$F = -10{,}000x \text{ dynes.}$$

$$\therefore \textit{ acceleration } \frac{d^2x}{dt^2} = -\frac{10{,}000x}{1} = -10{,}000x = -\left(\frac{C}{m}\right)x = -\omega_0^2 x,$$

*i.e.,* $\quad \omega_0^2 = 10{,}000$ and therefore, $\omega_0 \sqrt{10{,}000} = 100$.

$$\therefore \textit{ frequency for small oscillations, } n_0 = \frac{\omega_0}{2\pi} = \frac{100}{2\pi} = 15.92/\text{sec.}$$

Now, $\omega = \omega_0 + \frac{3}{8}\beta\frac{B_1^2}{\omega_0}$ (Equation VII, 1.9).

And, here, $\beta = 800$, $B_1 = 3$cm.

So that, $\omega = 100 + \frac{3}{8} \times 800 \times \frac{(3)^2}{100} = 127.$

$\therefore$ frequency, when amplitude of fundamental component is 3 cm is

$$n = \frac{\omega}{2\pi},$$

whence, $$n = \frac{127}{2\pi} = 20.21/\text{sec}.$$

Clearly, the *first overtone* here has thrice the frequency of the fundamental component, *i.e.*, a frequency n', say,

$$= \frac{3\omega}{2\pi} = 20.21 \times 3 = 60.63/\text{sec}.$$

and *its amplitude* is

$$\frac{\beta B_1^3}{32\omega_0^2} = \frac{800 \times (3)^3}{32 \times (100)^2} = \frac{27}{400} = 0.0675 \text{ cm}.$$

***Example 8:***

*A vertical U-tube of uniform cross-section contains water to a height of 30 cm. Show that if the water on one side is depressed and then released, its motion up and down the two sides of the tube is simple harmonic, and calculate its time-period.*

***Solution:***

Let AA′ be the initial level of water in the U-tube and let the column on the *left* be depressed through distance y to B. (Fig. 1.20). Then, obviously, the column on the right will rise up through the same distance y (liquids being incompressible) to the level C, so that the *difference of level between the two columns* = B′C = 2y, where B′ is in a level with B.

The *weight* of this column of water = 2y × a × ρ × g, where a is the internal area of cross-section of the tube (or of the water column), ρ, the density of water and g, the acceleration due to gravity at the place.

This is then the force acting on the *total mass of water*, m, say, = 2haρ = 2 × 30 × a × ρ = 60 aρ gm in the two limbs of the U-tube.

$\therefore$ *acceleration of the mass of water* = *force*

*mass of water, m*, say, = 2haρ = 2 × 30 × a × ρ = negative sign indicating that it is directed *opposite to the direction of displacement.*

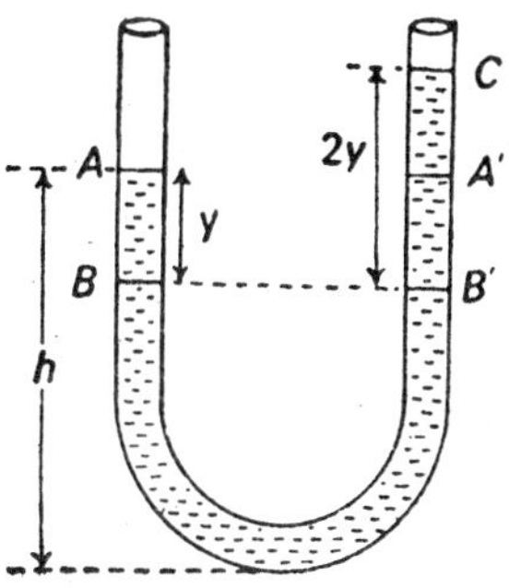


Fig. 1.20

Thus, the *acceleration of the mass of water is proportional to its displacement and is directed opposite to it.* It, therefore, executes a S.H.M. and its *time-period* is given by

$$T = 2\pi\sqrt{\frac{1}{\mu}} = 2\pi\sqrt{\frac{1}{\frac{g}{30}}} = 2\pi\sqrt{\frac{30}{g}} = 2\pi\sqrt{\frac{30}{981}} = 1.098 \text{ sec.}$$

***Example 9:***

*(a) A hyrdrogen atom has a mass of 1.68 × 10 ²⁴ gm. When attached to a certain massive molecule, it oscillates as a classical oscillator with a frequency of $10^{14}$ cycles per second and with an amplitude of 10 ⁹ cm. Calculate the force acting on the hydrogen atom.*

*(b) A test tube of weight 6 gm and of external diameter 2 cm is floated vertically in water by placing 10 gm of mercury at the bottom of the tube. The tube is depressed by a small amount and then released. Find the time of oscillation.*

***Solution:***

(a) We know that the frequency of a panicle executing S.H.M. is given by

$$n = \frac{\omega}{2\pi}. \text{ Here, } n = 10^{14}/\text{sec.}$$

So that, $10^{14}$/sec. So that, $10^{14} = \dfrac{\omega}{2\pi}$, whence, $\omega = 2\pi \times 10^{14}$.

Now, acceleration of the particle $= -\omega^2 x$, where x is its displacement from its mean or equilibrium position, the –ve sign indicating that it is directed towards that position, *i.e.* , opposite to displacement.

Here, since *displacement* = *amplitude* a, we have *acceleration of the particle, or the hydrogen atom*

$$= -\omega^2 a = -(2\pi \times 10^{14})^2 \times 10^{-9} \text{ cm/sec}^2$$

$\therefore$ *force acting on the hydrogen atom* = *mass* × *acceleration*

$$= 1.68 \times 10^{-24} \times (2\pi \times 10^{14})^2 \times 10^{-9}$$

$$= 6.63 \times 10^{-4} \text{ dynes.}$$

(b) Here *mass of the tube* + *mercury* = 6 + 10 = 16 gm, *external radius of the tube* = 2/2 = 1 cm and, therefore, its *area of cross section* $= \pi r^2 = \pi(1)^2 = \pi$ sq cm.

If the tube be depressed into water through a distance y cm, clearly, *volume of water displaced* = y × π c.c. Hence, *upthrust on the tube due to displaced water*

$$= y \times \pi \times \rho \times g = y\pi g \ (\because \rho \text{ for water} = 1 \text{ gm/c.c.}).$$

$$\therefore \textit{ Acceleration of the tube} = \frac{\text{force}}{\text{mass}} = -\frac{\pi g}{16} y = -\mu y,$$

where $$\frac{\pi g}{16} = \mu.$$

The –ve sign, as usual, indicates that the acceleration is directed oppositely to displacement.

Thus, the acceleration $\propto$ displacement and is directed oppositely to it. The tube, therefore, executes a S.H.M. and its *time-period* is given by

$$T = 2\pi\sqrt{\frac{1}{\mu}} = 2\pi\sqrt{\frac{1}{\frac{\pi g}{16}}} = 2\pi\sqrt{\frac{16}{\pi g}} = \sqrt{4\pi \times \frac{16}{g}} = 0.4527 \text{ sec.}$$

***Example 10:***

*If the earth were a homogeneous sphere of radius R and a straight hole were bored in it through its centre, show that a particle dropped into the hole will execute a simple harmonic motion and find out its time-period.*

***Solution:***

We know that the acceleration due to gravity on the surface of the earth is given by g = MG/R$^2$, where M is the *mass of the earth*, R, its *radius* and G, the *gravitational constant*,

If ρ be the density of the earth (assumed to be a homogeneous solid sphere), we have

$$M = \frac{4}{3}\pi R^3 \rho.$$

So that, $g = \frac{4}{3}\pi R^3 \rho G/R^2 = \frac{4}{3}\pi R\rho G/R^2$

At a depth r below the surface of the earth, (Fig. 1.21), similarly, the qcceleration due to gravity will be, say

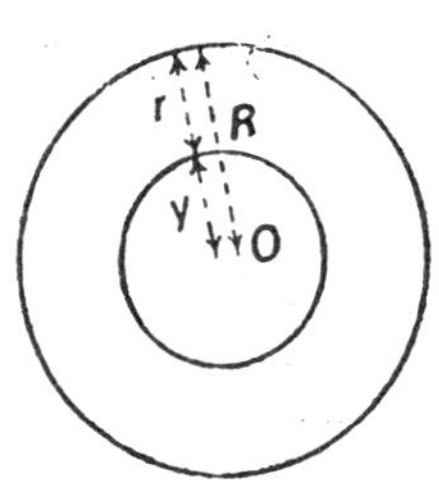

Fig. 1.21

$$g' = \frac{4}{3}\pi (R - r)\rho g.$$

$$\therefore \quad \frac{g'}{g} = \frac{(R-r)}{R}.$$

Or, $$g' = \left[\frac{(R\ \ r)}{R}\right]g = \left(\frac{g}{R}\right)(R-r)$$

If we put (R – r) = y = *distance from the centre of the earth,* we have

$$g' = -\frac{g}{R}\ y = -\ \mu y,$$

the –ve sign indicating its opposite direction to y.

Thus, the *acceleration α displacement* from the centre of the earth and is directed towards it, *i.e., opposite to displacement.*

*The particle, therefore, executes a S.H.M. about the centre of the earth.*

And, its time-period is given by

$$T = 2\pi\sqrt{\frac{1}{\mu}} = 2\pi\sqrt{\frac{1}{\frac{g}{R}}} = 2\pi\sqrt{\frac{R}{g}}.$$

***Example 11:***

*Show that the time-period for the swing of a magnet in the earth's field is given by* $t = 2\pi\sqrt{I/MH}$, *where M is the magnetic moment of the magnet, I, its moment of inertia about the axis of suspension and H, the earth's field.*

***Solution:***

Let NS be a magnet of *pole-strength* m and *magnetic moment* M, suspended at an angle θ with the earth's field H, as shown in Fig. 1.22,

Clearly, forces mH and mH acting on the two poles of the magnet, being equal, opposite and parallel, constitute a couple of moment mH × ST = mHNS sin θ = MH sin θ, tending to rotate the magnet back so as to lie parallel to the earth's field (H).

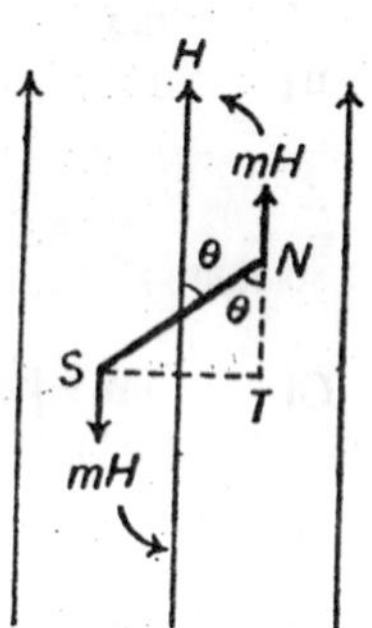


Fig. 1.22

Since θ is small, we may take sin θ = θ and, therefore,

*couple acting on the magnet* = MHθ.

If I be the *moment of inertia* of the magnet about the suspension thread and $d^2\theta/dt^2$, its angular acceleration, the couple on it is also equal to $Id^2\theta/dt^2$.

So that, we have $Id^2\theta/dt^2 = MH\theta$, whence, $d^2\theta/dt^2 = -(MH/I)\theta = -\mu\theta$, the – ve sign indicating that the acceleration is directed oppositely to the angular displacement θ from its equilibrium position parallel to the field.

The angular acceleration of the magnet is thus proportional to its angular displacement from its mean or equilibrium position and is directed oppositely to displacement. It, therefore, executes a S.H.M. and its *time-period* is given by

$$T = 2\pi\sqrt{\frac{1}{\mu}} = 2\pi\sqrt{\frac{1}{\frac{MH}{I}}} = 2\pi\sqrt{\frac{I}{MH}}.$$

***Example 12:***

*A particle is moving with S.H.M. in a straight line. When the distance of the particle from the equilibrium position has the values $x_1$ and $x_2$, the corresponding values of the velocity are $u_1$ and $u_2$. Show that the period is $2\pi[(x_2^2 - x_1^2)/(u_1^2 - u_2^2)]^{1/2}$.*

*Find also the maximum velocity and amplitude of the particle.*

***Solution:***

As we know, the velocity of a particle, executing S.H.M., at a distance x from its equilibrium position is given by $u = \omega\sqrt{a^2 - x^2}$, where $\omega$ is its *angular velocity* and a, its *amplitude*. So that, we have here

$$u_1 = \omega\sqrt{a^2 - x_1^2} \text{ and } u_2 = \omega\sqrt{a^2 - x_2^2}.$$

$$\therefore\ u_1^2 - u_2^2 = \omega^2\left(x_2^2 - x_1^2\right).$$

Or,
$$\omega = \left[\frac{\left(u_1^2 - u_2^2\right)}{\left(x_2^2 - x_1^2\right)}\right]^{1/2} \quad ...(i)$$

And, therefore, *time-period of the particle,* $T = \dfrac{2\pi}{\omega}$.

$$= 2\pi\left[\frac{\left(x_2^2 - x_1^2\right)}{\left(u_1^2 - u_2^2\right)}\right]^{1/2}$$

Now, we have, from above, $u_1^2 = \omega^2\left(a^2 - x_1^2\right)$.

Or,
$$\omega^2 = \frac{u_1^2}{\left(a^2 - x_1^2\right)} \quad ...(ii)$$

From relations (i) and (ii), therefore, we have

$$\frac{u_1^2 - u_2^2}{x_2^2 - x_1^2} = \frac{u_1^2}{\left(a^2 - x_1^2\right)}.$$

Or,
$$a^2 - x_1^2 = u_1^2\left(\frac{x_2^2 - x_1^2}{u_1^2 - u_2^2}\right)$$

$$\therefore\ a^2 = \frac{u_1^2 x_2^2 - u_1^2 x_1^2}{u_1^2 - u_2^2} + x_1^2 = \frac{u_1^2 x_2^2 - u_1^2 x_1^2 + u_1^2 x_1^2 - u_2^2 x_1^2}{u_1^2 - u_2^2}$$

whence, *amplitude of the particle,*

$$a = \left(\frac{u_1^2 x_2^2 - u_2^2 x_1^2}{u_1^2 - u_2^2}\right)^{1/2}.$$

Finally, we know that the maximum velocity of a particle executing S.H.M, is given by $v_{max} = a\omega$. So that here, *maximum velocity of the particle,*

$$v_{max} = \left(\frac{u_1^2 x_2^2 - u_2^2 x_1^2}{u_1^2 - u_2^2}\right)^{1/2} \left(\frac{u_1^2 - u_2^2}{x_2^2 - x_1^2}\right)^{1/2}$$

$$= \left(\frac{u_1^2 x_2^2 - u_2^2 x_1^2}{x_2^2 - x_1^2}\right)^{1/2}$$

***Example 13:***

*Two cities on the surface of the earth are joined by a straight, smooth underground tunnel of length 640 km. A body is released into the tunnel from one city. How much time will it take to reach the other city? Derive the formula used. $G = 6.67 \times 10^{-8}$ c.g.s. units and $\rho = 5.52$ gm/c.c. Calculate the velocity when the body would be nearest to the centre of the earth in its journey.*

***Solution.***

Taking the earth to be a solid and *homogeneous sphere of centre O, radius R* and density ρ, let A and B be the two cities on its surface, joined by a *straight, smooth* underground tunnel AB, 640 km long, as shown in Fig. 1.23. Then, a perpendicular dropped from O on to AB meets it in Q where Q is the *mid- point* of AB.

Let the body of mass m, released at city A be at a point P inside the tunnel at a given instant such that it is at a distance x from Q (the mid-point of the tunnel) and OP = r.

Then, clearly, the body at P lies on the surface of a solid sphere, concentric with the earth, of radius r and density ρ, as shown dotted in the Figure. It, therefore, experiences a gravitational force of attraction towards the centre O, given by F = (*mass of the sphere*) × m × G/r². Since the mass of the sphere = 4πr³ρ/3, we have

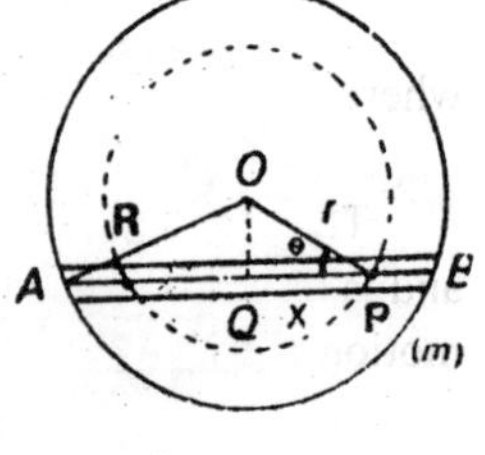


Fig. 1.23

$$F = \frac{\left(\frac{4}{3}\right)\pi r^3 \rho \times m}{r^2} \cdot G = \frac{4}{3}\pi r \rho m G \cdot$$

Multiplying and dividing by $R^3$, we have

$$F = \left(\frac{4}{3}\right)\pi r\rho mGr^3/R^3.$$

Clearly $\left(\frac{4}{3}\right)\pi R^3\rho = M$,

the *mass of the earth.* So that, *force acting on the mass,*

$$F = \frac{MmrG}{R^3}, \text{ directed along PO.}$$

The *component of this force along the tunnel*

$$= \frac{Mmr}{R^3} G \cos\theta, \text{ where } \theta = \angle OPQ.$$

Or, since $\cos\theta = \frac{x}{r}$, we have

*force acting on the mass, along the tunnel, towards*

$$Q = -\frac{MmrG}{R^3}\cdot\frac{x}{r}$$

$$= -\frac{MmG}{R^3}x,$$

the –ve sign indicating that its direction is opposite to that of displacement x.

$\therefore$ Acceleration of the body towards

$$Q = \frac{MmG}{R^3 m}x = -\frac{MG}{R^3} = -\omega^2 x = -\mu x,$$

where $\frac{MG}{R^3} = \omega^2 = \mu$, *a constant.*

The acceleration of the body is thus proportional to its displacement and is directed oppositely to it. It, therefore, executes a simple harmonic motion along AB about the point Q and its *time-period* is given by

$$T = 2\pi\sqrt{\frac{1}{\mu}} = 2\pi\sqrt{\frac{1}{\frac{MG}{R^3}}} = 2\pi\sqrt{\frac{R^3}{Mg}}.$$

Since the body is released at the end A, it will take time T to reach S and come back to A, completing one full oscillation.

$\therefore$ *time taken in reaching from A to B*

$$= \frac{T}{2} = \pi\sqrt{\frac{R^3}{MG}} = \pi\sqrt{\frac{R^3}{\frac{4}{3}\pi R^3 \rho G}}$$

$$= \sqrt{\frac{\pi^2 R^3}{\frac{4}{3}\pi R^3 \rho G}} = \sqrt{\frac{3\pi}{4\rho G}} = \sqrt{\frac{3\pi}{4\times 5.52\times 6.67\times 10^{-8}}}$$

$$= 2529 \text{ sec.}$$

Now, it is obvious that the body would be nearest to the centre of the earth (O) in its journey at the point Q, OQ being the perpendicular from O on to AB. Its velocity here will clearly be the maximum, this being its mean position or its position of equilibrium.

So that, *velocity of the body at Q (nearest to the centre of the earth)*, say, $v = \omega a$, where a is its amplitude equal to

$$OA = OB = \frac{AB}{2}.$$

Now, $\mu = \omega^2 = \dfrac{MG}{R^3}$

and therefore, $\omega = \sqrt{\dfrac{MG}{R^3}}$.

So that, *velocity of the body at Q, nearest to the centre of the earth, i.e.,*

$$v = \sqrt{\frac{MG}{R^3}}\,a = \sqrt{\frac{4\pi R^3 \rho G}{3R^3}}\,.a = \sqrt{\frac{4}{3}\pi\rho G}\times\frac{AB}{2}$$

$$= \sqrt{\frac{4}{3}\pi\times 5.52\times 6.67\times 10^{-8}}\times\frac{640}{2}\times 10^5$$

$$= 3.973 \times 10^4 \text{ cm/sec.}$$

***Example 14:***

*The total energy of a particle executing a simple harmonic motion of period $2\pi$ sec is 10240 ergs. $\pi/4$ sec after the particle passes the mid-*

*point of the swing, its displacement is* $8\sqrt{2}$ *cm. Calculate the amplitude of the motion and the mass of the particle.*

***Solution:***

We know that *total energy of a particle executing S.H.M.* = $2\pi^2$ $ma^2/T^2$, where the symbols have their usual meanings.

Here, T = $2\pi$ sec and, therefore, *total energy of the particle* = $2\pi^2 ma^2/T^2$, where the symbols have their usual meanings.

Here, T = $2\pi$ sec and, therefore, *total energy of the particle* = $2\pi^2ma^2/(2\pi^2)^2$ = 10240,

whence, $\frac{1}{2}ma^2 = 10240.$

Or, $ma^2 = 20480.$ ...(i)

Since the time here is counted from the instant the particle passes through its mean or equilibrium position, we have $\phi = 0$. Hence its *displacement* is given by the relation $x = a \sin \omega t$.

Now, $T = \frac{2\pi}{\omega} - 2\pi$ (given).

So that, $\omega = 1$ radian/sec,

$x = 8\sqrt{2}$ cm and $t = \frac{\pi}{4}$ sec.

So that, $8\sqrt{2} = a \sin\left(1 \times \frac{\pi}{4}\right) = \frac{a \sin \pi}{4} = a \times \frac{1}{\sqrt{2}},$

whence, $a = 8\sqrt{2} \times \sqrt{2} = 8 \times 2 = 16$ cm.,

*i.e., amplitude of the particle* is 16 cm.

Substituting this value of a in expression (i) above, we have

$m \times (16)^2 = 20480,$

whence, $m = 20480/16 \times 16 = 80$ gm.

Thus, the *mass of the particle* is 80 gm.

***Example 15:***

*(a) What is the frequency of a simple pendulum 2.0 metres long?*

*(b) Assuming small amplitude, what would its frequency be in an elevator accelerating upward at a rate of 2.0 metres/sec*$^2$*?*

*(c) What would its frequency be in free fall?*

***Solution:***

(a) We know that the *time-period* of a simple pendulum is given by $T = 2\pi\sqrt{\frac{l}{g}}$. Hence, its *frequency* $n = \frac{1}{T} = \frac{1}{2\pi}\sqrt{\frac{g}{l}}$.

Here, $l$ = 2.0 metres. Therefore, taking g = 9.8 m/sec$^2$, we have *frequency of the pendulum,*

$$n = \frac{1}{2\pi}\sqrt{\frac{9.8}{2}} = \frac{1}{2\pi}\sqrt{4.9} = 0.3524/\text{sec.}$$

(b) In the elevator going up with an acceleration a, the *effective weight* of the pendulum is $mg\left(1+\frac{a}{g}\right)$ = m(g + a). So that, here, the effective value of g is, say, g' = (g + a) = (9.8 + 2.0) = 11.8 m/sec$^2$.

∴ *frequency of the pendulum in the elevator,* say,

$$n' = \frac{1}{2\pi}\sqrt{\frac{g'}{l}} = \frac{1}{2\pi}\sqrt{\frac{11.8}{2}} = \frac{1}{2\pi}\sqrt{5.9} = 0.3867/\text{sec.}$$

(c) In *free fall* of the pendulum, the *effective weight* of the pendulum is $mg\left(1-\frac{g}{g}\right)$ = 0, *i.e.*, the effective value of g, *i.e.*, g' = 0.

∴ *time-period of the pendulum* = $2\pi\sqrt{\frac{l}{g'}}$ = *infinite, i.e., there is no oscillation of the pendulum at all. Its frequency (n) is, therefore, zero.*

***Example 16:***

*A simple pendulum of length l and mass m is suspended in a car that is travelling with a constant speed v around a circle of radius R. If the pendulum undergoes small oscillations about Its equiiibrium position, what will its frequency of oscillation be?*

***Solution:***

Here, in addition to g, the acceleration due to gravity, the pendulum will also be subject to the centripetal acceleration $v^2/R$ in a direction perpendicular to that of g on account of the car going round with speed v in a circle of radius R. Thus, *resultant acceleration to which the pendulum is subjected*

$$= \left[g^2 + \left(\frac{v^2}{R}\right)^2\right]^{1/2} = \left(g^2 + \frac{v^4}{R^2}\right)^{1/2}$$

On a small angular displacement θ from its mean or equilibrium position, therefore, the *restoring couple acting on the pendulum*

$$= m\left(g^2 \times \frac{v^4}{R^2}\right)^{1/2} l \sin\theta$$

$$= m\left(g^2 + \frac{v^4}{R^2}\right)^{1/2} l\theta. \qquad [\because \text{ is small}].$$

If $d^2\theta/dt^2$ be the acceleration of the pendulum and I, its moment of inertia about the suspension thread, we have the couple on the pendulum also equal to $I.d^2\theta/dt^2$.

Or equal to $ml^2d^2\theta/dt^2$, because $I = ml^2$.

We, therefore, have $ml^2d^2\theta/dt^2 = -m\left(g^2 + \frac{v^4}{R^2}\right)^{1/2} l\theta$

whence,
$$\frac{d^2\theta}{dt^2} = -\frac{\left(g^2 + \frac{v^4}{R^2}\right)^{1/2}\theta}{l} = -\mu\theta,$$

where
$$\mu = \frac{\left(g^2 + \frac{v^4}{R^2}\right)^{1/2}}{l}.$$

The pendulum thus executes a S.H.M. and, therefore, *frequency of the pendulum*

$$= n = \frac{1}{T} = \frac{1}{2\pi}\sqrt{\mu} = \frac{1}{2\pi}\sqrt{\frac{\left(g^2 + \frac{v^4}{R^2}\right)^{1/2}}{l}}.$$

***Example 17:***

*(a) Snow that the maximum tension in the string of a simple pendulum, when the amplitude $\theta_m$, is small, is mg $(1 + \theta_m^2)$. At what position of the pendulum is the tension a maximum?*

*(b) Using conservation of energy, show that the angular speed of a simple pendulum is given by* $\frac{d\theta}{dt} = \left[\frac{2}{ml^2}\{E - mgl(1-\cos\theta)\}\right]^{1/2}$ *where E is the total energy of oscillations, l and m are length and mass of the pendulum and θ is angular displacement from the vertical.*

**Solution:**

(a) Let SO be a simple pendulum of length $l$, with S, as its point of suspension and O, its mean or equilibrium position, (Fig. 1.24). If its angular amplitude be $\theta_m$ so that it is in the position SP, the whole of its energy is present in the potential form and is equal to mg × OP', where OP' is the vertical distance through which the bob rises with respect to its mean position O.

Or, since OP' = SO – SP'

$$= l - l\cos\theta_m$$

$$= l(1 - \cos\theta_m), \text{ we have}$$

*total energy of the pendulum*

$$= mgl(1 - \cos\theta_m)$$

Or, because $\theta_m$ is small,

we may take $\cos\theta_m = 1 - \frac{1}{2}\theta_m^2$.

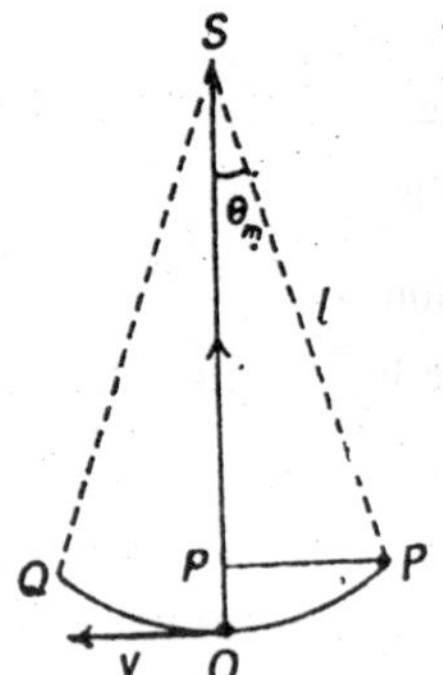


Fig. 1.24

So that, total energy of the pendulum

$$= mgl\left[1 - \left(1 - \frac{1}{2}\theta_{m2}\right)\right] = \frac{1}{2}mgl\theta_m^2$$

When the bob arrives at O, the whole of the energy of the pendulum is converted into the kinetic form, so that if v be its velocity at O tangential to the arc PO, its

$$\text{K.E.} = \frac{1}{2}mv^2.$$

Thus, $\frac{1}{2}mv^2 = \frac{1}{2}mgl\theta_m^2$,

whence, $v^2 = gl\theta_m^2$.

Or, $v = (gl\theta_m^2)^{1/2}$.

Now, the *centripetal force* necessary to keep the bob moving in an arc of radius $l = mv^2/l$ towards S and it must obviously be equal to T – mg, where T is the tension in the string along OS and mg, the weight of the bob, *i.e.*, $T - mg = mv^2/l$

Or, substituting the value of v from above, we have

$$T - mg = mgl\theta_m^{\ 2}/l = mg\theta_m^{\ 2},$$

whence, $T = mg + mg\theta_m^{\ 2} = mg\ (1 + \theta_m^{\ 2})$.

The tension is clearly the maximum when the bob passes through its mean or equilibrium position, because only in this position is the velocity of the bob, and hence the centripetal force acting upon it, the maximum, as also the entire weight of the bob effective.

(b) If the angular displacement of the pendulum bob be θ, *its potential energy in its displaced position* = $mgl\ (1 - \cos\theta)$, where, as we have just seen in part (a) above, $l(1 - \cos\theta)$ is the vertical distance through which the bob rises relative to its mean position.

Since θ is less than the amplitude of the pendulum, the bob in this position possesses also kinetic energy = $\frac{1}{2}mv^2$, where v is the velocity of the bob in the displaced position. Therefore,

*total energy of the pendulum,*

$$E = \frac{1}{2}mv^2 + mgl(1 - \cos\theta)$$

which throughout remains constant in accordance with the *law of conservation of energy.*

We thus have $\frac{1}{2}mv^2 = E - mgl\ (1 - \cos\theta)$.

Or, $$mv^2 = 2E - 2mgl\ (1 - \cos\theta).$$

Now, $v = l\omega = l d\theta/dt$, where $\omega = d\theta/dt$ is the angular velocity of the bob.

So that, $$ml^2\left(\frac{d\theta}{dt}\right)^2 = 2E - 2mgl\ (1 - \cos\theta).$$

Or, $$\left(\frac{d\theta}{4t}\right)^2 = \left(\frac{2}{ml^2}\right)[E - mgl(1 - \cos\theta)].$$

And, therefore, $$\frac{d\theta}{dt} = \left[\frac{2}{ml^2}\{E - mgl(1 - \cos\theta)\}\right]^{1/2}$$

***Example 18:***

*A. simple pendulum of length 100 cm has an energy equal to $2 \times 10^6$ ergs when its amplitude is 4 cm. Calculate its energy when (i) its length is doubled (ii) its amplitude is doubled.*

***Solution:***

We have *energy of a simple pendulums* = its max. P.E. = its max.

$$\text{K.E.} = \frac{1}{2}mv^2 = \frac{1}{2}m\omega^2a^2.$$

Now, $\omega = \frac{2\pi}{T}$, where T is the *time-period* of the pendulum

$$= 2\pi\sqrt{\frac{l}{g}}. \text{ Or, } \omega = \sqrt{\frac{g}{l}}.$$

Initially, $l$ = 100 cm.

$$\therefore \qquad \omega = \sqrt{\frac{g}{100}}.$$

So that, initially, with amplitude 4 cm, the energy of the pendulum

$$= E = \frac{1}{2}m\left(\frac{g}{100}\right)(4)^2$$

$$= 8mg/100 = 2 \times 10^6 \text{ ergs.}$$

(i) *If the length of the pendulum is doubled*, we have $l$ = 200 cm and

$$\therefore \qquad \omega = \sqrt{\frac{g}{200}}.$$

Hence, *energy of the pendulum,*

$$E' = \frac{1}{2}m\left(\frac{g}{200}\right)(4)^2 = \frac{8mg}{200}.$$

So that,
$$\frac{E'}{E} = \frac{\frac{8mg}{200}}{\frac{8mg}{100}} = \frac{1}{2}.$$

Or,
$$E' = \frac{1}{2}E.$$

Since the *initial energy of the pendulum*, E = 2 × $10^6$ ergs (given), we have *energy of the pendulum* ndw

$$= \frac{1}{2} \times 2 \times 10^6 = 10^6 \text{ ergs.}$$

(ii) *If the amplitude of the pendulum be doubled*, a = 8 cm, its length remaining 100 cm; we have energy of the pendulum,

$$E'' = \frac{1}{2} m \left(\frac{g}{100}\right) (8)^2 = 32 \text{ mg}/100.$$

$$\therefore \qquad \frac{E''}{E} = \frac{\frac{32mg}{100}}{\frac{8mg}{100}} = 4$$

Or, $\quad E'' = 4E.$

Or, $\quad E'' = 4 \times 2 \times 10^6 = 8 \times 10^6$ ergs.

Thus, *the energy of the pendulum is now* 8 × $10^6$ ergs.

***Example 17:***

*A uniform circular disc of radius R oscillates in a vertical plane about a horizontal axis. Find the distance of the axis of rotation from the centre for which the period is minimum. What is the value of this period?*

***Solution:***

The circular disc here oscillates as a compound or a physical pendulum of length *l*, whose time-period is given by

$$T = 2\pi\sqrt{\frac{\left(\frac{K^2}{l}\right) + l}{g}},$$

where k its *radius of gyration* about an axis through its *e.g.*, **parallel to** the axis of suspension.

Now, as we know, the *time-period of a compound* ***pendulum*** *is the minimum, when its length is equal to its radius of gyration* ***about*** *its e.g., i.e.*, when *l* = k.

So that, $$T_{min.} = 2\pi\sqrt{\frac{\left(\frac{k^2}{k}\right)+k}{g}} = 2\pi\sqrt{\frac{2k}{g}}.$$

Since the moment of inertia of a disc about an axis perpendicular to its plane and passing through its centre is equal to

$$I = Mk^2 = \frac{1}{2}MR^2$$

where M is the *mass* of the disc and R, its *radius*, we have

$$k^2 = \frac{R^2}{2} \text{ and}$$

$$\therefore \quad k = \frac{R}{\sqrt{2}}.$$

Thus, *the disc will oscillate with the minimum time-period when the distance of the axis of rotation from the centre is* $\frac{R}{\sqrt{2}}$.

*And the value of this minimum time-period will be*

$$T_{min} = 2\pi\sqrt{\frac{\frac{2R}{\sqrt{2}}}{g}} = 2\pi\sqrt{\frac{\sqrt{2}R}{g}} = 2\pi\sqrt{\frac{1.414\,R}{g}}.$$

***Example 20:***

*Three particles of the same mass in are fixed to a uniform circular hoop of mass M and radius a at the corners of an equilateral triangle. The hoop is free to move in a vertical plane about the point on the circumference opposite to one of the masses m. Prove that the equivalent simple pendulum is equal in length to the diameter of the hoop.*

***Solution:***

Let the three equal masses (m) be fixed to the hoop of mass M and radius a, as shown in Fig. 1.25, so as to lie at the corners of an equilateral triangle. Since all of them are equidistant from the centre, the *e.g.*, of the triangle lies at O, the centre of the hoop. *The whole arrangement is thus equivalent to a hoop of mass* (M + 3m) *and radius a, with its centre at gravity at its centre O.*

∴ M.I. of this loaded hoop about an axis passing through its centre and perpendicular to its plane = $(M + 3m)a^2$ and hence, by the principle of parallel axes, its M.I. about a parallel axis through its point of suspension

$$S = (M + 3m)\ a^2 + (M + 3m)\ a^2$$
$$= 2(M + 3m)\ a^2.$$

∴ its *time-period about S, i.e.,*

$$T = 2\pi\sqrt{\frac{I}{(M+3m)ga}},$$

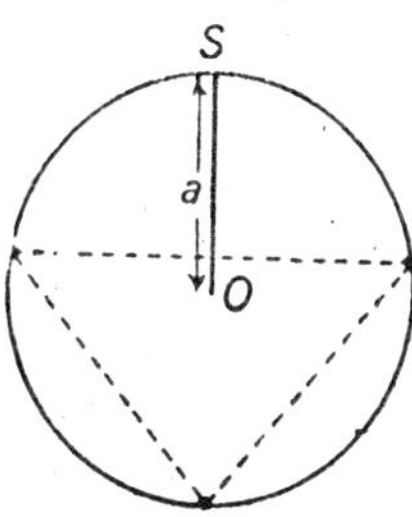


**Fig. 1.25**

because, here, *mass of the pendulum* = (M + 3m) and its length l = a. where it is shown that

$$T = 2\pi\sqrt{\frac{I}{mgl}}.$$

$$\text{Or,}\ T = 2\pi\sqrt{\frac{2(M+3m)a^2}{(M+3m)ga}} = \sqrt{\frac{2a}{g}},$$

the same as that of a simple pendulum of length 2a.

So that, *the length of the equivalent simple pendulum is equal to 2a or the diameter of the hoop.*

***Example 21:***

*A solid sphere of radius 0.3 metre executes torsional oscillations of time-period* $2\pi\sqrt{12}$ *sec at the end of a suspension wire whose upper end is fixed in a rigid support. If the torque constant of the wire be 6 ×* $10^{-3}$ *N-m/radian, calculate the mass of the sphere.*

***Solution:***

We know that the M.I. of a solid sphere about any diameter, *i.e.*, about any axis passing through its centre is 2/5 $MR^2$, where M is its mass and R, its radius.

We, therefore, have

$I = \frac{2}{5}MR^2$; *torque constant or torque per unit twist of the suspension wire, i.e.,*

$C = 6 \times 10^{-2}$ N-m/radian and $T = 2\pi\sqrt{12}$ sec.

Substituting these values in the relation $T = 2\pi\sqrt{\frac{I}{C}}$ for the time-period of a torsional oscillation, we have

$$2\pi\sqrt{12} = 2\pi\sqrt{\frac{I}{C}},$$

whence, $\sqrt{12} = \sqrt{\frac{I}{C}}$.

Or, $\frac{I}{C} = 12.$

Or, $I = 12C,$

*i.e.*, $\frac{2}{5}\,MR^2 = 12 \times 6 \times 10^{-3} = 72 \times 10^{-3}.$

Or, $MR^2 = 72 \times 10^{-3} \times \frac{5}{2}$

$= 180 \times 10^{-3} = 18 \times 10^{-2}.$

$$\therefore\ M = \frac{18\times10^{-2}}{R^2} = \frac{18\times10^{-2}}{(0.3)^2} = \frac{18\times10^{-2}}{9\times10^{-2}} = 2\,kg.$$

Thus, *the mass of the sphere is 1 kg.*

***Example 22:***

*The tune-period of a rod of mass 120 gm and length 10 cm, executing torsional vibration about a suspension wire passing through its centre and perpendicular to its length is 2 sec. The period of a flat plate of the shape of an equilateral triangle suspended from the suspension wire, passing through its centre of mass and perpendicular to its plane is found to be 4 sec. What is the moment of inertia of the triangular plate about the suspension wire as axis?*

***Solution:***

Let $I_1$ be the M.I. of the rod about the suspension wire as axis and $I_2$, that of the triangular plate about the same axis. Then, if $T_1$ and $T_2$ be the respective time-periods of the two torsional oscillations, we have

$$T_1 = 2\pi\sqrt{\frac{I_1}{C}}$$

and $T_2 = 2\pi\sqrt{\frac{I_2}{C}}$.

So that, $\frac{T_2}{T_1} = \sqrt{\frac{I_2}{I_1}}$.

Or, $$\frac{I_2}{I_1} = \frac{T_2^2}{T_1^2},$$

whence, $$I_2\left(\frac{T_2^2}{T_1^2}\right)I_1 = \left(\frac{T_2}{T_1}\right)^2 I_1.$$

Now, *M.I. of the rod about the suspension wire, i.e.*

$$I_1 = \frac{Ml^2}{12} = 120 \times (10)^2/12 = 10^3 \text{ gm-cm}^2,$$

$$T_1 = 2 \text{ sec and } T_2 = 4 \text{ sec.}$$

∴ *M.I. of the triangular plate about the suspension wire,*

$$I_2 = \left(\frac{4}{2}\right)^2 \times 10^3 = 4 \times 10^3 \text{ gm-cm}^2.$$

***Example 23:***

*The balance wheel of a watch vibrates with an angular amplitude of π radians and a period of 0.5 sec. Find (a) the maximum angular speed of the wheel, (b) the angular speed of the wheel when its displacement is π/2 radians and (c) the angular acceleration of the wheel when its displacement is π/4 radians.*

***Solution:***

(a) Just as in *linear* S.H.M., so also, here in torsional vibration, we have

$$\theta = \theta_0 \sin(\omega t + \phi),$$

where θ is the angular displacement at the given instant and $\theta_0$, the *maximum angular displacement or amplitude.*

∴ angular speed, $\frac{d\theta}{dt} = \theta_0\omega \cos(\omega t + \phi)$ the maximum value of which will clearly be $\theta_0$, *i.e.*, when $\cos(\omega t + \phi) = 1$.

Here, $\theta_0 = \pi$ radians

and $\omega = 2\pi n = 2\pi \times \frac{1}{T} = 2\pi \times \frac{1}{0.5} = 4\pi$.

∴ *maximum angular speed of the balance wheel* $= \theta_0\omega = \pi \times 4\pi = 4\pi^2 \approx 39.48$ *radians/sec.*

(b) Again, as in linear S.H.M., so also here, *angular speed at angular displacement* $\theta = \omega\sqrt{\theta_0^2 - \theta^2}$. So that, *angular speed of the wheel at displacement π/2 radian*

$$= 4\pi\sqrt{\pi^2 - \left(\frac{\pi}{2}\right)^2} = 4\pi\sqrt{\frac{3\pi^2}{4}}$$

$$= 2\pi^2\sqrt{3} = 34.18 \text{ radians/sec.}$$

(c) Again, acceleration of the balance wheel at displacement θ is given by

$$\frac{d^2\theta}{dt^2} = -\omega^2\theta.$$

Therefore, *Acceleration of the balance wheel at displacement* $\theta = \frac{\pi}{4}$ *radian* is clearly equal to $-(4\pi)^2\left(\frac{\pi}{4}\right) = -4\pi^3 = -124$ radians/sec².

***Example 24:***

*The scale of a spring balance reading from 0 to 32 lb is 4.0 in long. A package suspended from the balance is found to oscillate vertically with a frequency of 2.0 oscillations per second. How much does the package weigh?*

***Solution:***

Here, clearly, a mass of 32 *l*b suspended from the spring balance extends the spring by 4 inches or $\frac{4}{12} = \frac{1}{3}$ ft.

∴ *force constant of the spring, i.e.,*

$$C = \frac{\textbf{force applied}}{\textbf{extension produced}} = \frac{32 \times 32}{1/3} = 32 \times 32 \times 3 \text{ poundals/ft.}$$

Now, *time-period of oscillation of the spring,* $T = 2\pi\sqrt{\frac{m}{C}}$.

$$\therefore\ T^2 = \frac{4\pi^2 m}{C}, \text{ whence, } m = \frac{CT^2}{4\pi^2}.$$

Since C = 32 × 32 × 3 *poundals/ft* and $T = \frac{1}{2}$ sec, we have

$$m = \frac{32 \times 32 \times 3 \times 1/4}{4\pi^2} = \frac{192}{\pi^2} = 19.45\ lb.$$

*The package thus weighs* 19.45 *l*b.

### *Example 25:*

*Two springs $S_1$ and $S_2$ are connected to a mass M, as shown in Figs. 1.26 (a) and (b), with the mass lying on a frictionless surface. If the respective force constants* of the spring be $C_1$ and $C_2$, obtain expressions for the time-periods of the mass in the two cases. What are their electrical analogues?

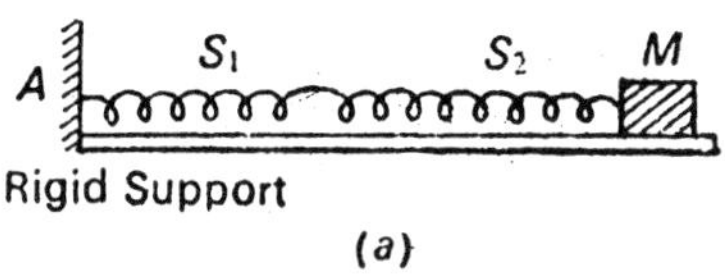


(a)

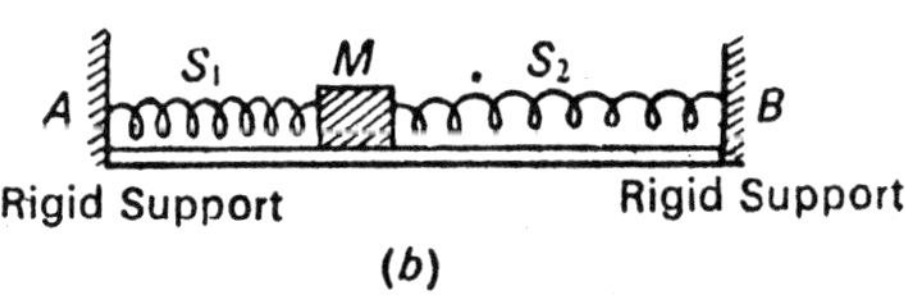


(b)

**Fig. 1.26**

### *Solution:*

(a) If $x_1$ and $x_2$ be the extensions produced in the two springs respectively as the mass (M) is displaced outwards, the *total extension* $x = x_1 + x_2$.

Since the restoring force due to each spring is the same, F, say, we have

$$F = -C_1x_1 = -C_2x_2;$$

so that, $x_1 = -\dfrac{F}{C_1}$ and $x_2 = -\dfrac{F}{C_2}$.

$$\therefore \qquad x = x_1 + x_2 = -F\left(\frac{1}{C_1} + \frac{1}{C_2}\right) = -F\left(\frac{C_1 + C_2}{C_1 C_2}\right),$$

whence, $F = -\left(\dfrac{C_1 C_2}{C_1 + C_2}\right)x,$

indicating that the effective force constant of the combination of the two springs, say,

$$C = \left(\frac{C_1 C_2}{C_1 + C_2}\right).$$

∴ *time-period of the oscillating mass,*

$$T = 2\pi\sqrt{\frac{M}{C}} = 2\pi\sqrt{\frac{M}{\left(\frac{C_1 C_2}{(C_1 + C_2)}\right)}}$$

$$= 2\pi\sqrt{\frac{(C_1 + C_2)M}{C_1 C_2}}.$$

(b) In this ease, if the mass (M) be; displaced to one side or the other, through a distance x, say, one spring gets extended and the other compressed, both exerting a restoring force on the mass *in the same direction*, tending to bring it back to its original position. Thus, if $F_1$ and $F_2$ be the restoring forces due to the two springs respectively, we have *resulting restoring force* $F = F_1 + F_2 = -C_1x - C_2x = -(C_1 + C_2)x = Cx$, indicating that the *effective force constant* C of the combination of the two springs is now equal to $(C_1 + C_2)$, the sum of their individual force constants.

*The time-period of the oscillating mass* is, therefore,

$$T = 2\pi\sqrt{\frac{M}{C}}$$

$$= 2\pi\sqrt{\frac{M}{(C_1 + C_2)}}.$$

*The electrical* ***analogues*** *of the two systems are a parallel and a series arrangement* ***of two capacitors*** *respectively.*

***Example 26:***

**A uniform** spring of normal length $l$ has a force constant C. It is cut into two pieces of lengths $l_1$ and $l_2$ such that $l_1 = nl_2$ where an is an integer. What are the force constants $C_1$ and $C_2$ of the two pieces respectively in terms of n and C?

***Solution:***

Here, clearly, $l = l_1 + l_2$.

Or, since $l_1 = nl_2$ and, therefore, $l_2 = l_1/n$, we have

$$l = l_1 + l_1/n = l_1(1 + l/n) = l_1(n + l)/n \quad \text{....(i)}$$

Also, $\quad l = nl_2 + l_2 = l_2(n + l) \quad$ ...(ii)

Now, for a given spring, its force constant is inversely proportional to its length, *i.e.*, $C \propto 1/l$ or $Cl$ = a constant. We, therefore, have

$$C_1 l_1 = Cl = Cl_1\ (n + 1)/n,$$

whence, $\quad C_1 = C(n + 1)/n,$

Similarly, $\quad C_2 l_2 = Cl = Cl_2(n + 1),$

whence, $\quad C_2 = C(n + 1).$

Thus, the force constants of the two pieces of the spring are respectively $C_1 = C\ (n + 1)/n$ and $C_2 = C(n + 1)$.

***Example 27:***

*Show that it a given body oscillates at the end of a spring whose mass is not negligible, its time-period is the same as though its mass were increased by one-third of that of the spring.*

***Solution:***

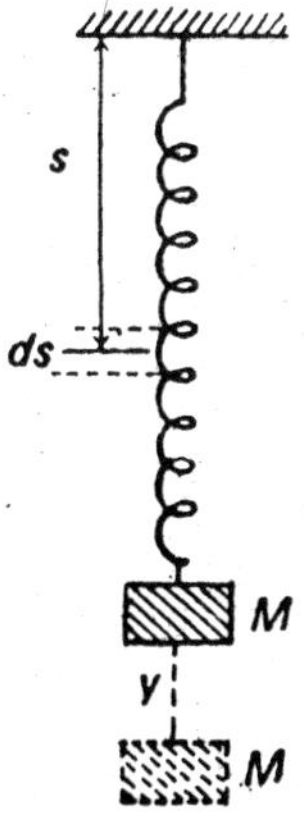

Let a uniform spring of length $l$ and mass m be suspended from a rigid support and carry a mass M at its lower free end, (Fig. 1.27). Then, clearly, *mass per unit length of the spring* = m/$l$, and therefore *mass of an element of length ds of it at a distances from the upper fixed end* = (m/$l$) ds.

If at a given instant, the velocity of the lower end of the spring (and, therefore, also of mass M) be v, the velocity of the element ds at a distance s from the fixed end = (v/$l$)s.

$\therefore$ *KE. of the element*

$$= \frac{1}{2}\left(\frac{m}{l}ds\right)\left(\frac{vs}{l}\right)^2 = \frac{1}{2}\frac{mv^2}{l^3}s^2 ds.$$

Fig. 1.27

Hence *K.E. of the whole spring at the given instant*

$$= \int_0^l \frac{1}{2} \frac{mv^2}{l^3} s^2 ds = \frac{1}{2} \frac{mv^2}{l^3} \cdot \frac{l^3}{3} = \frac{mv^2}{6}.$$

And *K.E. of the suspended mass (M)* $= \frac{1}{2} Mv^2$.

$\therefore$ *KE. of the system, as a whole* $= \frac{1}{2} Mv^2 + \frac{mv^2}{6}$

$$= \frac{1}{2}\left(M + \frac{m}{3}\right)v^2.$$

If at the instant considered, y be the displacement of mass M from its mean or equilibrium position, we have, substituting dy/dt for v,

*K.E. of the system, as a whole* $= \frac{1}{2}\left(M + \frac{m}{3}\right)\left(\frac{dy}{dt}\right)^2$. ...(i)

If C be the *force constant* of the spring, we have *restoring force set up in the spring* = – Cy and, therefore,

*P.E. of the system* $= \int_0^y Cy\,dy = \frac{1}{2} Cy^2$ ...(ii)

$\therefore$ *total energy of the system,*

$$E = K.E. + P.E. = \frac{1}{2}\left(M + \frac{m}{3}\right)\left(\frac{dy}{dt}\right)^2 + \frac{1}{2} Cy^2.$$

Since the total energy of the system must remain conserved, we have

$$\frac{dE}{dy} = 0,$$

*i.e.* $$\frac{d}{dy}\left[\frac{1}{2}\left(M + \frac{m}{3}\right)\left(\frac{dy}{dt}\right)^2 + \frac{1}{2} Cy^2\right] = 0,$$

*i.e.,* $$\left(M + \frac{m}{3}\right)\frac{d^2y}{dt^2} + Cy = 0.$$

Or, $$\frac{d^2y}{dt^2} = -\frac{C}{\left(M + \frac{m}{3}\right)} y,$$

indicating that *the spring executes a S.H.M. of time-period*

$$T = 2\pi\sqrt{\frac{\dfrac{1}{C}}{\dfrac{1}{M+\dfrac{m}{3}}}} = 2\pi\sqrt{\frac{M+m/3}{C}}$$

which is clearly the same as though the value of the suspended mass has been increased by one-third of the mass of the spring.

***Example 28:***

*An elastic cord of length l is suspended from a rigid support. A mass m, attached to its lower end, stretches it by a length a. An additional mass m', attached to it, stretches it farther by a length b. When the system has come to rest, mass m' gets detached. Show that the distance of mass m from the upper end of the cord is, at any instant* $l + a + b\cos\sqrt{g/a}\,t$.

***Solution:***

In Fig. 1.28, let SA be the elastic cord of length $l$, suspended from a rigid support at S. When a mass m is attached to its lower end, let it extend by a length AB = a and, on attaching another mass m' to it, let it further extend by a length BC = b.

When the system has come to rest at the point C, mass m' is suddenly detached, so that mass m starts oscillating up and down about its mean position B, the equation of motion of the mass being $d^2y/dt^2 = -(C/m)y$, where C is the force constant of the cord.

Clearly, before the mass m was attached to it, the position of the lower end of the cord was at A. The addition of mass m, *i.e.*, the application of a force mg to it, brought it down to B or produced in it an extension AB = a. So that, mg = Ca, whence, C = mg/a. And, therefore, $d^2y/dt^2 = -(g/a)y$, *i.e.*, mass m executes a S.H.M. about B as its mean or equilibrium position, its displacement at any instant t being given by $y = a\sin(\omega t + \phi)$, where a is the *amplitude* of the oscillation

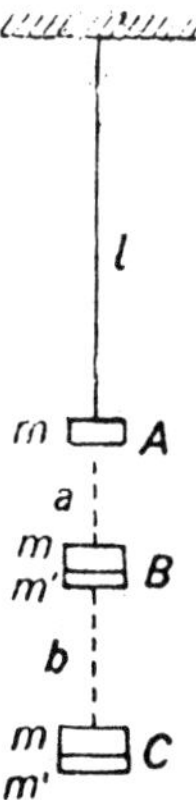


Fig. 1.28

and $\omega^2 = \dfrac{g}{a}$ or $\omega = \sqrt{\dfrac{g}{a}}$.

∴ *velocity of the mass at the given instant*

$$= \frac{dy}{dt} = \omega\sqrt{a^2 - y^2}.$$

At t = 0, *i.e.*, at the instant that the mass *m' just* gels detached, y = b and dy/dt = 0. So that,

$$0 = \omega\sqrt{a^2 - b^2}.$$

Or, $a^2 - b^2 = 0.$

Or, a = b.

Substituting these values in the expression for y above, we have

$$b = b \sin \phi, \text{ whence, } \sin \phi = 1 \text{ or } \phi = \frac{\pi}{2}.$$

We therefore have $y = b\sin\left(\omega t + \frac{\pi}{2}\right) = b \cos\omega t = b\cos\sqrt{\frac{g}{a}}t.$

∴ *distance of mass m from tile point of suspension (or the upper m' end of the cord) at instant*

$$t + l + a + y = l + a + \cos\sqrt{\frac{g}{a}}t.$$

***Example 29:***

*A spherical Helmholtz resonator of capacity 10 litres has a narrow neck of length 1 cm and radios 1 cm. If the velocity of sound in air be 340 m/sec, calculate the frequency and the wavelength of the note to which it will sharply respond.*

***Solution:***

We know that the frequency of the note to which a spherical resonator responds is given by $n = \left(\frac{v}{2\pi}\right)\sqrt{\frac{\alpha}{Vl}}$, where v is the *velocity of sound in air, l, the length and* α, *the area of cross-section of the neck* and V, the volume (or *capacity*) of the resonator.

Here, v = 340 m/sec = 34000 cm/sec,

$\alpha = \pi\ (1)^2 = \pi$ sq cm, $l$ = 1 cm

and V = 10 litres = 10000 c.c.

∴ *frequency of the note,* $n = \frac{34000}{2\pi}\sqrt{\frac{\pi}{1000 \times 1}} = \frac{170}{\sqrt{\pi}} = 95.92/\text{sec}.$

And, *wavelength of the note,* $\lambda = \frac{v}{n} = \frac{34000}{95.92} = 345.5$ cm.

**Example 30:**

*Calculate the frequency of an L-C circuit in which the inductance and capacitance are respectively 5 millihenry and 2μF. If the maximum potential difference across the capacitor be 10 volts, what is the energy of the system?*

**Solution:**

For the frequency of the oscillations in an L-C circuit, we have the relation

$$n = \frac{1}{2\pi\sqrt{LC}},$$

where L and C are the *inductance* and *capacitance* in the circuit respectively.

Here, $L = 5$ *milli henry* $= 5 \times 10^{-3}$ henry

and $C = 2\mu F = 2 \times 10^{-6}$ farad

$\therefore$ *frequency of the L.C circuit,*

$$n = \frac{1}{2\pi\sqrt{5\times10^{-3}\times2\times10^{-6}}} = \frac{10^4}{2\pi} = 1592/\text{sec}.$$

And, with $V = 10$ volts, *energy of the system*

$$= \frac{1}{2}CV^2 = \frac{1}{2} \times 2 \times 10^{-6} \times (10)^2 = 10^{-4} \text{ joule}.$$

**Example 31:**

*(a) What is the reduced mass of each of the following diatomic molecules: $O_2$, HCl, CO? Express your answers in atomic mass units, the mass of a hydrogen atom being approximately 1.00 amu.*

*(b) An HCl molecule is known to vibrate at a fundamental frequency of $8.7 \times 10^{13}$ cycles/sec. What is the effective 'force constant' C for the coupling forces between the atoms?*

**Solution:**

(a) We know that a diatomic molecule behaves like a system of two masses connected by a small massless spring. So that, its reduced mass is given by

$$\mu = \frac{m_1 \times m_2}{m_1 + m_2},$$

where MI and mn are the masses of the two atoms respectively in amu.

Now, in the case of $O_2$, $m_1$ = 16 amu

and $m_2$ = 16 amu

$$\therefore \quad \mu = \frac{16 \times 16}{16+16} = \frac{256}{32} = 8 \text{ amu.}$$

*In the case of* HCl, $m_1$ = 1 amu

and $m_2$ = 35.5 amu

$$\therefore \quad \mu = \frac{1 \times 35.5}{1+35.5} = \frac{35.5}{36.5} = 0.97 \text{ amu.}$$

And, *in the case of* CO, $m_1$ = 12 amu and $m_2$ = 16 amu

$$\therefore \quad \mu = \frac{12 \times 16}{12+16} = \frac{192}{28} = 6.857 \text{ amu.}$$

(b) The frequency of a two-body harmonic oscillator, as we know, is given by

$$n = \frac{1}{2\pi}\sqrt{\frac{C}{\mu}},$$

where C is the '*force constant*' for the coupling forces and μ, the *reduced mass of the system.*

Here, the reduced mass μ for the HCl molecule = 0.97 amu [See above under (a)], *i.e.*, equal to 0.97 × *mass of a hydrogen atom*

$$n = 8.7 \times 10^{23} \text{ cycles/sec}$$

So that, $$8.7 \times 10^{23} = \frac{1}{2\pi}\sqrt{\frac{C}{0.97 \times 1.67 \times 10^{-24}}},$$

whence, $C = 4\pi^2 (8.7 \times 10^{13})^2 \times 0.97 \times 1.67 \times 10^{-24}$

$= 485200$ dyne/cm $= 485$ N/m.

***Example 32:***

*(a) Two discs $D_1$ and $D_2$ of masses 1 kg and 2 kg respectively are connected to the two ends of a massless spring and placed on a table, as shown in Fig. 1.29, with the spring vertical and the heavier disc ($D_2$)*

*resting on the table. Calculate the frequency of oscillation of the system, if the force constant of the spring be $10^6$ dynes/cm.*

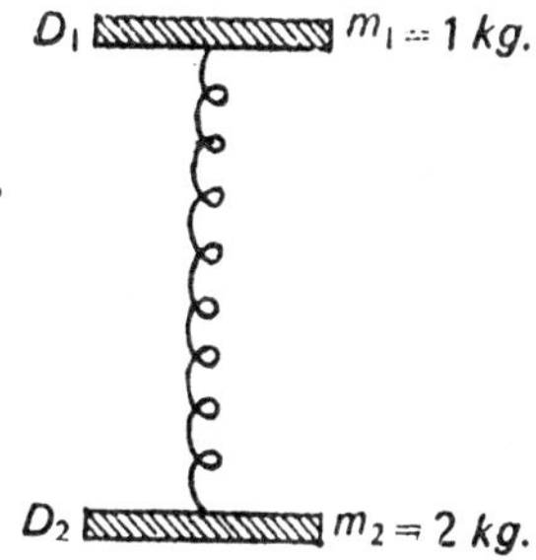


**Fig. 1.29**

*(b) After the system has come to rest, the table is suddenly removed from under disc $D_2$. Calculate :*

*(i) the frequency of the system,*

*(ii) the instantaneous acceleration of the two discs and of the centre of mass of the system.*

**Solution:**

(a) Obviously, here, the system functions as an ordinary *one-body harmonic oscillator*, with the lower end of the spring fixed and the upper one carrying disc $D_1$ of mass $m_1$ = 1kg.

If, therefore, $n_1$ be the frequency of the system, we have

$$n_1 = \frac{1}{2\pi}\sqrt{\frac{C}{m_1}} = \frac{1}{2\pi}\sqrt{\frac{10^6}{1000}} = 5.025 \text{ or } 5.0/\text{sec}.$$

(b) (i) When the table is suddenly removed from under disc $D_2$, the system oscillates as a two-body harmonic oscillator, with both the discs oscillating simultaneously.

If $n_2$ be the frequency of the system, we have

$$n_2 = \frac{1}{2\pi}\sqrt{\frac{C}{\mu}},$$

where $\mu$ is the *reduced mass of the system*, equal to

$$\frac{m_1 m_2}{m_1 + m_2} = \frac{1\times 2}{1+2} = \frac{2}{3} \text{ kg}.$$

$$\therefore \quad n_2 = \frac{1}{2\pi}\sqrt{\frac{10^6}{\left(\frac{2}{3}\right)\times 1000}} = \frac{1}{2\pi}\sqrt{\frac{3\times 10^6}{2000}} = \frac{1}{2\pi}\sqrt{\frac{3\times 10^3}{2}}$$

$= 6.165/\text{sec}.$

(ii) Clearly, with the sudden removal of the table from under disc $D_2$, the latter is subjected to an instantaneous force equal to (1 kg + 2 kg) = 3kg ωt = $3 \times 10^3 \times g$ dynes.

If its acceleration be a, the force on it is clearly

$$m_2\, a = 2 \times 10^3\, a = 3 \times 10^3 \times g,$$

whence, $\quad a = 3g/2.$

Thus, the *instantaneous acceleration of disc* $D_2 = \frac{3}{2}$ g cm/sec².

As to acceleration of disc $D_1$, we note that the system was at rest before removal of the table from under disc $D_2$. Since disc $D_1$ was at rest on top of the spring, its weight was being supported by the elastic reaction of the spring and it continues to be so supported at the instant the table is just removed. So that, *the instantaneous acceleration of disc* $D_1$ *is zero.*

The acceleration of the centre of mass of the system will naturally be the same as that of a body falling freely under the action of gravity, *viz.*, g. This may be shown as follows:

If y be the instantaneous displacement of the centre of mass (taking y-direction along the vertical) and $y_1$ and $y_2$, the corresponding displacements of the two discs of masses $m_1$ and $m_2$ respectively, we have $(m_1 + m_2)y = m_1y_1 + m_2y_2$. Differentiating with respect to t, we have

$$\frac{(m_1 + m_2)d^2y}{dt^2} = \frac{m_1 d^2y_1}{dt^2} + \frac{m_2 d^2y_2}{dt^2},$$

where, as we know, $\frac{d^2y_1}{dt^2} = 0$ and $\frac{d^2y_2}{dt^2} = \frac{3g}{2}$.

If, therefore, the acceleration of the centre of mass, *i.e.*,

$\frac{d^2y}{dt^2}$ be a′ cm/sec², we have

$$(1+2)\,a' = 0+2\times\frac{3}{2}g.$$

Or, $\quad 3a' = 3g$, whence, $a' = g$.

*i.e., the acceleration of the centre of mass is g.*

### *Example 33:*

*If a particle moves in a potential energy field $U = U_0 - ax + bx^2$, where a and b are positive constants, obtain an expression for the force acting on it as a function of position. At what point does the force vanish? Is this a point of stable equilibrium?*

*Calculate the force constant, time-period and frequency of the particle.*

### *Solution:*

(i) As we know, *force acting on the particle* is given by

$$F = -\frac{dU}{dx} = -\frac{d}{dx}\left(U_0 - ax + bx^2\right),$$

whence, $F = a - 2bx$.

(ii) The force vanishes at the point where

$$\frac{dU}{dx} = 0, \; i.e.,$$

where $\quad a - 2bx = 0$. So that, $2bx = a$

Or, $$x = \frac{a}{2b},$$

which gives the position of the point where the force vanishes.

(iii) We have $\dfrac{d^2U}{dx^2} = 2b$.

So that, b being *positive,* the point , $x = \dfrac{a}{2b}$

represents the point of *minimum potential energy* on the energy curve of the particle. *It is, therefore, a point of stable equilibrium.*

(iv) From the expression for force F in (i) above, it is clear that it is a *linear restoring force* and *the force constant C* is therefore equal to 2b.

(v) The *time-period of the particle,*

$$T = 2\pi\sqrt{\frac{1}{\mu}},$$

where, as know, $\mu = \frac{C}{m} = \frac{2b}{m}$.

So that,

$$T = 2\pi\sqrt{\frac{1}{\frac{2b}{m}}} = 2\pi\sqrt{\frac{m}{2b}}.$$

And, therefore, *frequency of the particle,*

$$n = \frac{1}{T} = \frac{1}{2\pi}\sqrt{\frac{2b}{m}}.$$

***Example 34:***

*(a) The potential energy of a harmonic oscillator of mass 2kg in its resting position is 5 joules, its total energy is 9 joules and its amplitude, 1cm. Calculate its time-period.*

*(b) A particle of mass 5gm lies in a potential field $V = (8x^2 + 200)$ ergs/gm. Calculate its time-period.*

***Solution:***

(a) Clearly, the *energy gained by the oscillator in moving from its resting position to its extreme position, i.e., in moving through a distance equal to its amplitude a, is equal to (9 – 5) = 4 joules.*

Since, *energy gained* $= \frac{1}{2}Cx^2$, where x is the *displacement* of the particle and C, the *force constant,* we have $\frac{1}{2}C(1)^2 = 4$,

whence, C = 4 × 2 = 8 joules/gm. [$\because$ x = a = 1 cm].

$\therefore$ *time period of the oscillator,*

$$T = 2\pi\sqrt{\frac{m}{C}} = 2\pi\sqrt{\frac{2}{8}} = 2\pi\sqrt{\frac{1}{4}}$$

$$= 2\pi \times \frac{1}{2} = \pi = 3.142 \text{ sec.}$$

(b) Here, P.E. *of the particle, i.e.,* U = mV = $5(8x^2 + 200)$ ergs.

∴ *force acting on the particle,*

$$F = -\frac{dU}{dx} = -5 \times 16x = 80x \text{ dynes.}$$

So that, its *equation of motion* is

$$\frac{md^2x}{dt^2} = -80x,$$

whence, $$\frac{d^2x}{dt^2} = -\frac{80x}{m} = -\frac{80x}{5} = -16x = -\mu x.$$

Or, $$\frac{d^2x}{dt^2} \propto x.$$

The particle thus executes a simple harmonic motion and its *time-period* is given by

$$T = 2\pi\sqrt{\frac{1}{\mu}} = 2\pi\sqrt{\frac{1}{16}} = 2\pi \times \frac{1}{4} = \frac{\pi}{2} = 1.571 \text{ sec.}$$

***Example 35:***

*A particle of mass 10gm moves under a potential $V_x$ given by $V_{(x)} = 8 \times 10^5x^2$ ergs/gm, where x is in centimetres. Deduce the time-displacement relation when the total energy is $8 \times 10^5$ ergs.*

***Solution:***

Clearly, *maximum energy acquired by the particle* = $mV_{(x)} = 10 \times 8 \times 10^5x^2 = 8 \times 10^6x^2$ ergs. The whole of this must be present in the form of potential energy (U) at the end of its maximum displacement and must be equal to the total energy of the particle, given to be $8 \times 10^5$ ergs. Thus, $8 \times 10^5x^2 = 8 \times 10^5$, where x is the *maximum displacement.*

Since the *maximum displacement* is the *amplitude* a of the particle, we may replace x by a.

So that, $8 \times 10^6a^2 = 8 \times 10^5$,

whence, $a^2 = 8 \times 10^5/8 \times 10^6 = 1/10$.

Or, $$a = \frac{1}{\sqrt{10}}.$$

Now, *force acting on the particle,*

$$F = -\frac{dU}{dx} = -\frac{d\left(8 \times 10^6 x^2\right)}{dx}$$
$$= -2 \times 8 \times 10^6 x = -16 \times 10^6 x.$$

∴ the *equation of motion of the particle* is $md^2x/dt^2 = -16 \times 10^6 x$,

whence, $$\frac{d^2x}{dt^2} = -\frac{16 \times 10^6 x}{m} = -\frac{16 \times 10^6 x}{10}$$

$$= -16 \times 10^5 x = \omega^2 x. \qquad ...(a)$$

So that, $w^2 = 16 \times 10^5$ and

$$\therefore \ \omega = \sqrt{16 \times 10^5} = 400\sqrt{10}\cdot$$

If, therefore, the solution of the differential equation of motion of the particle, (which is clearly a simple harmonic motion) be $x = a \sin(\omega t + \phi)$, we have

$$x = \frac{1}{\sqrt{10}} \sin\left(400\sqrt{10}t + \phi\right).$$

## EXERCISES

1. Is there any connection between the F versus x relation at the molecular level and the macroscopic relation between F and x in a spring? Explain briefly.

2. Draw a typical potential energy curve for a diatomic molecule, clearly pointing out and explaining its peculiarities, if any. Show that, for small amplitudes, its oscillations are simple harmonic and deduce their time-period.

3. (a) Two masses mi and mg of 1.0 kg and 3.0 kg respectively are connected by a spring of force constant 250 N/M and are placed on a smooth table. They are slightly pulled apart and released. What is the frequency of the two-body system?

**Ans.** 2.90/sec.

(b) What is the ratio between the kinetic energies of the two masses of the system?

**Ans.** 3:1.

[**Hint.** Since linear momentum remains conserved during the oscillations, we have $m_1v_1 = m_2v_2$, or $m_1/m_2 = v_2/v_1$ Hence *K.E.*

*of mass* $m_1$/K.E. *of mass* $m_2 \frac{1}{2} m_1 v_1^2 / \frac{1}{2} m_2 v_2^2 = (m_1/m_2)\left(v_1^2/v_2^2\right)$

$= (m_1/m_2)\left(m_2^2/m_1^2\right) = m_2/m_1 = 3/1/.$]

4. Draw a characteristic potential energy diagram of a diatomic molecule and mark on it (i) the binding energy, (ii) equilibrium internuclear-distance.

5. Draw the energy-level diagram of a diatomic molecule and explain its salient points. What is meant by the terms: vibrational quantum number and zero-point energy?

6. In an HCl molecule, the force required to alter the distance between the atoms from its equilibrium value is $5.4 \times 10^5$ dynes per cm. What is the fundamental frequency of the vibration of the molecule, assuming the vibration to be simple harmonic and the mass of the Cl atom to be infinite compared to that of the H atom, which is $1.66 \times 10^{-24}$ gm? **Ans.** $9.078 \times 10^{13}$.

7. The separation of energy levels in HCl molecules made from $H^1$ and $Cl^{25}$ isotopes of atomic weights 1 and 35 is 0.36 electron volts. What will be the value of the separation levels is HCl molecules made from $D^2$ and $Cl^{37}$ isotopes. **Ans.** 0.258 eV.

8. Discuss a large amplitude pendulum as an anharmonic oscillator.

9. Assuming the solution of the differential equation $md^2x/dt^2 = -Cx + C_1x^2$ for an anharmonic oscillator to be $x = A \cos \omega t + B \cos^2 \omega t + x_1$ and neglecting all terms involving products of A and B and the second and higher powers of B, show that the average value of x is $x_1 = \alpha A^2/z\omega_0^2$, where $\omega^2 = \omega_0^2 = C/m$ and $\alpha = C_1/m$.

10. A particle oscillates in a potential field $U = \frac{1}{2}Cx - \frac{1}{2}C_1x^2$, where $C_1$ is positive and very much smaller them C. Will its potential well be symmetrical? If not, how will it affect the mean position of the particle?

11. What is meant by a harmonic oscillator? Obtain expressions for (i) its *displacement* and velocity at a given instant, (ii) time-period and frequency.

12. What are the important properties of a simple harmonic oscillator? Why is a simple harmonic motion considered to be a fundamental

periodic motion? Show that the time-period of a simple harmonic oscillation is given by

$$T = 2\pi\sqrt{\frac{\text{displacement}}{\text{acceleration}}}.$$

13. Explain the variation of the kinetic and potential energies of a simple harmonic oscillator. Illustrate your answer with suitable graphs.

14. Show that for a simple harmonic oscillator, mechanical energy remains conserved and that its energy is, *on an average*, half kinetic and half potential in form. At what particular displacement is this *exactly* so? And what is the ratio between its kinetic and potential energies at a displacement equal to half its amplitude?

    **Ans.** *At displacement* $a\sqrt{2}$, where a is the *amplitude*: 3:1.

15. It is said that all oscillations, if their amplitudes be infinitely small, may be regarded as simple harmonic. Do you agree? If so, explain why. Is it possible to have an oscillator for which even with small amplitudes, the oscillations are not simple harmonic?

16. Solve the differential equation $\frac{d^2x}{dt^2} + \omega^2 x = 0$ to obtain the expression $x = a \sin(\omega t + \phi)$ for the displacement of a particle executing S.H.M.

17. Show that a simple harmonic motion may be expressed as either a sine or a cosine function, there being only a difference of initial phase in the two cases.

18. (a) A particle of mass 10 gm lies in a potential field $V = 32x^2 + 200$ ergs/gm. What is its frequency of oscillation?

    **Ans.** 1.27 $sec^{-1}$.

    (b) A block executes a S.H.M. on a horizontal surface. If its frequency be 2 oscillations per second and the coefficient of static friction between the block and the surface, 0.50, how great can the amplitude be if the block does not slip on the surface?

    **Ans.** 3.1 cm.

19. A quantity of gas is enclosed in a cylinder, fitted with a smooth, heavy piston. The axis of the cylinder is vertical. The piston is

thrust downwards to compress the gas and then let go. Is the ensuing motion of the piston a simple harmonic motion? If so, what is its time-period?

**Ans.** Yes; $T = \frac{2\pi}{a}\sqrt{\frac{Vm}{P}}$, where V and P are the *initial volume and pressure* of the gas respectively, a, the *area of cross-section of the piston* and m, its mass.

20. Prove that in a simple harmonic motion, the average potential energy equals the average kinetic energy $= \frac{1}{4}Ca^2$ *when the average is taken with respect to time over one period of the motion*, and that, if the average is taken with respect to *position over one cycle*, it is equal to $\frac{1}{2}$ $Ca^2$. Explain why.

21. (a) A particle executing a S.H.M. has a maximum displacement of 4 cm and its acceleration at a distance of 1 cm from its mean position is 3 cm/sec$^2$. What will its velocity be when it is at a distance of 2 cm from its mean position? **Ans.** 6 cm/sec.

    (b) A particle, moving in a straight line with S.H.M. of period $2\pi/\omega$ about a fixed point O, has a velocity $\sqrt{3b\omega}$ when at a distance b from O. Show that its amplitude s 2b and that it will cover the rest of its distance in time $\pi/3\omega$.

22. A particle executes linear harmonic motion about the point x = 0; at t = 0, it has displacement x = 0.37 cm and zero velocity. If the frequency of the motion is 0.25/sec, determine (a) the period, (b) the amplitude, (c) the maximum speed and (d) the maximum acceleration.

    **Ans.** (a) 4.0 we, (b) 0.37 cm,

    (c) 0.58 cm/sec. (d) 0.91 cm/sec$^2$.

23. A light string is suspended vertically from a point and carries a heavy mass at its lower free end which stretches it through distance *l* cm. Show that the vertical oscillations of the system are simple harmonic in nature and of a time-period equal to that of a simple pendulum of length *l* cm.

24. Two particles execute S.H.M. of the same amplitude and frequency along adjacent straight lines. They pass one another

when going in opposite directions each time their displacement is half their amplitude. What is the phase difference between them? **Ans.** $2\pi/3$.

25. A particle is dropped down in a deep hole which extends to the centre of the earth. Calculate its velocity at a depth of 1 kilometer from the surface of the earth. Assume that g =1000 cm/sec$^2$ and radius of the earth = 6400 kilometers.

**Ans.** $1.414 \times 10^4$ cm/sec.

26. The displacement of an oscillating particle at an instant r is given by $x = a \cos \omega t + b \sin \omega t$. Show that it is executing a S.H.M.

If a = 5 cm, b = 12 cm and $\omega$ = 4 radians/sec. calculate (i) the amplitude, (ii) the time-period, (iii) the maximum velocity and (iv) the maximum acceleration of the particle.

**Ans.** (i) 13 cm, (ii) 1.57 sec, (iii) 52 cm/sec, and (iv) 208 cm/sec$^2$.

27. A smooth, straight tunnel is bored through the moon along any line taken at random. Show that if a smooth ball is dropped inside the tunnel, it executes S.H.M. In how much lime will it reach one end of the tunnel from the other end? Given: mass of the moon = $7.35 \times 10^{25}$, radius of the moon = $1.75 \times 10^8$ cm and G = $6.67 \times 10^{-8}$ dynes-cm$^2$-gm$^2$. **Ans.** 54 win. 44 sec.

28. The positions of a particle executing S.H.M. along the x = axis are x = A x = B at time I and time 2t respectively. Show that its period of oscillation is given by $T = 2\ \pi t/\cos^{-1}$ (B/2A).

[**Hint.** Here, $A = a \sin \omega t$ and $B = a \sin 2\omega t = 2\ a \sin \omega t \cos \omega t$. If at t = 0, x = 0, we have $\cos \omega t$ =B/2A, *i.e.*, $w = \cos^{-1}$ (B/2A)/t. $\therefore\ T = 2\pi/\omega = 2\pi t/\cos^{-1}$ (B/2A).]

29. If the potential energy curve is $U = -A/x + B/x^2$, where A and B are positive constants, obtain the position of stable equilibrium and the force constant for small oscillations. What is the time-period for such oscillations?

**Ans.** $\dfrac{2B}{A}$; $\dfrac{A^4}{8B^3}$; $4\pi B\sqrt{\dfrac{2mB}{A^2}}$.

30. A thin uniform bar of length 120 cm is made to oscillate about an axis through its end. Find the period of oscillation and other

points about which it can oscillate with the same time-period.

**Ans.** 1.795 sec; at 40 cm, 80 cm and 120 cm from either end.

31. The ends of one of the prongs of a tuning fork, which executes S.H.M. of frequency 1000/sec, has an amplitude of 0.40 mm. Calculate (a) its maximum acceleration and maximum speed (b) its acceleration and its speed when it has a displacement equal to half its amplitude.

**Ans.** (a) $1.6 \times 10^4$ m/sec$^2$; 2.5 m/sec.

(b) $7.9 \times 10^3$ m/sec$^2$; 2.2 m/sec.

32. In a reversible pendulum, the periods about the two knife-edges are t and (t + T), where T is a small quantity. The knife-edges are distant $l$ and $l'$ from the centre of gravity of the pendulum. Prove that $l + l' = \frac{gl}{4\pi^2}\left(t + \frac{2l'}{l'-l}T\right)$.

33. Obtain an expression for the time-period of a compound pendulum and show that (i) there are four points collinear with the *e.g.*, about which the time-period is the same, (ii) its time-period remains unaffected by the fixing of a small additional mass to it at its centre of suspension.

34. What is meant by an equivalent simple pendulum? If the periods of a Kater's pendulum in the erect and inverted positions are equal, prove that the distance between the knife-edges is equal to the length of the equivalent simple pendulum.

35. A body of mass 200 gm oscillates about a horizontal axis at a distance of 20 cm from its centre of gravity. If the length of the equivalent simple pendulum be 35 cm, find its moment of inertia about the axis of suspension. **Ans.** $1.4 \times 10^6$ gm-cm$^2$.

36. A uniform rod AB of mass 100 gm and length 120 cm can swing in a vertical plane about A as a pendulum. A particle of mass 200 gm is attached to the rod at a distance x from A. Find x such that the period of vibration is a minimum.

**Ans.** x = 2.748 cm.

37. A uniform circular disc of radius 25 cm oscillates in a vertical plane about a horizontal axis. Find the distance of the axis of

rotation from the centre for which the period is the minimum. What is the value of this period?

**Ans.** $25\sqrt{2}$ cm; $2\pi\sqrt{1.414\times25/g}$ sec.

38. A uniform cube is free to turn about one edge which is horizontal. Find in terms of a seconds pendulum the length of the edge, so that it may execute a complete oscillation in 2 sec. **Ans.** $3\sqrt{2l}$

39. At a certain place the value of g is 980 cm/sec$^2$ and the length of the pendulum is so adjusted that the period is 1 sec. When the period of the pendulum is measured on an elevator, undergoing uniform acceleration, it is found to be 1.025 sec. (i) For an acceleration a << g of the elevator, show that the period measured on the elevator is given by $T = T_0\,(1 - a/2g)$, where $T_0$ is the period of the unaccelerated pendulum and upward acceleration is assumed to be positive, (ii) What is the acceleration of the elevator? **Ans.** 49 cm/sec$^2$ downward.

40. An annular ring of internal and outer radii r and R, respectively oscillates in the vertical plane about a horizontal axis perpendicular to its plane and passing through a point on its outer edge. Calculate its time-period and show that the length of an equivalent simple pendulum is 2 R as $r \to 0$ and 3R as $r \to R$. **Ans.** $T = 2\pi\sqrt{r^2 + 3R^2/Rg}$.

41. What is a torsional pendulum? Obtain an expression for its time-period. Explain why, unlike a simple or a compound pendulum, the time-period in this case remains unaffected even if the amplitude be large.

42. A disc of 10 cm radius and mass 1 kg is suspended in a horizontal plane by a vertical wire attached to its centre. If the diameter of the wire is 1 mm and its length is 1.5 metres and the period of torsional vibration of the disc is 5 sec, find the rigidity (n) of the material of the wire. **Ans.** $1.206 \times 10^{12}$ dynes-cm$^2$.

43. A solid cylinder of 2 cm radius weighing 200 gm is rigidly connected, with its axis vertical, to the lower end of a fine wire. The period of oscillation of the cylinder under the influence of the torsion wire is 2 seconds. Calculate the couple necessary to twist it through four complete turns.

**Ans.** $9.9 \times 10^4$ dyne-cm$^2$.

44. A body, suspended by means of a wire, executes torsional/ vibrations about the wire as axis, of time-period 5.0 sec. If the moment of inertia of the body about the suspension wire as axis be $\pi/2$ gm-cm$^2$, calculate the maximum angular velocity of the body if its amplitude be $\pi/2$ radian. What would then be its average kinetic energy? **Ans.** 1.973 radian.sec; $7.30 \times 10^3$ ergs.

[**Hint.** Since $\theta = \theta_0 \sin(\omega t + \phi)$, we have *angular velocity* $d\theta/dt = \omega\theta_0 \cos(\omega t + \phi)$, where $\theta_0$ is the amplitude. $\therefore$ maximum value of $d\theta/dt = \omega\theta_0 = 2\pi n\theta_0$ ($\because \omega = 2\pi n$).

Here, a = 1/T = 1/5 per sec and $\theta_0 = \pi/2$ radian. Hence, maximum angular velocity of the body $= 2\pi \times (1/5) \times (\pi/2) = \pi^2/5 = 1.973$ radian/sec.

$$\text{And, } \textit{average} \text{ K.E.} = \frac{1}{2}$$

$$\textit{total energy} = \frac{1}{2}\left(\frac{1}{2}C\theta_0^2\right) = \frac{1}{4}C\theta_0^2.$$

$$\text{Since } T = 2\pi\sqrt{\frac{I}{C}} \text{ and } \backslash\ C = \frac{4\pi^2 I}{I^2}, \text{ we have}$$

$$\textit{average K.E.} = \frac{1}{4} \times \frac{4\pi^2 I}{T^2}\left(\frac{\pi}{2}\right)^2\Bigg].$$

45. An unstressed spring has a force constant C. It is stretched by a weight hung from it to an equilibrium length well within the elastic limit. Does the spring have the same force constant C for displacements from this new equilibrium position?

46. Explain how we can determine the period of oscillation of a spring-mass system from the extension produced in the spring on suspending the mass from it, even though we neither know the magnitude of the mass nor the force constant of the spring.

47. A spring has a force constant C. If it be cut into two pieces whose lengths are in the ratio 1:2, will the force constant continue to be the same (C) for each piece? If not, obtain the value of the force constant of each piece in terms of C.

**Ans.** 3C, 3C/2.

48. A massless spring, as we know, is purely a myth. Any real spring must have some mass. If this fact be taken into account, explain

what change, if any, will it bring about in the expression for the time-period of a spring-mass system.

49. A spring of force constant 19.6 N/m hangs vertically. A body of mass 0.20 kg is attached to its free end and then released. Assume that the spring was unstretched before the body was released and find how far below the initial position the body descends. Find also the frequency and amplitude of the resulting S.H.M. **Ans.** 0,1 m; 1.576/w; 0.1 m.

50. A 2 kg mass hangs from a spring. A 300 gm body hung below the mass stretches the spring 2 cm further. If the 300 gm body is removed and the mass set into oscillation, find the period of motion. **Ans.** 0.73 sec.

51. (a) Show that 'gravity' has no effect upon the force constant or the period of oscillation of a spring-mass system.

    (b) Calculate the ratio of K.E. to P.E. when a mass oscillating at the lower end of a spring has covered half its amplitude.

    **Ans.** 3:1

52. An 8 *l*b block is suspended from a spring with a force constant of 3.0 *l*b/in. A bullet weighing 0.10 *l*b is fired into the black from below with a velocity of 500 ft/sec and comes to rest in the block, (a) Find the amplitude of the resulting S.H.M. (b) What fraction of the original K.E. of the bullet is stored in the harmonic oscillator? **Ans.** (a) 6.2 in., (b) 1.2%.

53. A mass M resting on a smooth table between two firm supports A and B and controlled by two massless springs. If the mass M is 30 gm and the force constants of the two springs are 2000 and 3000 dynes/cm, deduce (i) the frequency of small oscillations of M, (ii) the energy of oscillations for amplitude 0.5 cm.

    **Ans.** (i) 1.592/sec., (ii) 375 ergs.

54. (a) Calculate the frequency for small oscillations of mass At in problem 53 above, if it were connected to the two springs.

    **Ans.** 0.75/MC.

    (b) A car may be considered to be mounted on a stout vertical spring so far as its vertical oscillations are concerned. Thus, if the frequency of such vertical oscillations of a car weighing 1700 kg be 4 when unoccupied, what would be the frequency if it were occupied by five persons, each weighing 60 kg on an average? **Ans.** 3.4/sec.

55. A spherical vessel of capacity 500 c.c. is fitted with a tube of 5 cm. length and $\sqrt{\pi}$ cm radius. Calculate the wavelength of sound for which it will serve as a resonator. If the velocity of sound be 340 m/sec, what will be the frequency of the sound to which it will respond? **Ans.** $\lambda$ = 100 cm; n = 340 cps.

$$\left[\text{Hint. } n = \frac{v}{2\pi}\sqrt{\frac{\alpha}{Vl}} \therefore \lambda = \frac{v}{n} = 2\pi\sqrt{\frac{Vl}{\alpha}}.\right]$$

56. A spherical resonator with a neck of length 4 cm and area of cross section $4\pi^2$ sq cm responds sharply to a note of frequency 112. If the velocity of sound in air be 336 m/sec, what is the capacity of the resonator? **Ans.** 10 litres.

57. It is desired to have an L.C. circuit of frequency 10 kilocycles/ sec. Calculate the capacitance that must be put in series with an inductance of 1 millihenry to achieve the purpose.

    **Ans.** 0.2533 μF.

58. An L.C. circuit oscillates with a frequency of 20 cps. The capacitance in the circuit is 5 μF. Calculate the value of inductance in the circuit. **Ans.** 12.67 henry.

59. What is meant by a two-body harmonic oscillator? Obtain an expression for the time-period of its oscillation and point out its similarities with a one-body oscillator. What is meant by reduced mass? Why is it so called?

60. Show that the K.E. of a two-body oscillator (*i.e.*, two bodies connected by a spring) is given by $\frac{1}{2}\mu v^2$, where $\mu$ is the reduced mass and v (= $v_1 - v_2$), the relative velocity.

61. Is it really possible to construct a truly simple pendulum? Show that as the angular amplitude of the pendulum approaches 180°, its time-period approaches infinity.

62. What are the drawbacks of a simple pendulum? The time-period of a simple pendulum for infinitely small amplitudes is given by $T = 2\pi\sqrt{l/g}$. What will it be for a finite amplitude?

63. A simple pendulum of length 100 cm has energy equal to 0.3 joule when its amplitude is 2 cm. What will be its energy if (i) its length is increased to 150 cm, (ii) its amplitude is increased to 3 cm? **Ans.** (i) 0.2 joule, (ii) 0.675 joule.

64. A hollow sphere is filled with water through a small hole in it. It is hung by a long thread and, as the water slowly flows out of the hole at the bottom, the period of oscillation first increases and then decreases. Why is it so?

65. Show that the time-period of a simple pendulum at depth h below the earths surface is inversely proportional to $\sqrt{R-H}$, where R is the radius of the earth. Calculate its time-period when h = R/2. **Ans.** 2.8 sec.

66. How does a compound pendulum differ from a simple pendulum? Obtain an expression for its time-period and mention its points of superiority over a simple pendulum.

67. How is the period of a pendulum affected when its point of suspension is (a) moved horizontally with acceleration a, (b) moved vertically upward with acceleration a, (c) moved vertically downward with acceleration $a < g$? Which case, if any, applies to a pendulum mounted on a cart rolling down an inclined plane?

68. Define centres of suspension and oscillation of a compound pendulum and show that they are interchangeable. What length of the pendulum has its minimum time-period?

69. Show that if a uniform stick of length $l$ is mounted so as to rotate about a horizontal axis perpendicular to the stick and at a distance d from its centre of mass, the period has the minimum value when $d = l/\sqrt{12} = 0.289\ l$.

70. Give the theory of Kater's pendulum and find an expression for acceleration due to gravity in terms of two nearly equal periods of oscillation about the two parallel knife-edges.

71. Explain how we can determine the period of oscillation of a spring-mass system from the extension produced in the spring on suspending the mass from it, even though we neither know the magnitude of the mass nor the force constant of the spring.

# 2

# HARMONIC OSCILLATOR *(Continued)*

## FRICTIONAL EFFECTS—(DAMPING)

An oscillator, in actual practice, almost always lies in a resisting medium, like *air*, *oil* etc., part of its energy is dissipated in overcoming the opposing frictional or viscous forces and its amplitude, therefore, goes on decreasing progressively. Such forces, which are non-conservative in nature, have thus a *damping effect* on the oscillations and are aptly referred to as *damping, resistive* or *dissipative* forces. In the absence of any such forces, the oscillations will continue, indefinitely, without any change in amplitude, as shown in Fig. 2.1.

Let us discuss in necessary detail the frictional effects on a harmonic oscillator.

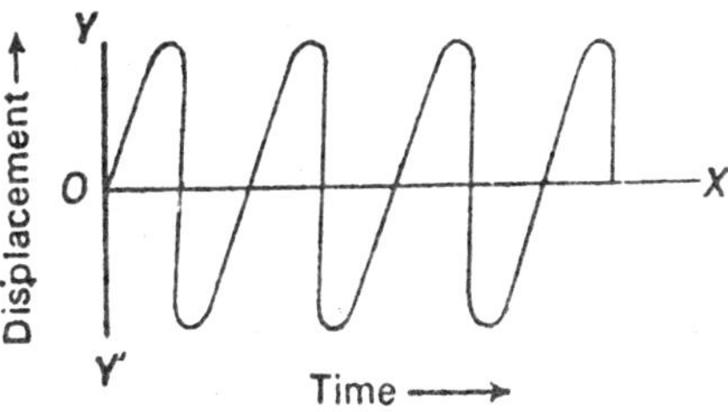


**Fig. 2.1**

It has been shown by *Mayevski* that at *ordinary velocities*—and most cases of interest to us fall in this category—*the opposing, resistive or damping force is, to a first approximation, proportional to velocity* and may thus be represented by

$$F = -\gamma v = -\frac{\gamma\, dx}{dt}, \qquad \ldots(i)$$

where $\gamma$ is a *positive constant*, called *damping coefficient* of the medium and may be looked upon as the *resistive force per unit velocity*.

So that, if there is no other force, other than this resistive or damping force, acting on the oscillating body or particle, of mass m, we have, in accordance with Newton's second law of motion,

$$F = m\frac{d^2x}{dt^2} = m\frac{dv}{dt} = -\gamma v.$$

Or, $$\frac{dv}{dt} + \frac{\gamma}{m}v = 0. \quad ...(ii)$$

Here, $\frac{m}{\gamma}$ is usually denoted by, a constant, having the dimensions of time and called *relaxation time,*

We, therefore, have $$\frac{dv}{dt} + \frac{1}{\tau}v = 0. \quad ...(iii)$$

The constant $\frac{1}{\tau} = \frac{\gamma}{m}$, or *the resistive force per unit mass per unit velocity*, is often denoted by 2k, where k is called the *damping constant* of the medium.

Now, rewriting differential equation (iii) in the form

$\frac{dv}{v} = -\frac{dt}{\tau}$, we have

$\int\frac{dv}{v} = -\frac{1}{\tau}\int dt$, which gives $\log_e v = -\frac{t}{\tau} + C$,

where C is a constant of integration to be determined from the initial conditions.

Thus, if at t = 0, v = $v_0$, we have $\log_e v_0 = C$. And, therefore,

$$\log_e v - \log_e v_0 = -\frac{t}{\tau}.$$

Or, $$v = \frac{v_0 e^{-t}}{\tau}, \quad ...(iv)$$

clearly showing that *the velocity decreases exponentially with time*, as shown by the curve in. Fig. 2.2 between the function $e^{-t/\tau}$ and t. We express this by saying that *the velocity is damped, with lime constant* τ.

As will be readily seen, at

$$t = \tau,\ v = v_0e^{-1} = v_0/e = v_0/2.718 = 0.368\ v_0.$$

This enables us to define the *time constant* (or the *relaxation time*) τ as, *the time in which the velocity of the oscillating particle falls to l/e th (i.e., 0.368 or, roughly, one-third) of its initial value.*

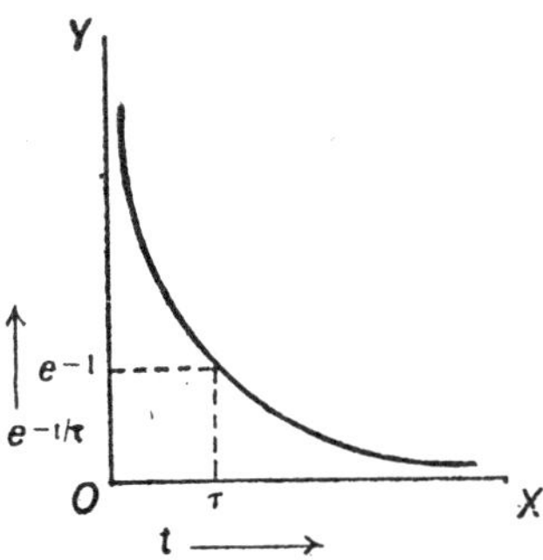


Fig. 2.2

And, since the kinetic energy of the oscillating particle is given by $T = \frac{1}{2}mv^2$, we have, on substituting the value of v from relation (iv) above, $T = \frac{1}{2}mv_0^2 e^{-2t}/\tau$. Or, representing the initial kinetic energy $\frac{1}{2}mv_0^2$ by $T_0$, we have $T = T_0e^{-2t/\tau}$ indicating that *the kinetic energy of the oscillating particle too falls exponentially with time, with a relaxation time half that for velocity, i.e.,* $\tau/2$, which is only to be expected since K.E. $\alpha$ (velocity)$^2$.

Putting dx/dt for v in relation (iv) above, we have $\frac{dx}{dt} = v_0e^{-t/t}$, which, on integration, gives $x = -v_0te^{-t/\tau} + C$, where C is a constant of integration.

At $t = 0$, $x = 0$, so that $C = v_0\tau$. And, therefore, $x = v_0\tau(1 - e^{-t/\tau})$

Now, as $t \to \infty$, $e^{-t/t} \to 0$ and, therefore, $x \to v_0\tau$.

Thus, *the maximum value of x is the distance that would be covered by the particle in time $\tau$ if its velocity remained constant at its initial value $v_0$.*

**Examples:** (i) As an example of the resistive or damping force of the type represented by relation (i) above may be cited the force experienced by a flat disc moving normally to its plane through a gas (or air), at very low pressure, at a speed very much smaller than that of its molecules.

(ii) Or, perhaps a more familiar example is that furnished by the ohmic resistance in an electrical circuit containing an inductance, such as the one shown in Fig. 2.3. On suddenly breaking the circuit, an induced emf, $-\frac{LdI}{dt}$, is set up across the inductance L, where I is the value of the current flowing through the circuit at the instant it is broken. Since there is now no external emf operative, we have

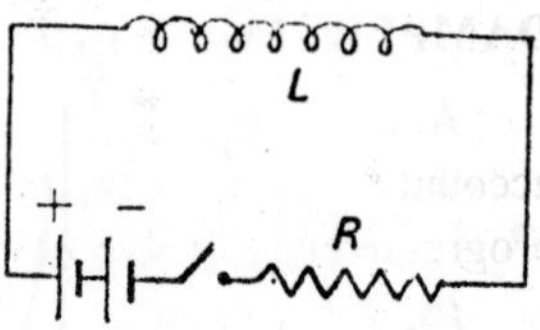

Fig. 2.3

$$RI = -\frac{LdI}{dt}.$$

Or, $$\frac{LdI}{dt} + RI = 0,$$

where RI is the potential drop across the resistance R.

This is a relation, identical in form with relations (ii) and (iii) above, with the time constant $\tau = \frac{L}{R}$, clearly indicating that the ohmic resistance plays the same part here as the damping force there, so that *the current in the circuit falls exponentially with time*. For, putting relation (v) in the form $\frac{dI}{I} = -\left(\frac{R}{L}\right)dt$ and integrating, we have $\log_e I = -\frac{R}{L}t + C$, where C is a constant of integration.

Since at t = 0, I = $I_0$, its initial (or maximum) value, we have C = $\log_e I_0$. And, therefore,

$$\log_e I = -\frac{R}{L}t + \log_e I_0.$$

Or, $$\log_e \frac{I}{I_0} = -\frac{R}{L}t, \text{ whence, } I = I_0 e^{-(R/L)t}.$$

It will he seen at once that if t = *time constant* L/R, we have

$$I = I_0 e^{-1} = \frac{I_0}{e}.$$

Thus, *the time constant here too is the time in which the value of the current in the circuit falls, to 1/e, or roughly, one-third, of its initiator maximum value.*

## DAMPED HARMONIC OSCILLATOR

A harmonic oscillator in which the oscillations are damped on account of resistive forces, as explained, above under, with its amplitude progressively decreasing to zero, is called a *damped harmonic* oscillator.

Obviously, in the case of such an oscillator, in addition to the restoring force $-Cx$, a resistive or damping force $\gamma dx/dt$ also acts upon it, where $dx/dt$ is its velocity at displacement x. We, therefore, have

$$m\frac{d^2x}{dt^2} = -\gamma\frac{dx}{dt} - Cx.$$

Or, $$\frac{d^2x}{dt^2} + \frac{\gamma dx}{m dt} + \frac{C}{m}x = 0.$$

Or, since $\frac{\gamma}{m} = 2k$ (where k is the *damping constant* of the resistive medium) and $\frac{C}{m} = \omega_0$, the *natural angular frequency of the oscillating particle, i.e.*, its frequency in the absence of damping, we have

$$\frac{d^2x}{dt^2} + 2k\frac{dx}{dt} + \omega_0^2 x = 0, \qquad \text{...(vi)}$$

which is called the *differential equation of a damped harmonic oscillator*.

*Solution of the equation:* The differential equation, which is a *homogeneous linear type, of the second order*, has at least one solution of the form $x = Ae^{\alpha t}$, where A and $\alpha$ are both *arbitrary constants*. We shall, therefore, use this as a trial solution.

Differentiating with respect to t, we have

$$dx/dt = \alpha.\ Ae^{\alpha t} \text{ and } \frac{d^2x}{dt^2} = \alpha^2 Ae^{\alpha t}.$$

Substituting these values in the differential equation (vi) above, we have

$$\alpha^2 Ae^{\alpha t} + 2k\alpha\ Ae^{\alpha t} + \omega_0^2 Ae^{\alpha t} = 0.$$

Or, dividing throughout by $Ae^{\alpha t}$, we have $\alpha^2 + 2k\alpha + \omega_0^2 = 0$. (vii)
This is clearly a *quadratic equation* in $\alpha$ and, therefore,

$$\alpha = -k + \sqrt{k^2 - \omega_0^2}.$$

The differential equation (vi) is thus satisfied by two values of x, *viz.*,

$x = Ae^{\left(-k+\sqrt{k^2-\omega_0^2}\right)t}$ and $x = Ae^{\left(-k+\sqrt{k^2-\omega_0^2}\right)t}$.

The equation being a linear one, the linear sum of the two linearly independent solutions of the equation is also a – and, indeed, the most general solution. Thus, the general solution is

$$x = A_1e^{\left(-k+\sqrt{k^2-\omega_0^2}\right)t} + A_2e^{\left(-k-\sqrt{k^2-\omega_0^2}\right)t}. \qquad \text{...(viii)}$$

Now, $k = \dfrac{1}{2\tau}$ and if we put $\sqrt{k^2 - \omega_0^2} = \beta$, we may also put the solution in the form

$$x = A_1e^{-t/2\tau+\beta t} + A_2e^{-t/2\tau-\beta t}$$

where the values of the *arbitrary constants* $A_1$ and $A_2$ may be determined as follows:

Differentiating expression (viii) with respect to t, we have

$$\frac{dx}{dt} = \left(-k+\sqrt{k^2-\omega_0^2}\right)A_1e^{\left(-k+\sqrt{k^2-\omega_0^2}\right)t}$$

$$+\left(-k-\sqrt{k^2-\omega_0^2}\right)A_2e^{\left(-k-\sqrt{k^2-\omega_0^2}\right)t} \qquad \text{...(ix)}$$

So that, if the displacement x be the *maximum*, equal to $x_{max} = a_0$, say, at t = 0 and, therefore, the velocity $\dfrac{dx}{dt} = 0$, we have from relation (viii) above,

$$x_{max} = a_0 = (A_1 + A_2) \qquad \text{...(A)}$$

$$\left(-k+\sqrt{k^2-\omega_0^2}\right)A_1 + \left(-k-\sqrt{k^2-\omega_0^2}\right)A_2 = 0.$$

Or, $\quad -k(A_1 + A_2) + \sqrt{k^2-\omega_0^2}\,(A_1 - A_2) = 0.$

Or, $\quad \sqrt{k^2-\omega_0^2}\,(A_1 - A_2) = k(A_1 + A_2) = ka_0,$

whence, $\quad (A_1 - A_2) = \dfrac{ka_0}{\sqrt{k^2-\omega_0^2}} \qquad \text{...(B)}$

Adding relations A and B, therefore, we have

$$2A_1 = a_0 + \frac{ka_0}{\sqrt{k^2-\omega_0^2}}$$

And ∴ $\quad A_1 = \dfrac{1}{2}\left(a_0 + \dfrac{ka_0}{\sqrt{k^2-\omega_0^2}}\right) = \dfrac{1}{2}a_0\left(1 + \dfrac{k}{\sqrt{k^2-\omega_0^2}}\right)$

$$= \frac{1}{2}a_0\left(1+\frac{1}{2\beta\tau}\right)$$

and $A_2 = (A_1 + A_2) - A_1 = a_0 - \frac{1}{2}a_0\left(1+\frac{k}{\sqrt{k^2-\omega_0^2}}\right)$

$$= \frac{1}{2}a_0\left(1-\frac{k}{\sqrt{k^2-\omega_0^2}}\right) = \frac{1}{2}a_0\left(1-\frac{1}{2\beta\tau}\right)$$

Substituting these values in expression (viii) above, we have

$$x = \frac{1}{2}a_0 e^{-kt}\left[\left(1+\frac{k}{\sqrt{k^2-\omega_0^2}}\right)e^{\sqrt{(k^2-\omega_0^2)}t} + \left(1-\frac{k}{\sqrt{k^2-\omega_0^2}}\right)e^{-\sqrt{(k^2-\omega_0^2)}t}\right]. \quad ...(x)$$

Or, since $k = \frac{1}{2\tau}$ and $\sqrt{k^2-\omega_0^2} = \beta$, we have

$$x = \frac{1}{2}a_0 e^{-t/2\tau}\left[\left(1+\frac{1}{2\beta\tau}\right)e^{\beta t} + \left(1-\frac{1}{2\beta t}\right)e^{-\beta t}\right] \quad ...(xi)$$

Now, *three important cases* arise:

**1. When k (or 1/2τ) > $\omega_0$ – (Case of overdamping):** In case of high damping such as this, clearly, $\sqrt{k^2-\omega_0^2}$ (or $\sqrt{1/4\tau^2-\omega_0^2}$) is a *real quantity* with a *positive value, less than k.* So that, each of the two terms on the right hand side of equation (x) or (xi) has an exponential term with a negative power. The displacement, after attaining its maximum value, therefore, dies off exponentially with time, without changing direction. There is thus no oscillation and the motion is, therefore, called *overdamped, aperiodic or dead beat*, as we have in the case of a *dead beat galvanometer* or that of a pendulum oscillating in a viscous fluid like oil.

**2. When k (or 1/2τ) = $\omega_0$ – (case of critical damping):** In this case, obviously, $\sqrt{k^2-\omega_0^2}$ (or $\sqrt{1/4\tau^2-\omega_0^2}$) = 0, so that each of the two terms on the right hand side of equation (x) or (xi) above, becomes infinite and the solution breaks down.

Let us, however; consider the case when $\sqrt{k^2 - \omega_0^2} = h$, *a very small quantity but not zero.* We shall then have, from relation (viii) above,

$$x = A_1 e^{(-k+h)t} + A_2 e^{(-k-h)t} = e^{-kt}(A_1 e^{ht} + A_2 e^{-ht}).$$

Or, $$x = e^{-kt}\left[A_1\left(1 + ht + \frac{h^2t^2}{2!} + \frac{h^3t^3}{3!} + \ldots\right) + A_2\left(1 - ht + \frac{h^2t^2}{2!} - \frac{h^3t^3}{3!} + \ldots\right)\right].$$

Or, neglecting terms containing the second and higher powers of h, we have

$$x = e^{-kt}[A_1(1 + ht) + A_2(1 - ht)]$$
$$= e^{-kt}[(A_1 + A_2) + (A_1 - A_2)ht]$$

Or, putting $(A_1 + A_2) = M$ and $(A_1 - A_2)h = N$, we have

$$x = e^{-kt}(M + Nt).$$

Now, taking $x = x_{max} = a_0$ and $dy/dt = 0$ at $t = 0$, we have

$$M = x_{max} = a_0 \text{ and } N = ka_0.$$

So that, $x = e^{-kt}(a_0 + ka_0t)$

$$= a_0e^{-kt}(1 + kt) = a_0e^{-t/2\tau}\left(1 + \frac{t}{2\tau}\right).$$

Here, the second term $a_0 + kte^{-kt}$ [or $(a_0t/2\tau)e^{-t/2\tau}$] decays less rapidly than the first term $a_0e^{-kt}$ (or $a_0e^{-t/2\tau}$) and the displacement of the oscillator first increases but it then returns back quickly to its equilibrium position. The motion of the oscillator thus becomes *just aperiodic or non-oscillatory, i.e., if just ceases to oscillate.* This is called the case of *critical damping*, the necessary condition for which, as we have just seen, is that $k \rightarrow \omega_0$. It finds an application in many pointer-type instruments like galvanometers where the pointer moves at once to, and stays at, the correct position, without any annoying oscillations.

**3. When k (or 1/2τ) < $\omega_0$ – (case of underdamping):** Here, clearly, the quantity $\sqrt{k^2 - \omega_0^2}$ will be an *imaginary* one, say, equal to no, where $i = \sqrt{-1}$ and $\omega = \sqrt{\omega_0^2 - k^2}$, a *real quantity*. Expression (viii) above thus becomes

$$x = A_1e^{(-k + i\omega)t} + A_2e^{(-k - i\omega)t}$$

Or, $$x = e^{-kt}[A_1 (\cos \omega t + i \sin \omega t) + A_2(\cos \omega t - i \sin \omega t)]$$

$$= e^{-kt}[(A_1 + A_2) \cos \omega t + i (A_1 - A_2) \sin \omega t]$$

Or, putting $(A_1 + A_2) = A$ and $i(A_1 - A_2) = B$, we have

$$x = e^{-kt} (A \cos \omega t + B \sin \omega t)$$

If A, B and $a_0$ be related as shown in Fig. 1.4, we have

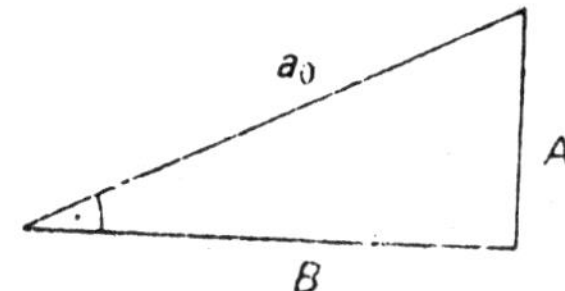


**Fig. 2.4**

$$x = e^{-kt}\left(a_0 \cos\omega t \frac{A}{a_0} + a \sin \omega t \frac{B}{a_0}\right)$$

$$= e^{-kt}(a_0 \cos \omega t \sin \phi + a_0 \omega t \cos \phi)$$

Or, $$x = a_0e^{-kt} \sin (\omega t + \phi)$$

$$= a_0e^{-t/2}\tau \sin (\omega t + \phi) \qquad \text{...(xii)}$$

This is the *equation of a damped harmonic oscillator,* with *amplitude* $e_0e^{-kt}$ (or $a_0e^{-t/2r}$) and *frequency* $\omega/2\pi = \sqrt{\omega_0^2 - k^2/2\pi}$. It is so called because the sine term in the equation suggests the oscillatory character of the motion and the exponential term, the gradual damping out of the oscillations.

Damping thus clearly produces two effects:

(*i*) *The frequency of the damped harmonic oscillator,* $\omega/2\pi$ *is smaller than its undamped or natural frequency* $\omega_0/2\pi$, *i.e.*, damping some what decreases the frequency or increases the time-period of the oscillator. In actual practice, in a majority of cases, particularly in the case of musical instruments, the damping is small and its effect on the frequency or the time-period of the oscillator, therefore, quite negligible.

(ii) *The amplitude of the oscillator does not remain constant at* $a_0$, *which represents the amplitude in the absence of damping, but decays exponentially with time, to zero, in* accordance with the term $e^{-kt}$, *called the damping factor.*

This is illustrated by the *time-displacement curve* of the damped harmonic oscillator, shown in Fig. 1.5

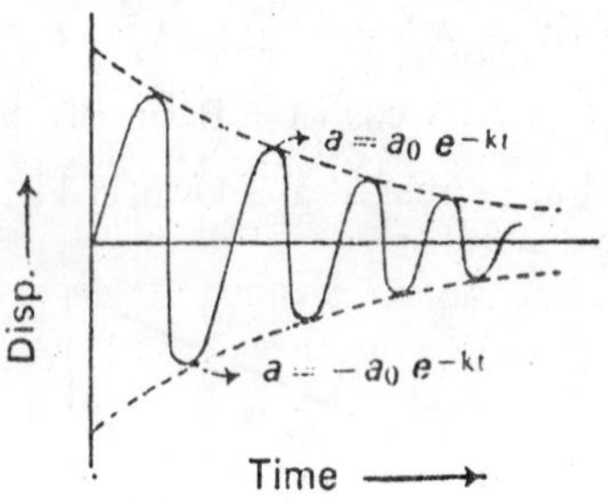


Fig. 2.5

Since the maximum values of sin $(\omega t + \phi)$ are +1 and −1 *alternately*, the time-displacement graph of the oscillating particle is bounded by the dotted curves

$$a = a_0 e^{-kt} \text{ (or } a = e_0 e^{-t/2\tau})$$

and $$a = -a_0 e^{-kt} \text{ (or } a = -a_9 e^{-t/2r}).$$

Thus, although its amplitude decreases exponentially with time, the underdamped harmonic oscillator does perform a sort of oscillatory motion. *The motion does not, of course, repeat itself and is thus not periodic in the usual sense of the term.* However, it has still a time-period $= 2\pi/\omega = 2\pi/\sqrt{\omega_0^2 - k^2}$, which is the time-interval between its successive passages in the same direction past the equilibrium point. It is obviously also the time-interval between successive maximum displacements on the same side of the equilibrium point.

N.B. Actually, as we shall see later, a damped oscillation may, in terms of *Fourier's theorem*, be imagined to result from the superposition of a very large number (*i.e.*, an infinite series) of undamped oscillations, with their frequencies varying continuously on either side of the main or principal frequency $\omega/2\pi$ and with their amplitudes diminishing in proportion to the departure of their frequencies from the principal one.

**Logarithmic Decrement**

As we have just seen above, the amplitude in the case of a damped harmonic motion goes on decreasing progressively. So that, if $a_n$ and

$a_{(n+1)}$ be the successive amplitudes of the oscillating particle on the two sides of the equilibrium position respectively, the time-interval between them is clearly half the time-period (T) of oscillation and is thus T/2. We thus have

$$a_n = a_0\, e^{-kt} \text{ and } a_{(n+1)} = a_0 e^{-k(t+T/2)}$$

and therefore, $\dfrac{a_n}{a_{(n+1)}} = \dfrac{e^{-kt}}{e^{-k(t+/t/2}} = e^{kT/2} = d$, a constant.

This constant d between two successive amplitudes of a given damped harmonic motion is called the *decrement* for that motion. The same naturally also applies to angular amplitudes, where we have

$$\frac{\theta_n}{\theta_{(n+1)}} = d.$$

The logarithm of the decrement, *i.e.*, $\log_e d = kT/2 = \lambda$. Or, $d = e^{\lambda}$.

This constant, λ, which is obviously the natural logarithm of the *decrement or the ratio between two successive amplitudes of the oscillation* is referred to as the *logarithmic decrement* for that motion or oscillation.

Use is made of this in applying the necessary correction for damping to the *first deflection* or *the first 'throw'* $\theta_1$ of a ballistic galvanometer as follows:

Each half oscillation, as we know, comprises one swing from $\theta_1$ to $\theta_2$ or from $\theta_2$ to $\theta_3$, *i.e.*, from the extreme position on one-side to the extreme position on the other, as shown in Fig. 1.6.

So that, $\dfrac{\theta_1}{\theta_2} = \dfrac{\theta_2}{\theta_3} = d = e^{\lambda}$.

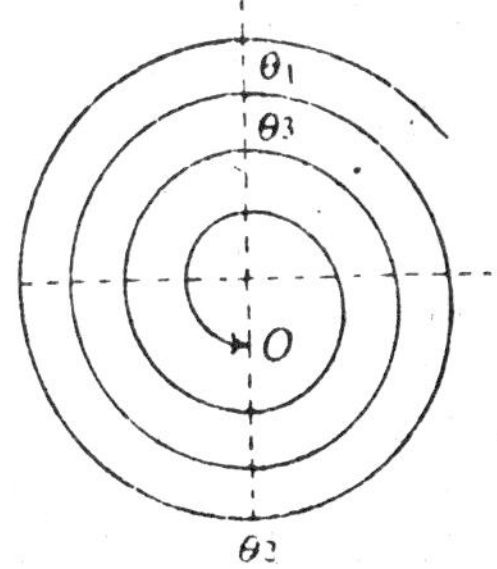


**Fig. 2.6**

Now, the first throw of a ballistic galvanometer, from the mean or equilibrium position to either extreme, constitutes only a *quarter oscillation* on a *half swing*. If, therefore, θ be the *true* value of this first throw, *i.e.*, its value in the absence of damping, and $\theta_1$, its *observed* value, we have

$$\frac{\theta}{\theta_1} = d^{1/2} = e^{\lambda/2} = \left(1 + \frac{\lambda}{2} + \frac{(\lambda/2)^2}{2!} + \dots\right) = 1 + \frac{\lambda}{2} \text{ very nearly.}$$

Hence, $\theta = \theta_1\left(1+\dfrac{\lambda}{2}\right)$.

Thus, knowing the logarithmic decrement ($\lambda$) for the given galvanometer, we can easily correct its first throw ($\theta_1$) for damping.

## POWER DISSIPATION

Since the amplitude of a damped harmonic oscillator goes on falling exponentially with time, on account of the resistive or damping forces it has to overcome, it is clear that its energy gets continuously dissipated during its oscillation. Let us calculate its *rate of dissipation of energy* or its *power dissipation*, as it is more commonly called (because *energy or work / time = power*).

Since the displacement of the oscillator is given by

$$x = a_0\, e^{-kt} \sin(\omega t + \phi),$$

we have *velocity of the particle at a given instant t, i.e.,*

$$\frac{dx}{dt} = a_0 e^{-kt}[-k \sin(\omega t + \phi) + \omega \cos(\omega t + \phi)]$$

$\therefore$ *K.E. of oscillation of the particle at the given instant t*

$$= \frac{1}{2}m\left(\frac{dx}{dt}\right)^2 = \frac{1}{2}ma_0^2 e^{-2kt}\,[-k \sin(\omega t + \phi) + \omega \cos(\omega t + \phi)]^2$$

$$= \frac{1}{2}ma_0^2 e^{-2kt}\,[k^2 \sin^2(\omega t + \phi) + \omega^2 \cos^2(\omega t + \phi) - 2k\omega \sin(\omega t + \phi) \cos(\omega t + \phi)]$$

If the damping be small (as it usually is), the amplitude does not change appreciably over a time-period, so that the factor $e^{-2kt}$ may be taken to be constant over the period. Further, since the average value over a time-period of both $\sin^2(\omega t + \phi)$ and $\cos^2(\omega t + \phi)$ is 1/2 and that of $2\omega \sin(\omega t + \phi) \cos(\omega t + \phi) = \sin 2(\omega t + \phi) = 0$, we have *average K.E. of the particle or the oscillator over one cycle at the given instant* $t = \frac{1}{2}ma_0^2 e^{-2kt}\left(\frac{1}{2}k^2 + \frac{1}{2}\omega^2 - 0\right) = \frac{1}{4}ma_0^2\omega^2 e^{-2kt}$, neglecting $k^2$ in comparison with $\omega^2$, in view of its very small value.

And, *P.E. of the particle or the oscillator at the given instant t (when the displacement is x)* $= \frac{1}{2}Cx^2 = \frac{1}{2}m\omega_0^2 x^2$, where $\omega_0$ is the natural (undamped) angular frequency of the particle.

$$\left[\because \text{P.E.} = \int_0^x Cx\,dx = \frac{1}{2}Cx^2 \text{ and } \frac{C}{m} = \omega_0.\right]$$

Or, substituting the value of x, we have

*P.E. of the oscillating particle* $= \frac{1}{2}m\omega_0^2\ [a_0\ e^{-kt} \sin(\omega t + \phi)]^2$

$$= \frac{1}{2}ma_0^2 e^{-2kt}\left[\omega_0^2 \sin^2(\omega t + \phi)\right]$$

Again, since the average value of $\sin^2(\omega t + \phi)$ over a time-period is 1/2. We have *average P.E. of the particle or the oscillator over one cycle, at the given instant* $t = \frac{1}{4}ma_0^2\omega_0^2 e^{-kt} = \frac{1}{4}ma_0^2\omega^2 e^{-2kt}$, because, damping being small, $\omega_0 \approx \omega$.

Hence, *average total energy of the oscillator at the given instant = (average K.E. + average PE.) at the instant*

$$= \frac{1}{4}ma_0^2\omega^2 e^{-2kt} + \frac{1}{4}ma_0^2\omega^2 e^{-2kt} = \frac{1}{2}ma_0^2\omega^2 e^{-2kt}$$

$$= E = E_0 e^{-2kt} = E_0 e^{-t/\tau},$$

where $\frac{1}{2}ma_0^2\omega^2 = E_0$,

*the average total energy of the undamped oscillator.*

$\therefore$ *average power dissipation (P) = rate of loss of energy*

$$= -\frac{dE}{dt} = ma_0^2\omega^2 ke^{-2kt} = 2kE = \frac{E}{\tau}. \qquad \left(\because 2k = \frac{1}{\tau}\right)$$

This loss of energy is obviously due to the work done against the damping or dissipative force $-\gamma\, dx/dt$ and usually appears in the form of heat.

*Alternatively,* we could obtain the same result as follows:

As in the case of a harmonic oscillator, so also here, the energy of an oscillator must be proportional to the square of its amplitude, *i.e.*,

$$E \propto (a_0 e^{-kt})^2.$$

Or, $$E = Ca_0^2\ e^{-2kt},$$

where C is a constant of proportionality.

After one full cycle (*i.e.*, one full time-period T), the energy is, say,

$$E_T = Ca_0^2\, e^{-2k(t+T)} = Ee^{-2kT}$$

∴ *loss of energy in one time-period*

$$T = E - Ee^{-2kT} = E - E\left(1 - 2kT + \frac{4k^2T^2}{2!}\cdots\right)$$

$$= E - E(1 - 2kT) = 2kET,$$

neglecting higher terms in k, since k is small.

And ∴ *rate of loss of energy over a time-period, i.e.*

$$P = \frac{2kET}{T} = 2kE = \frac{E}{\tau}.$$

## QUALITY FACTOR Q

As its very name indicates, it is a factor which measures the *quality* of a harmonic oscillator in so far as damping is concerned. *The less the damping, the better the quality of the harmonic oscillator as an oscillator and, therefore, the higher its quality factor Q.*

It is also referred to as the *figure of merit* of a harmonic oscillator and is defined as $2\pi$ *times the ratio between the energy stored and the energy lost per period.* Being thus a mere number, it is a *dimensionless quantity.*

Thus, $$Q = 2\pi \frac{\text{energy store}}{\text{energy lost per period}} = \frac{2\pi E}{PT},$$

where P is the *average loss of energy over a period* $= \dfrac{E}{\tau}$.

And, since $\dfrac{2\pi}{T} = \omega$, we have $Q = \dfrac{E\omega}{P} = \dfrac{E\omega}{E/\tau} = \omega\tau$.

In the case of low damping, $\omega = \omega_0$ and, therefore, $Q = \omega_0\tau$.

But, as we know, $\omega_0 = \sqrt{\dfrac{C}{m}}$ and $\tau = \dfrac{m}{\gamma}$. So that,

$$Q = \sqrt{\frac{C}{m}} \cdot \frac{m}{\gamma} = \frac{\sqrt{Cm}}{\gamma},$$

clearly indicating that the lower the value of $\gamma$, *i.e.*, the lower the damping, the higher the value of Q, *i.e.*, as $\gamma \to 0$, $Q \to \infty$.

Again; we know that the energy of a damped harmonic oscillator is given by $E = E_0e^{-t/\tau}$. Hence, if $t = \tau$, we have $E = E_0e^{-1} = E_0/e$, *i.e.*, the energy of the oscillator falls 1/eth of its initial value in time $t = \tau$. In this interval of time, the oscillator executes $(\omega_0/2\pi)\tau$ or $Q/2\pi$ oscillations, so that *the phase of the oscillator changes by Q.*

This gives us another method of defining Q, *viz.* as the phase *change brought about in the oscillator in the lime taken by its energy to fall to 1/eth of its initial value.*

## IMPORTANT EXAMPLES OF DAMPED HARMONIC OSCILLATORS

### (I) Dead Beat and Ballistic Galvanometers

In the case of a moving coil galvanometer, the turning of the coil under the *deflecting couple* is opposed, as we know; by a *restoring couple* C0, where C is the torsional couple per unit twist in the suspension fibre (equal to $\pi nr^4/2l$) and $\theta$, the deflection of the coil at the given instant t. In addition to this, there are two other clamping couples acting on the coil, *viz*., (i) a couple $-\gamma\, d\theta/dt$ due to the *mechanical damping* on account of viscosity of air and the elastic hysteresis of the suspension fibre, where $\gamma$ is the *damping coefficient* for this type of damping and (ii) a couple $-(\wedge/R)d\theta/dt$ due to *electromagnetic damping, i.e.*, damping on account of the induced current set up in the coil, which apart from being directly proportional to its velocity $d\theta/dt$, is also inversely proportional to its resistance and directly proportional to the magnetic flux generated through it and its area etc., all included in the constant $\wedge$.

If, therefore, I be the moment of inertia of the coil or the suspended system about the suspension fibre, the equation of motion of the system is

$$I\frac{d^2\theta}{dt^2} = -\gamma\frac{d\theta}{dt} - \frac{\wedge d\theta}{R\,dt} - C\theta.$$

Or, $$\frac{d^2\theta}{dt^2} + \left(\frac{\gamma + \wedge/R}{I}\right)\frac{d\theta}{dt} + \frac{C\theta}{I} = 0.$$

Now, $(\gamma + \wedge/R)/I = 2k = 1/\tau$ and $C/I = \omega_0^2$, where $w_0$ is the undamped angular frequency of the system.

So that, $$\frac{d^2\theta}{dt^2} + 2k\frac{d\theta}{dt} + \omega_0^2\theta = 0.$$

Or, $$\frac{d^2\theta}{dt^2} + \frac{1}{\tau}\frac{d\theta}{dt} + \omega_0^2\theta = 0.$$

This differential equation of motion of the coil, it will be readily seen, is the same as that of a damped harmonic oscillator, discussed under, except that we have here an angular, instead of a linear, displacement. And, therefore, its solution, as explained there, is given by

$$\theta = A_1 e^{\left(-k+\sqrt{k^2-\omega_0^2}\right)t} + A_2 e^{\left(-k-\sqrt{k^2-\omega_0^2}\right)t}$$

where the values of the arbitrary constants $A_1$ and $A_2$, obtained in the same manner, are

$$\frac{1}{2}\theta_0\left(1+\frac{k}{\sqrt{k^2-\omega_0^2}}\right) \text{ and } \frac{1}{2}\theta_0\left(1-\frac{k}{\sqrt{k^2-\omega_0^2}}\right)$$

respectively, with $\theta_0$ as the initial or maximum deflection.

So that, $$\theta = \frac{1}{2}\theta_0 e^{-kt}\left[\left(1+\frac{k}{\sqrt{k^2-\omega_0^2}}\right)e^{\left(\sqrt{k^2-\omega_0^2}\right)t} + \left(1-\frac{k}{\sqrt{k^2-\omega_0^2}}\right)e^{\left(-\sqrt{k^2-\omega_0^2}\right)t}\right].$$

As before, *three cases* arise:

(i) When $k > \omega_0$. This is the case of overdamping or heavy damping and $k^2 > \omega_0^2$ or $\left(\frac{\gamma + \wedge/R}{4I^2}\right)^2 > \frac{C}{I}$. So that, both the terms in the expression for $\theta$ above have an exponential term with a negative power and $\theta$, therefore, decays exponentially without changing sign. *The motion is thus non-oscillatory, aperiodic or dead beat.*

Hence, *for a galvanometer to be dead beat*, $k = 1/2\tau = (\gamma + \wedge/R)/2I$ *must be large, i.e.*, (a) *I, the M.I. of the coil or the suspended system must be small, (b) R, the resistance of the coil should be small, (c) ^ should be large, indicating that the induced currents set up in the coil and hence the magnetic flux through it should be large, i.e., the coil must be wound on a conducting frame and (d) the damping coefficient γ for mechanical damping must be large.*

(ii) **When** $k = \omega_0$, or $(\gamma + \wedge/R)^2/4I^2 = C/I$. In this case, the damping is critical and the coil comes back to its original position quickly.

(iii) **When** $k < \omega_0$ or $(\gamma + \wedge/R)^2/4I^2 < C/I$. This is the *case of underdamping* when, as explained under (3), the solution of the differential equation of motion of the coil is of the form $\theta = \theta_0 e^{-kt}$ sin

$(\omega t + \phi)$, where $\omega = \sqrt{\omega_0^2 - k^2} = \sqrt{\frac{C}{I} - (\gamma + \wedge/R)^2/4I}$

and, therefore, *frequency* $n = \frac{\omega}{2\pi}$.

The amplitude here goes on decreasing exponentially with time.

A galvanometer used for measuring charges must have the minimum damping and is called a *ballistic galvanometer.* Thus, in a ballistic galvanometer, $k = 1/2\tau = (\gamma + \wedge/R)/2I$ must be as small as possible and, therefore, the conditions for it are just the opposite of these for a dead beat galvanometer, *i.e., (a) I must be large, (b) R must be large, (c) the mechanical damping coefficient y must be small and (d) A should be small, i.e., the magnetic flux through the coil should be small and it should, therefore, be wound on a non-conducting frame of ebonite, ivory, bamboo etc.*–all these conditions being the very opposite of those required for a dead beat type of galvanometer.

The first throw of the ballistic galvanometer is corrected for any damping that may still be present, with the help of the logarithmic decrement of the instrument.

## (II) The L.C.R. Circuit

Any electrical circuit containing inductance (L), capacitance (C) and resistance (R) is an excellent example of a damped harmonic, oscillation, with the resistance, by itself alone, playing the part of a resistive or dissipative force analogous to that of friction or viscosity in the case of mechanical oscillations.

Thus, suppose we have such an electrical circuit, connected up as shown in Fig. 2.7, such that on pressing the knob of the Morse Key, M.K., the capacitor gets charged by the battery and on releasing it, the battery is thrown out of the circuit and the capacitor (C) connected in series with the inductance (L) and the resistance (R) through which it then discharges itself.

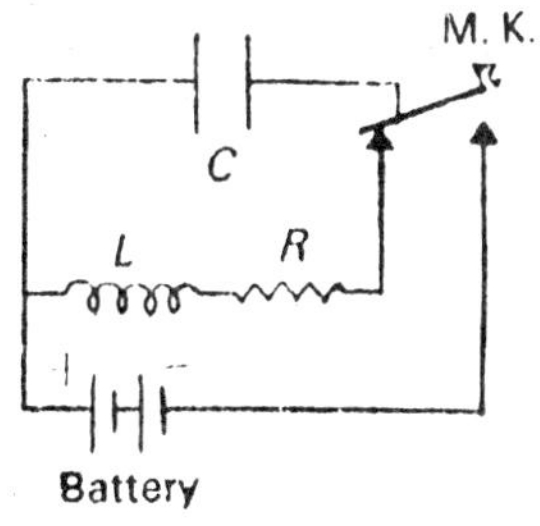


Fig. 2.7

Suppose further that at a given instant t, the charge on the capacitor is Q, the current flowing through the circuit (consisting of capacitor, inductance and resistance, with the battery thrown out) is I, so that the *induced emf* across the inductance is LdI/dt and the potential difference across the resistance is RI. Then, since there is now no external emf in the circuit (the battery being out), we have, in accordance with Kirchoff's law,

$$RI + \frac{Q}{C} = -L\frac{dI}{dt}.$$

Or, $$L\frac{dI}{dt} + RI + \frac{Q}{C} = 0.$$

But $I = \frac{dQ}{dt}$ and, therefore, $L\frac{d^2Q}{dt^2} + R\frac{dQ}{dt} + \frac{Q}{C} = 0.$

Or, $$\frac{d^2Q}{dt^2} + \frac{R}{L}\frac{dQ}{dt} + \frac{Q}{LC} = 0. \qquad \text{...(I)}$$

This equation is of the same form as that for a damped harmonic oscillator with the difference that displacement x there is replaced by charge Q here, 2k (or $1/\tau$) by R/L and $\omega_0^2$ by I/LC.

Proceeding in the same manner, therefore, we find the solution of the equation to be

$$Q = A_1 e^{\left(-k+\sqrt{k^2-\omega_0^2}\right)t} + A_2 e^{\left(-k-\sqrt{k^2-\omega_0^2}\right)t},$$

where the values of the arbitrary constants $A_1$ and $A_2$ respectively are

$$\frac{1}{2}Q_0\left(1+\frac{k}{\sqrt{k^2-\omega_0^2}}\right) \text{ and } \frac{1}{2}Q_0\left(1-\frac{k}{\sqrt{k^2-\omega_0^2}}\right).$$

Thus, $$Q = \frac{1}{2}Q_0 e^{-kt}\left[\left(1+\frac{k}{\sqrt{k^2-\omega_0^2}}\right)e^{\left(\sqrt{k^2-\omega_0^2}\right)t} + \left(1-\frac{k}{\sqrt{k^2-\omega_0^2}}\right)e^{\left(-\sqrt{k^2-\omega_0^2}\right)t}\right] \qquad \text{...(II)}$$

Again, the following *three cases* arise:

(i) **When the damping is large,** *i.e.*, when $k^2 > \omega_0^2$

or $$\left(\frac{R^2}{4L^2}\right) > \frac{1}{LC} \text{ or } R^2 > \frac{4L}{C}.$$

In this case, both the terms in expression II for Q above have exponential terms with a negative power, so that the charge decays exponentially with time, without changing sign, as shown by curve (i) in Fig. 2.8, and there is thus .no oscillation, *i.e., the discharge of the capacitor is non-oscillatory or aperiodic.*

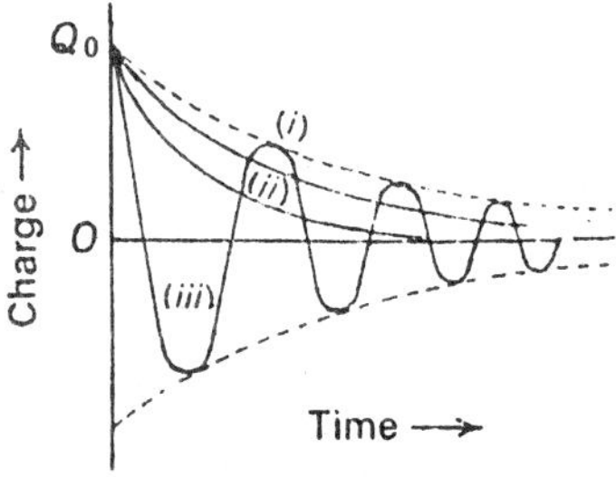


**Fig. 2.8**

(ii) **When the damping is critical,** *i.e.*, when $k^2 = \omega_0^2$ or $\frac{R^2}{4L^2} = \frac{1}{LC}$ or $R^2 = \frac{4L}{C}$. Here, the discharge becomes just aperiodic and dies down in the shortest possible time, as shown by curve (ii) in Fig. 2.8.

(iii) **With the damping is small,** *i.e.*, when $k^2 < \omega_0^2$ or $\frac{R^2}{4L^2} < \frac{1}{LC}$ or $R^2 < \frac{4L}{C}$. This is the case of interest to us, since here the solution of the equation of motion of the charge takes the form $Q = Q_0 e^{-kt} \sin (\omega t + \phi)$, *viz*, that of a *damped harmonic oscillator. The charge on the capacitor thus repeatedly acquires positive and negative values before ultimately decaying to zero. The discharge curve* of the capacitor is of the from shown by curve (iii) in Fig. 2.8 above, which, it will be recalled, is identical in form with the displacement curve of a damped harmonic oscillator, shown in Fig. 2.5 and is similarly bound by the curves $Q = Q_0 e^{-kt}$ and $Q = -Q_0 e^{-kt}$.

Thus, *the discharge of the capacitor is oscillatory in character, the oscillations being damped as. in the case of a damped harmonic oscillator.*

Differentiating the expression for Q with respect to t, we have

$$\frac{dQ}{dt} = \frac{dQ_0}{dt} e^{-kt} \sin(\omega t + \phi).$$

Or, since $dQ/dt = I$ and $dQ_0/dt = I_0$, the maximum value or the amplitude of the current, we have

$$I = I_0 e^{-kt} \sin(\omega t + \phi).$$

The *frequency of the oscillations* is $n = \frac{\omega}{2\pi} = \frac{\sqrt{\omega_0^2 - k^2}}{2\pi}$

$$= \frac{1}{2\pi}\sqrt{\frac{1}{LC} - \frac{R^2}{4L^2}}.$$

That the resistance R alone is responsible for the damping of the oscillations in this case (the energy dissipated appearing as heat) is clear from the fact-that in the absence of R, *i.e.*, if $R = 0$, we shall have $n = \frac{1}{2\pi}\sqrt{\frac{1}{LC}}$, the same as that of an L–C circuit, with no damping of the oscillations.

The *quality factor of the circuit,*

$$Q = \omega\tau = \frac{\omega}{2k} = \frac{L\omega}{R}, \text{ where}$$

$$\omega - \sqrt{\omega_0^2 - k^2} = \sqrt{\omega_0^2 - \frac{R}{4}l^2}.$$

Since R is small, $\omega \approx \omega_0$ and, therefore, *quality factor* $Q = \frac{L\omega_0}{R}$.

It will be readily seen that *in a purely inductive circuit, i.e., with* $R = 0$, *the quality factor Q will be infinite.*

## DRIVEN HARMONIC OSCILLATOR

When a harmonic oscillator oscillates in a medium like air, its oscillations, as we know, get damped, *i.e.*, its amplitude falls exponentially with time to *zero.*

If, however, we apply an *external periodic force* to the oscillator of a frequency *not necessarily the same as its own natural frequency*, a sort of tussle ensues between the damping force tending to retard its motion and the applied force tending to continue it. So that, after some initial erratic movements, it ultimately succumbs to the applied or the driving force and *settles down to oscillating with the forcing or the driving frequency (i.e., the frequency of the applied or the driving force) and a constant amplitude and phase so long as the applied force remains operative.*

An oscillator, thus compelled to oscillate with a frequency other than its own natural frequency, is called a *driven harmonic oscillator* and the oscillations executed by it are, therefore, called *driven* or *forced oscillations.*

In the *particular case* of forced oscillations *when the frequency of applied force is the same as the natural frequency of the oscillator itself, we have the phenomenon of resonance or resonant oscillations.*

Let us first deal with the *theory of a forced or driven oscillator*, in general.

Suppose the periodic force to which a damped harmonic oscillator is subjected is $F = F_0 \sin pt$ which is obviously a sinusoidal force, of *amplitude* (or *maximum value*) $F_0$ and frequency $\pi/2\pi$.

Since the damping and the restoring forces acting on the oscillator are respectively $-\gamma\, dx/dt$ its *equation of motion* becomes

$$m\frac{d^2x}{dt^2} = -\gamma\frac{dx}{dt} - Cx + F.$$

Or,
$$\frac{d^2x}{dt^2} + \frac{\gamma}{m}\frac{dx}{dt} + \frac{C}{m}x = \frac{F_0}{m}\sin pt.$$

Now, $\dfrac{\gamma}{m} = 2k$ and $\dfrac{C}{m} = \omega_0^2$, where $\omega_0$ is the *natural angular frequency* of the oscillator. So that, putting $\dfrac{F_0}{m}\sin pt = f_0 \sin pt$, representing the *applied force per unit mass*, we have

$$\frac{d^2x}{dt^2} + 2k\frac{dx}{dt} + \omega_0^2 x = f_0 \sin pt. \quad ...(I)$$

When the *steady state* has been attained, *i.e.*, when, after the tussle between the damping and the applied forces, the oscillator has settled down to oscillate with the forcing frequency $p/2\pi$ and a constant amplitude, let us try $x = A \sin(pt - \theta)$ as a particular solution of its equation of motion (I, above), where $\theta$ is the possible phase difference between the applied force and the displacement of the oscillator.

Then, clearly, $dx/dt = A\, p \cos (pt - \theta)$

and $d^2x/dt^2 = -A\, p^2 \sin (pt - \theta)$.

Substituting these values in equation (I), we have

$-\ Ap^2 \sin(pt - \theta) + 2k\ Ap \cos(pt - \theta) + \omega_0^{\ 2} A \sin(pt - \theta)$

$= f_0 \sin[(pt - \theta) + \theta]$

$= f_0 \sin(pt - \theta) \cos\theta + f_0 \cos(pt - \theta) \sin\theta.$

Or, $A(\omega_0^{\ 2} - p^2) \sin(pt - \theta) + 2k\ Ap \cos(pt - \theta)$

$$= f_0 \cos\theta \sin(pt - \theta) + f_0 \sin\theta \cos(pt - \theta) \quad ...(II)$$

If this solution is to hold good for all values of t, the coefficients of sin(pt – θ) and cos(pt – θ) on either side of equation II must respectively be equal, *i.e.*, we must have $A(\omega_0^{\ 2} - p^2) = f_9 \cos\theta$ ...(i)

and $\qquad 2k\ Ap = f_0 \sin\theta.$ $\qquad$ ...(ii)

Squaring and adding relations (i) and (ii), we have

$$A^2(\omega_0^{\ 2} - p^2)^2 + 4k^2 A^2p^2 = f_0^{\ 2} \cos^2\theta + f_0^{\ 2} \sin^2\theta.$$

whence,
$$A^2 = \frac{f_0^2}{\left(\omega_0^2 - p^2\right)^2 4k^2p^2}$$

∴ *amplitude of the driven or forced oscillator,*

$$A = \frac{f_0}{\sqrt{\left(\omega_0^2 - p^2\right)^2 + 4k^2p^2}}, \quad ...(III)$$

taking only the positive value of the square root. Its negative value will mean opposite phase but then 6 too will change by w and there would, therefore, be no effect on the value of A.

And, $\tan\theta = \dfrac{f_0 \sin\theta}{f_0 \cos\theta} = \dfrac{2kp}{\left(\omega_0^2 - p^2\right)}.$

Or, the *phase difference* between the driven or forced oscillator and the applied force is

$$\theta = \tan^{-1} \frac{2kp}{\left(\omega_0^2 - p^2\right)} \quad ...(IV)$$

Since sin θ is positive, it follows that θ *lies within the range 0 to π.*

Substituting these values in the relation x = A sin (pt – θ), we have

$$x = \frac{f_0}{\sqrt{\left(\omega_0^2 - p^2\right)^2 + 4k^2p^2}} \sin\left(pt - \tan^{-1} \frac{2kp}{\omega_0^2 - p^2}\right) \quad ...(V)$$

which represents a S.H.M. of frequency $p/2\pi$, *i.e.*, the same as that of the driving force, but lagging behind it in phase by $\theta = \tan^{-1} [2kp/(\omega_0^2 - p^2)]$, where $\theta$ lies between 0 and $\pi$ ($\therefore$ sin $\theta$ is positive).

Now, $x = A \sin (pt - \theta)$ is not really the complete solution of equation I, which is an *inhomogeneous differential equation* because of the presence of the term $f_0 \sin pt$ which contains neither the variable (x) nor its derivative. The solution will, therefore, be complete only if we add to it a complementary function which is a solution of the related homogeneous equation

$$\frac{d^2x}{dt^2} + \frac{2k\,dx}{dt} + \omega^2 x = 0.$$

One such solution, as we know, is $x = a_0 e^{-kt} \sin (\omega t + \phi)$, representing a damped harmonic oscillation, where $\omega = \sqrt{\omega_0^2 - k^2}$. The addition of this term $[x = a_0 e^{-kt} \sin (\omega t + \phi)]$ to the solution $x = A \sin (pt - \theta)$ does not in any way impair the validity of the latter, since the term, taken by itself, reduces the left hand side of equation I to zero.

The *complete solution of equation* I is thus $x = a_0 e^{-kt} (\omega t + \phi) + A \sin (pt - \theta)$, where, obviously, the first term on the right hand side represents an initial damped oscillation of frequency $\omega/2\pi$, with its amplitude decaying *exponentially* to zero, and the second represents a forced or driven oscillation of the forcing (or driving) frequency $p/2\pi$ and a *constant amplitude* A. The former oscillation dies out quickly and the latter alone then remains effective, so that we are left with $x = A \sin(pt - 0)$ as the *equation of motion of the forced or driven oscillation*, with its *amplitude A* and *phase angle* $\theta$ given by relations III and IV above respectively.

Now, when k has a finite value, greater than zero, the value of A will obviously be the maximum when the denominator in expression III (for A) has its minimum value, *i.e.*, when

$$\frac{d}{dp}\left[\left(\omega_0^2 - p^2\right)^2 + 4k^2p^2\right] = 0,$$

Or, $\quad - 4(\omega_0^2 - p^2)\, p + 8k^2p = 0,$

*i.e.*, when $p^2 = \omega_0^2 - 2k^2$ and, therefore, $p = \sqrt{\omega_0^2 - 2k^2}$.

*Alternatively,* we can obtain the same result by putting expression III for A in the form

$$A = \frac{f_0}{\sqrt{\left(p^2 + 2k^2 - \omega_0^2\right)^2 + 4k^2\omega_0^2 - 4k^4}},$$

from which also, A will be maximum when $p^2 + 2k^2 - \omega_0^2 = 0$, or, when

$$p^2 = \omega_0^2 - 2k^2,$$

or

$$p = \sqrt{\omega_0^2 - 2k^2}.$$

Thus, the amplitude of the driven oscillator will be the maximum when the driving frequency is $\sqrt{\omega_0^2 - 2k^2}/2\pi$, which we may denote by pr/π.

This phenomenon where for a particular driving frequency, the response or the amplitude of the driven oscillator is the maximum is called *amplitude resonance* and the particular driving frequency is referred to as the *resonant frequency.*

It may be noted that *this resonant frequency*

$$\left(pR/2\pi = \sqrt{\omega_0^2 - 2k^2}/2\pi\right)$$

is *smaller than both the natural, undamped frequency* $\omega_0/2\pi$ *and the natural damped frequency*

$$\frac{\omega}{2\pi} = \sqrt{\left(\omega_0^2 - k^2\right)}/2\pi \textit{ of the oscillator.}$$

Substituting pR for p in the relation for A above therefore, we have

$$\textit{maximum amplitude, } A_{max} = \frac{f_0}{2k\left(\omega_0^2 - k^2\right)^{1/2}} \qquad \text{...(i)}$$

Or, because $\omega_0^2 - 2k^2 = p^2$

or, $\omega_0^2 = p^2 + 2k^2$, we have

$$A_{max} = \frac{f_0}{2k\left(p^2 + k^2\right)^{1/2}}, \qquad \text{...(ii)}$$

showing that *the smaller the value of k, the greater the value of* $A_{max}$

(i) It follows, therefore, that *in case me damping be low,* $p \approx \omega_0$ and the maximum amplitude in that case is $A_{max} = f_0/2k\,\omega_0 = f_0\tau/\omega_0$, ($\because$ $1/2k = \tau$). Obviously then, *in the absence of damping* (*i.e.*, k = 0), the amplitude should become *infinite*. Since, however, damping is never actually zero, this never happens.

(ii) Again, if *the driving frequency be negligibly small or zero and the damping low*, we have

$$A = \frac{f_0}{\omega_0^2}.$$

Thus, the ratio of the response (or amplitude) of the oscillator when the driving frequency is equal to the resonant frequency to its response when the driving frequency is zero or negligible is clearly

$$\frac{f_0 \tau/\omega_0}{f_0/\omega_0^2} = \omega_0 \tau = Q, \text{ the } \textit{quality factor of the oscillator,}$$

which thus actually controls its response or amplitude at this frequency.

It will also be easily seen that $\frac{f_0}{\omega_0^2} = \frac{F_0/m}{C/m} = \frac{F_0}{C}$,

showing that if *the driving frequency be negligible or zero and the damping low, the amplitude or the response of the oscillator is controlled neither by its mass nor any damping but simply by the force constant C.*

(iii) Finally, if *the damping be low and the driving frequency high*, so that $\omega_0$ is negligible in comparison, the amplitude of the driven oscillator is given by

$$A = \frac{f_0}{p^2},$$

indicating that *the response in this case falls as p increases.*

It is thus clear from the above discussion, as indeed from the expression for A itself, that the magnitude of A depends upon the relative values of p and $\omega_0$ and that it is, intact, controlled by the factor $(\omega_0^2 - p^2)$. The natural consequence is that its value diminishes both when $p < \omega_0$ and when $p > \omega_0$. This will be amply clear from Fig. 2.9, where a number of curves are drawn showing the relation between and the amplitude (A) of the driven oscillator for different cases of damping. It will be noted that

(a) *The peak value of amplitude*, $A_{max} = \frac{f_0}{2k\omega_0} = \frac{f_0}{2kp}$ *occurs when* $(\omega_0^2 - p^2) = 0$, *i.e., when* $p = \omega_0$ *which represents the condition for amplitude resonance.*

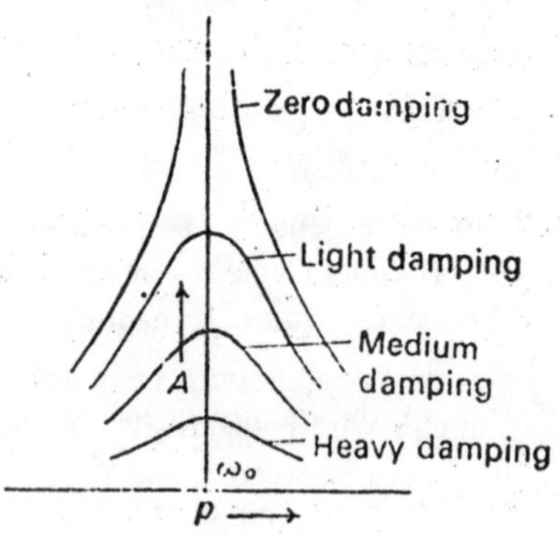


**Fig. 2.9**

And, if k = 0, *i.e.*, if there be no damping, $A_{max}$ becomes infinite for this value of p, the curve being asymptotic to the y or the amplitude axis. Since damping is never zero, we do not obtain such a curve in actual practice but only those of the type shown for light and medium damping.

(b) *The peak value of A (i.e., $A_{max}$) is naturally different for different cases of damping, becoming less and less as damping increases, but always occurs at, or very nearly at, $(\omega_0^2 - p^2) = 0$ provided the damping is not too large, i.e., the resonant frequency ($p_R$) in all these cases is equal to, or very nearly equal to, $\omega_0$.*

*Only in the case of heavy damping does the peak value occur at a frequency less than $\omega_0$, (being equal to $\sqrt{\omega_0^2 - 2k^2}$ ).*

*(c) The fall in the curves on either side of $(\omega_0^2 - p^2) = 0$ is steeper in the case of smaller than in the case of heavier damping.*

## SHARPNESS OF RESONANCE

*We have just seen in Fig. 2.9 above how the curves between p and A fall more steeply on either side of the respective peak values of A when the damping is low than when it is high. Thus, in the case of high damping, the amplitude remains more or less at its peak value over an appreciable range on either side of $(\omega_0^2 - p^2) = 0$ i.e., even when $p \neq \omega_0$. The oscillator thus responds to a number of frequencies near about $\omega_0$ on either side of it. The resonance in this case is, therefore, said to be flat.*

On the other hand, *if the damping be low*, the **steep fall** of curve on either side of the peak value of A shows that *the oscillator responds only to the frequency exactly equal to its natural frequency* $\omega_0$ *and to none others. The resonance here is, therefore, said to be sharp.*

Thus, sharpness of resonance is, in a way, a measure of the rate of fall of amplitude from its maximum value at resonant frequency, on either side of it. *The sharper the fall in amplitude, the sharper the resonance.*

Familiar examples of *flat* and *sharp* resonance are the case of air column and the sonometer wire respectively. On account of the large value of k for air, the curve between p and A is comparatively flat at the peak Value of A. The air column thus continues to respond to frequencies over an appreciable range "n either side of its natural frequency $\omega_0$, so much so that it becomes difficult to determine the coned length of the air column that exactly responds to the oscillator. This is, therefore, a case of *flat resonance.*

A sonometer wire, on the other hand, with little or no damping, responds only to one particular frequency, *viz.*, its own natural frequency $\omega_0$ and to none others. *The resonance here is, therefore, sharp.*

Thus, *the smaller the damping, the sharper the resonance.* In fact, it can be shown that *sharpness of resonance is inversely proportional to the square of the damping constant k.*

## PHASE OF THE DRIVEN OSCILLATOR

We have seen above how a forced or driven oscillation is represented by the equation $x = A \sin(pt - 0)$, indicating that it always lags a phase angle $0 = \tan^{-1}[2kp/(\omega_0^2 - p^2)]$ behind the driving force $f_0 \sin pt$.

Obviously, this phase angle ($\theta$) depends upon the damping and the relative values of $\omega_0$ and p. *If the damping be negligible, the following three cases arise:*

(i) When $p < \omega_0$, *i.e., when the forcing or driving frequency is smaller than the natural, undamped frequency of the oscillator.* In this case, clearly, $\tan\theta = 2kp/(\omega_0^2 - p^2)$ be a small positive quantity and *the driven oscillator will, therefore, very nearly be in phase with the driving force.*

(ii) When $p > \omega_0$, *i.e., when the driving frequency is greater than the natural, undamped frequency of the oscillator*. Here, obviously, tan

$\theta$ will be a small negative quantity and the *driven oscillator will thus be out of phase with the driving force, i.e.*, the phase difference between the two will be $\pi$ (or T/2).

(iii) When $p = \omega_0$, *i.e., when the driving frequency is equal to the natural, undamped frequency of the oscillator.* In this case, $\tan\theta = \infty$ and, therefore, $\theta = \pi/2$, *i.e.., the driven or forced oscillator will differ in phase from the driving force by $\pi/2$ (or T/4). Thus, its displacement will be the maximum when the driving force is zero and vice versa.*

It is clear from the above that *the phase angle $\theta$ changes from 0 to $\pi$ but remains positive throughout.*

The general nature of the phase lag of a driven oscillator as the driving frequency increases, passing through the value (do (the natural or undamped frequency of the oscillator) will be clear from the curves in Fig. 2.10, drawn for different cases of damping. The following points emerge;

(i) When the damping is zero, the curve runs along the frequency axis from 0 to $\omega_0$ and again parallel to it from ay to $2\omega_0$ but removed from the first part by $\pi$. This means, in other words, that for $p << \omega_0$, $\theta = 0$ but when $p >> \omega_0$, $\theta$ suddenly jumps to $\pi$.

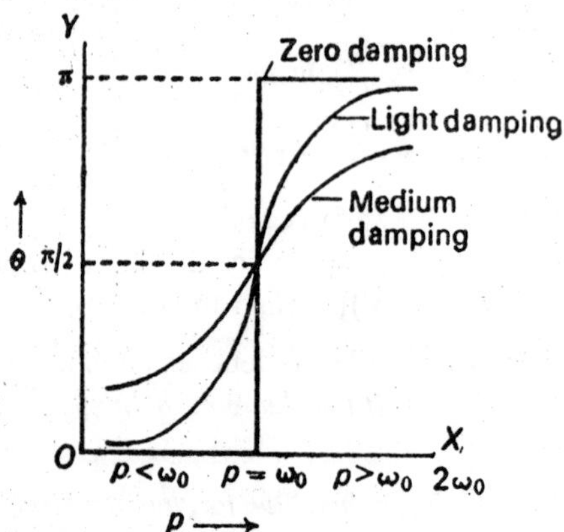


Fig. 2.10

(ii) With damping present, the phase lag ($\theta$) increases from 0 to $\pi/2$ as the driving frequency p increases from 0 to $\omega_0$ and then approaches $\pi$ as p continues to increase beyond $\omega_0$.

(iii) The rate of change of the phase angle is more rapid when the damping is low than when it is high.

(iv) Except at very high damping, all curves pass through $\pi/2$ when $p = \omega_0$, *i.e.*, amplitude resonance occurs when the driving frequency is equal to the natural, undamped frequency of the oscillator, or the resonance frequency in all these cases remains $\omega_0$.

## VELOCITY RESONANCE

The displacement of a driven oscillator at a given instant t, is, as we know, given by

$$x = A \sin(pt - \theta) = \frac{f_0}{\sqrt{\left(\omega_0^2 - p^2\right) + 4k^2p^2}} \sin(pt - \theta). \quad \text{...(I)}$$

$\therefore$ the *velocity of the driven oscillator at that instant*, is

$$v = \frac{dx}{dt} = pA \cos(pt - \theta) = v_0 \cos(pt - \theta), \text{ because}$$

$$pA = p\frac{f_0}{\sqrt{\left(\omega_0^2 - p^2\right)^2 + 4k^2p^2}} = v_0.$$

Or, $$v = v_0\left(pt + \theta + \frac{\pi}{2}\right),$$

showing that the velocity leads the displacement in phase by $\pi/2$.

Clearly, the velocity amplitude

$$v_0 = pA = \frac{f_0 p}{\sqrt{\left(\omega_0^2 - p^2\right)^2 + 4k^2p^2}}$$

and 0, as we know, equal to

$$\tan^{-1}\left[\frac{2kp}{\left(\omega_0^2 - p^2\right)}\right].$$

Thus, *the velocity amplitude* ($v_0$) *varies with p, being zero when p = 0 and the maximum when* $p = p_R = \omega_0$, this maximum value being $f_0/2k = f_0\tau$.

This *acquiring of its maximum velocity by a driven oscillator is spoken of as velocity resonance and obviously occurs when the driving frequency (p) is equal to the resonant frequency* ($p_R$), *equal to the natural, undamped frequency of the oscillator*, $\omega_0 = C/m$.

For all other values of p, greater or smaller than $p_R = \omega_0$, the velocity amplitude is naturally smaller, because the displacement amplitude (A) is then much smaller.

Now, as we know, the displacement at resonance lags in phase by $\pi/2$ behind the driving force and, as we have just seen, the velocity, then, leads the displacement in phase by $\pi/2$, *i.e.*, the displacement also lags in phase by $\pi/2$ behind the velocity. Obviously, therefore, *at resonance, the velocity of the driven oscillator is in phase with the driving force.* This is, therefore, the most favourable circumstance for the transference of energy from the driving force (F ) to the driven oscillator, for both F and v being in the same phase, with v at its maximum, the rate, of transference of energy, Fv, has its highest positive value.

## HALF WIDTH OF RESONANCE CURVE

The frequency—amplitude curve of a driven oscillator, (Fig. 2.9), is, as we know, symmetrical (for small damping) about the resonant frequency $p_R = \omega_0$, indicating that its amplitude, which is the maxi- mum ($A_{max}$) at $p_R$, falls on either side of it. There must, therefore, be a frequency $p_H$, say, for which the amplitude will be half of the maximum, *i.e.*, equal to $A_{max}/2$, as shown in Fig. 2.11.

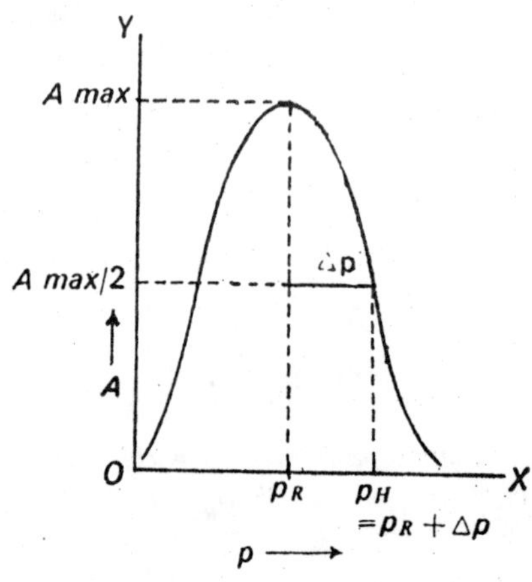


**Fig. 2.11**

*The change $\Delta p$, (from $p_R$ to $p_H$), in the value of the driving frequency p for which the amplitude of the driven harmonic oscillator falls from* *its maximum value $A_{max}$ to half this maximum value $A_{max}/2$ is called the half width of the resonance curve.*

Thus, *half width of the resonance curve,* $\Delta p = |p_H - p_R|$.

Now, as we know,

$$A_{max} = \frac{f_0}{2k\left(\omega_0^2 - k^2\right)^{1/2}} = \frac{f_0}{\sqrt{4k^2\omega_0^2 - 4k^4}}.$$

$$\therefore \quad A_{max}/2 = \frac{f_0}{2\sqrt{4k^2\omega_0^2 - 4k^4}} \qquad \text{...(i)}$$

Now,
$$A = \frac{f_0}{\sqrt{\left(\omega_0^2 - p^2\right)^2 + 4k^2p^2}}$$

$$= \frac{f_0}{\sqrt{\left(\omega_0^2 - 2k^2 - p^2\right)^2 + 4k^2\omega_0^2 - 4k^4}}$$

So that, $A_{max}/2$ also
$$= \frac{f_0}{\sqrt{\left(\omega_0^2 - 2k^2 - pH^2\right)^2 + 4k^2\omega_0^2 - 4k^4}} \qquad \text{...(iii)}$$

[$\because$ p = pH when A = $A_{max}/2$].

From the two values of $_{Amax}/2$, as given by relations (i) and (ii), we, therefore, have $(\omega_0^2 - 2k^2 - pH^2)^2 + 4k^2\omega_0^2 - 4k^4 = 4(4k^2\omega_0^2 - 4k^4)$.

But, as we know, $\left(\omega_0^2 - 2k^2\right) = p_R^2$. And, therefore,

$(p_R^2 - p_H^2) = 3(4k^2\omega_0^2 - 4k^4 = 3(4k^2 p_R^2 - 4k^4)$ [$\because = \omega_0 = p_R$].

Since with low damping, k is small and, therefore, $4k^4$ quite negligible, we have

$(p_R^2 - p_H^2) = 3(4k^2 p_R^2)$, whence, $p_R^2 - p_H^2 = \pm\sqrt{3(2kp_R)}$.

Or,
$$p_H^2 = p_R^2 \mp \sqrt{3(2kp_R)} = p_R^2\left(1 \mp \sqrt{3}\frac{2k}{p_R}\right).$$

Or,
$$p_H = p_R\left(1 \mp \sqrt{3}\frac{2k}{p_R}\right)^{1/2} \approx p_R\left(1 \pm \frac{\sqrt{3}k}{p_R}\right) = p_R \mp \sqrt{3}k.$$

$\therefore$ *for low damping, half width of the resonance curve,* $\Delta p = (p_H - p_R) \approx \sqrt{3}k$.

It is thus possible to estimate the value of the damping constant, as also, of course, that of Mg, from the frequency-amplitude curve of a driven harmonic oscillator.

## POWER ABSORPTION

In many an important problem on driven harmonic oscillators, as for example, in discussing the driven oscillations of an electron about a point of stable equilibrium, we are interested not so much in its displacement from the equilibrium position—which, in the case of an election, at any rate, we cannot even hope to observe directly—as in the energy absorbed by the oscillator to keep itself in motion.

We shall, therefore, calculate the average power absorbed per cycle by the driven oscillator to offset its loss of power in overcoming the frictional or resistive forces and thus to maintain its oscillations. *When the oscillator has settled down to a steady state of oscillation, the average power absorbed will obviously be just equal to the average power dissipated.*

If $F = F_0 \sin pt$ be the applied or the driving force and dx, the displacement of the oscillator in time dt, we have *work done, or energy supplied by the applied force to the oscillator, is*

*given by* $$dE = Fdx = F\frac{dx}{dt}dt,$$

whence, $$\frac{dE}{dt} = F\frac{dx}{dt}. \qquad \text{...(i)}$$

Now, $F = F_0 \sin pt = mf_0 \sin pt$

and $$\frac{dx}{dt} = \frac{pf_0}{\sqrt{(\omega_0^2 - p^2)^2 + 4k^2p^2}}\cos(pt - \theta) \qquad \text{...(ii)}$$

Substituting these values in relation (i) above, we have

*power absorbed by the oscillator*, $P = \frac{dE}{dt} = F\frac{dx}{dt}$

$$= mf_0 \sin pt \frac{pf_0 \cos(pt - \theta)}{\sqrt{(\omega_0^2 - p^2)^2 + 4k^2p^2}}.$$

Or, $$P = F\frac{dx}{dt} = \frac{mf_0^2 p}{\sqrt{(\omega_0^2 - p^2)^2 + 4k^2p^2}}\sin pt \cos(pt - \theta).$$

To obtain the *average power*, $P_{(av)}$, *over a full cycle or time-period*, we note that the average value of sin pt cos (pt – θ)

$$= \frac{1}{2}\left[\sin(2p-\theta)-\sin\theta\right] \text{ over a cycle or time-period T is}$$

$$\frac{1}{T}\int_0^T \frac{1}{2}\left[\sin(2pt-\theta)-\sin\theta\right]dt = \frac{1}{2}\sin\theta.$$

And, since 2k Ap = $f_0$ sin θ and $A = \dfrac{f_6}{\sqrt{\left(\omega_0^2-p^2\right)^2+4k^2p^2}}$, we have

$$\sin\theta = \frac{2k\,Ap}{f_0} = \frac{2kp}{f_0}\cdot\frac{f_0}{\sqrt{\left(\omega_0^2-p^2\right)^2+4k^2p^2}}.$$

$$= \frac{2kp}{\sqrt{\left(\omega_0^2-p^2\right)^2+4k^2p^2}}.$$

So that, average value of sin pt cos (pt – θ) over a full cycle or time-period

$$= \frac{1}{2}\sin\theta = \frac{1}{2}\frac{2kp}{\sqrt{\left(\omega_0^2-p^2\right)^2+4k^2p^2}} = \frac{kp}{\sqrt{\left(\omega_0^2-p^2\right)^2+4k^2p^2}}$$

∴ *average power,*

$$P_{(av)} = \frac{mf_0^2p}{\sqrt{\left(\omega_0^2-p^2\right)^2+4k^2p^2}}\cdot\frac{kp}{\sqrt{\left(\omega_0^2-p^2\right)^2+4k^2p^2}}$$

$$= \frac{mk\,f_0^2p^2}{\left(\omega_0^2-p^2\right)+4k^2p^2} \quad \text{...(iii)}$$

But $\dfrac{f_0^2p^2}{\left(\omega_0^2-p^2\right)^2+4k^2p^2} = v_0^2,$

where $v_0$ is the *velocity amplitude.*

Therefore, *average power absorbed by the driven oscillator per cycle or time-period,*

$$P_{(av)} = mk\,v_0^2 = mv_0^2/2\tau = \gamma\,v_0^2/2 \quad \text{...(iv)}$$

[∵ k = 1/2τ and m/τ = γ].

It is clear from relation (iii) above that the average power absorbed will have its *maximum value*, $P_{max}$, when p = $\omega_0$. So that, *maximum power absorbed,* $P_{max} = \frac{1}{4}\frac{mf_0^2}{k} = \frac{1}{2}mf_0^2\tau.$ ...(v)

Thus, $\omega_0 = p_R$, *the resonant frequency for velocity also turns out to be the resonant frequency for the average power absorbed.*

*It may be noted that the maximum power is absorbed at the frequency of velocity resonance,* $p_R = \omega_0$, *and not at the frequency of amplitude resonance,* $\sqrt{\omega_0^2 - 2k^2}$.

*Half width of average power absorbed versus frequency carve.* As we have just seen above (expression v), the maximum value of average power absorbed is $P_{max} = mf_0^2/4k$ at resonance frequency $p_R = \omega_0$. So that, half its maximum value of average power absorbed = $P_{max}/2 = mf_0^2/8k$. Let us calculate the frequency at which this is the average power absorbed.

We have relation (iii) above for average power absorbed ($P_{(av)}$). So that, if $P_{(av)} = P_{max}/2 = mf_0^2/8k$, we have

$$P_{(av)} = \frac{mkf_0^2 p^2}{\left(\omega_0^2 - p^2\right)^2 + 4k^2p^2} = \frac{mf_0^2}{8k},$$

whence, $$\frac{kp^2}{\left(\omega_0^2 - p^2\right)^2 + 4k^2p^2} = \frac{1}{8k}.$$

Or, $$(\omega_0^2 - p^2)^2 + 4k^2p^2 = 8k^2p^2.$$

Or, $$\left(\omega_0^2 - p^2\right)^2 = 4k^2p^2,$$

whence, $$\omega_0^2 - \omega_0^2 = \pm\ 2kp$$

Or, $$\omega_0 - p = \pm\frac{2kp}{\omega_0 + p}$$

$$= \frac{2k}{\omega_0/p + 1}$$

$$= \pm\frac{2k}{2} = \pm k. \qquad [\because\ \omega_0 \approx p.]$$

Thus, $p = \omega_0 \pm k = p_H$, say.

Or, since $\omega_0 = p_R$, we have $|p_H - p_R| = \Delta p = k$.

This change $\Delta p$, (from $p_R$ to $p_H$), in the value of the driving frequency p for which the average power absorbed by the driven oscillator falls from its maximum value $P_{max}$ to half this value $P_{max}/2$ is called the half width of the average power absorbed versus driving frequency curve, shown in Fig. 2.12.

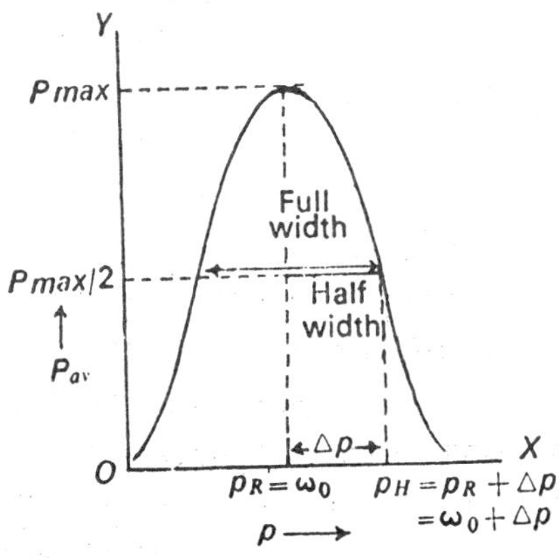


**Fig. 2.12**

So that, *half width of the average power absorbed versus driving frequency curve*, $\Delta p = k = 1/2\tau$.

*Quality factor Q of the driven oscillator.* We know that $Q = w_0\tau$.

Here, $\Delta p = k = \dfrac{1}{2\tau}$ and $\therefore\ \tau = \dfrac{1}{2}\Delta p$.

So that, $Q = \omega_0\tau = \dfrac{\omega_0}{2\Delta p} = \dfrac{\text{resonance frequency}}{\text{full width of power frequency curve at half maximum power}}$.

It may, however, also be obtained directly as follows:

By definition, $Q = 2\pi \dfrac{\text{average energy stored}}{\text{energy dissipated per cycle or per time time - period}} = \dfrac{E_{(av)}}{P_{(av)} \times T}$

Now, at any instant, E = K.E. + P.E.

$$= \frac{1}{2} mp^2 A^2 \cos^2(pt - \theta) + \frac{1}{2} m\omega_0^2 A^2 \sin^2(pt - \theta).$$

Since the average value of $\cos^2(pt-\theta)$ and $\sin^2(pt-\theta)$ over a whole time-period is 1/2 each and the *average power absorbed,*

$$Q = 2\pi \frac{\frac{1}{4}mp^2A^2 + \frac{1}{4}m\omega_0^2A^2}{\left(\frac{mA^2p^2}{2\tau}\right)\times T} = 2\pi \frac{\frac{1}{4}mA^2\left(p^2+\omega_0^2\right)}{\left(\frac{mA^2p^2}{2\tau}\right)\times\left(\frac{2\pi}{p}\right)}$$

$$= \frac{1}{2}\frac{\left(p^2+\omega_0^2\right)}{p}\tau \quad [\because T = 2\pi/p].$$

Or, $$Q = \frac{1}{2}\left(p+\frac{\omega_0^2}{p}\right)\tau = \frac{1}{2}\left(1+\frac{\omega_0^2}{p^2}\right)p\tau.$$

Since at, or near, resonance, $p = \omega_0$, we have

$$Q = \frac{1}{2}(2)p\tau = p\tau = \omega_0\tau = \frac{\omega_0}{2\Delta p}, \text{ as obtained above.}$$

At low damping, obviously $k = 1/2\tau$ will be small or $\tau$ large, so that $Q = \omega_0\tau$ *will also be large, making for sharpness of resonance.* Thus, the *quality factor Q is a measure of sharpness of resonance in the case of a driven harmonic oscillator.*

## SUPERPOSITION-PRINCIPLE

This principle is a general one, applicable to all cases where the equations governing motion or displacement are linear. It merely states that *solutions in all such cases are additive, i.e.*, if $x_1$ be the displacement due to a force $F_1$ and $x_2$ due to another force $F_2$, the displacement due to the combined force $(F_1 + F_2)$ is equal to $(x_1 + x_2)$. This process of vector addition of displacements is termed *super-position.*

It is thus possible to obtain the displacement under the combined force $(F_1 + F_2)$ if the displacement under each of the two forces individually be known.

The principle is also applicable to electromagnetic waves because the relationship between electrical and magnetic fields is linear. It, however, fails to apply the moment the displacement's or disturbances are large and the ordinary linear laws of mechanical action cease to hold. Thus, in the case of sound, it certainly applies to ordinary sound waves but not to shock waves caused by 'violent explosions.

In our present case of harmonic oscillators it applies because of the linearity of their equations of motion. It ceases to apply when this linearity is lost. Thus, it does not apply in the case of an harmonic oscillators or even in the case of harmonic oscillators with large displacements when the linear relationship given by Hooke's law no longer holds.

In view of this principle, therefore, we can look upon the response of a harmonic oscillator to a sinusoidal force as being the sum of its responses to a number of components of various frequencies into which the given force may be imagined to be resolved.

## DRIVEN L.C.R CIRCUIT

An LCR circuit, *i.e.*, a circuit containing an inductance (L), a capacitance (C) and a resistance (K) in series, (Fig. 2.13), functions as a driven oscillator if it be connected to an external source of alternating emf to supply the necessary energy to maintain the oscillations.

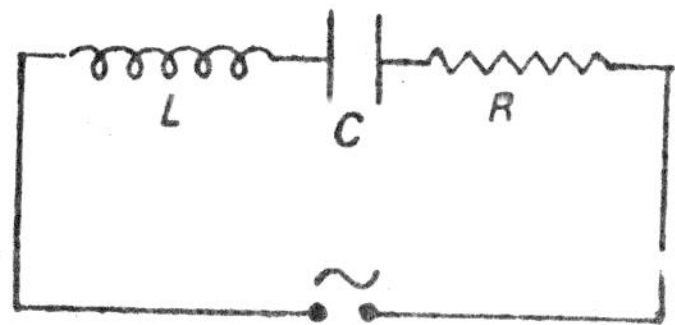


**Fig. 2.13**

If I be the current in the circuit at a given instant, with Q as the charge on the capacitor at that instant, we have potential difference across the inductance = – LdI/dt, potential difference across the capacitor = Q/C and potential difference across the resistance = RI.

If, therefore, $E = E_0 \sin pt$ be the external emf applied to the circuit, we have.

$$E - L(dI/dt) = \left(\frac{Q}{C}\right) + RI.$$

Or, $$L\left(\frac{dI}{dt}\right) + RI + \left(\frac{Q}{C}\right) = E = E_0 \sin pt \qquad ...(i)$$

Or, $$\frac{dI}{dt} + \left(\frac{R}{L}\right)I + \frac{Q}{LC} = \left(\frac{E_0}{L}\right)\sin pt.$$

Or, since $I = \frac{dQ}{dt}$, we have

$$\frac{d^2Q}{dt^2} + \left(\frac{R}{L}\right)\left(\frac{dQ}{dt}\right) + \frac{Q}{LC} = \left(\frac{E_0}{L}\right)\sin pt.$$

This is an equation of an identical form with that for a driven harmonic oscillator, discussed earlier with Q replacing x and R/L, 1/LC and $E_0/L$ *replacing* 2k, $\omega_0^2$ and $f_0$ respectively,

It steady state solution too is similarly

$$Q = \frac{\frac{E_0}{L}}{\sqrt{\left(\frac{1}{LC} - p^2\right) + \left(\frac{pR}{L}\right)^2}}\sin(pt - \theta) \qquad \text{...(ii)}$$

where $$\theta = \tan^{-1}\frac{\frac{pR}{L}}{\frac{1}{LC} - p^2}.$$

We can easily obtain the value of the current flowing through the circuit by differentiating this expression for Q with respect to t, just as we can obtain the value of the velocity v of the oscillator-by differentiating the earlier expression for x with respect to t.

*Alternatively,* we can proceed directly with a trial solution $I = I_0 \sin (pt - \phi)$.

So that, $$\frac{dI}{dt} = I_0 p\cos(pt - \phi)$$

and $$Q = \int I\,dt = \int I_0 \sin(pt - \phi)dt = \frac{I_0}{p}\cos(pt - \phi).$$

Substituting these values in equation (i) above, we have

$$LpI_0(\cos pt - \phi) + RI_0 \sin(pt - \phi) - \frac{I_0}{Cp}\cos(pt - \phi) = E_0 \sin pt.$$

Or, $I_0[R \sin (pt - \phi) + \left(Lp - \frac{1}{Cp}\right) \cos (pt - \phi) = E_0 \sin pt$ ...(iii)

Now, putting $R = a \cos \phi$ and $\left(Lp - \frac{1}{Cp}\right) = a\sin\phi$, we have

$$a = \sqrt{\left(R^2 + \left(Lp - \frac{1}{Cp}\right)^2\right)} \text{ and } \tan\phi = \frac{\left(Lp - \frac{1}{Cp}\right)}{R}.$$

Substituting these values of R and $\left(Lp - \frac{1}{Cp}\right)$ in relation (iii) above, we have

$$I_0 a[\cos\phi \sin(pt - \phi) + \sin\phi \cos(pt - \phi)] = E_0 \sin pt.$$

Or, $\quad I_0\, a \sin[(pt - \phi) + \phi] = I_0\, a \sin pt = E_0 \sin pt.$

Or, $$I_0 \sqrt{R^2 + \left(LP - \frac{1}{Cp}\right)^2} \sin pt = E_0 \sin pt,$$

whence, $$I_0 \sqrt{R^2 + \left(Lp - \frac{1}{Cp}\right)^2} = E_0.$$

Or, $$I_0 = \frac{E_0}{\sqrt{R^2 + \left(Lp - \frac{1}{Cp}\right)^2}} \qquad \text{...(iv)}$$

$$\therefore I = I_0 \sin(pt - \phi) = \frac{E_0}{\sqrt{R^2 + \left(Lp - \frac{1}{Cp}\right)^2}} \sin(pt - \phi) \qquad \text{...(v)}$$

where $\phi = \tan^{-1} \frac{\left(Lp - \frac{1}{Cp}\right)}{R}$.

It will be readily seen that in expression (iv) for $I_0$, the denominator $\sqrt{R^2 + (Lp - 1/Cp)^2}$ functions as the *effective resistance* in the circuit. It is called *impedance* of the circuit, denoted by the letter Z and measured in *ohms*.

Clearly, the impedance is made up of two parts, the *ohmic* resistance R and the quantity $(Lp - 1/Cp)$, called *reactance,* usually denoted by X and also measured in *ohms* (though L and C are individually measured in *henry* and *farad* respectively).

The relation between *resistance, reactance, impedance* and $\phi$ is best illustrated as in Fig. 2.14.

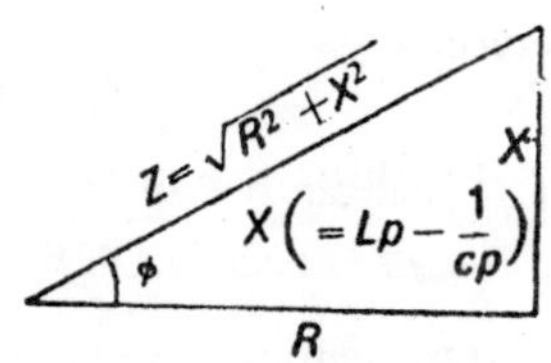


**Fig. 2.14**

The *reactance* too, as can readily be seen, .is made up of two parts: Lp, the *reactance due to inductance*, usually denoted by $X_L$ and 1/Cp, the *reactance due to capacitance*, denoted by $X_C$, so that,

$$X = X_L + X_C.$$

We thus have $Z = \sqrt{R^2 + X^2}$.

Or, $\quad$ impedance $= \sqrt{\text{Resistance}^2 + \text{Reactance}^2}$

And, clearly, *peak value or amplitude* of the current, *i.e.*,

$$I_0 = \frac{E_0}{Z} = \frac{E_0}{\sqrt{R^2 + X^2}}.$$

And, therefore, $I = \frac{E_0}{Z} \sin(pt - \phi) = I_0 \sin(pt - \phi)$.

Obviously, the *current* and emf in the circuit differ in phase by $\phi$,

where $\phi = \tan^{-1} \frac{Lp - 1/Cp}{R} = \tan^{-1}\left(\frac{\text{Reactance}}{\text{Resistance}}\right)$

$$= \tan^{-1}\left(\frac{X_L - X_C}{R}\right).$$

*Now, whereas the ohmic resistance (R) has no effect on the phase angle* (because it is quite independent of the frequency), *the inductance in a circuit makes the emf lead the current (in phase), and the capacitance in a circuit makes the current lead the emf. When both are present in a circuit, therefore, the lead or lag of the emf or the current depends upon the relative values of $X_L$ or Lp and Xc or 1/Cp and the following three cases arise:*

(i) **When the value of p is such that $X_L = X_C$.** In this case, obviously, Lp = 1/Cp, *i.e.*, *the reactance in the circuit*, $X = X_L - X_C = 0$. So that, *impedance in the circuit = resistance* R, and, therefore, the least.

∴ *amplitude or peak value of the current,* $I_0 = E_0/R$, the *maximum,* which approaches ∞ as R → 0.

Obviously, here, tan f $= \dfrac{X_L - X_C}{R}$ and

∴ $\phi = 0$, *i.e., the current in the circuit is in phase with the applied emf.*

This is, therefore, the case of resonance, with $p^2 = 1/LC$ and ∴ $p = 1/\sqrt{LC} = \omega_0$.

Thus, *resonance frequency* $= \dfrac{p}{2\pi} = \dfrac{\omega_0}{2\pi} = \dfrac{1}{2\pi\sqrt{LC}}$.

This circuit is, therefore, called a *series resonance circuit.*

It may be noted that current resonance here is analogous to velocity resonance in the case of mechanical oscillators.

(ii) **When p has, a value such that $X_L > X_C$.** Here, since Lp > l/Cp, the *net reactance in the circuit is inductive, tan $\phi$, a positive quantity and, therefore, $\phi = \pi/2$. The applied emf thus leads the current by $\pi/2$ and the value of p is greater than $\omega_0$.*

(iii) **When the value of p is such that $X_L < X_C$.** In this case, Lp being less than 1/Cp, *the net reactance in the circuit is capacitive,* tan $\phi$, a *negative quantity* and, therefore, $\phi = -\dfrac{\pi}{2}$. *The current thus leads the applied emf by $\pi/2$ and the value of p is less than $\omega_0$.*

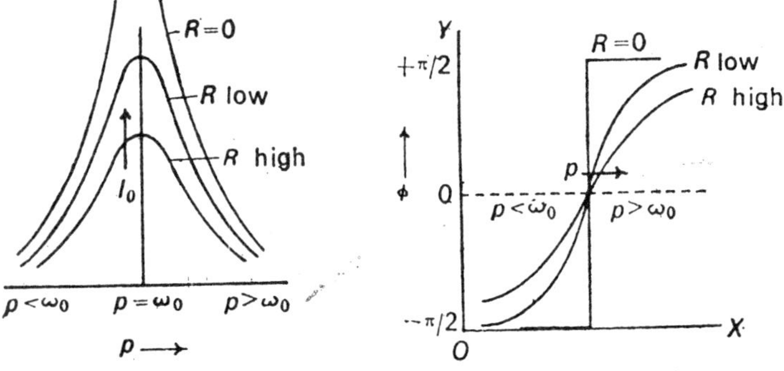


Fig. 2.15

These changes in the peak value of the current $I_0$ and in the value of the phase angle $\phi$ with p are shown in Figs. 2.15 (a) and (b) respectively, for different values of R, which corresponds to k in the case of mechanical oscillators.

It will be seen at once that *the lower the value of R, the sharper the resonance and that the current lags or leads the emf according as p is greater than or less than $\omega_0$.*

## PARALLEL RESONANCE CIRCUIT

When the *inductance* (L) and the *capacitance* (C) in a circuit are placed in parallel (instead of in series), as shown in Fig. 2.16, the applied voltage is the same across both inductance and capacitance.

Now, as we know, the current $I_L$ through the inductance *lags*, and the current $I_C$ through the capacitance *leads*, the emf by $\pi/2$, *i.e.*,

$$I_L = \frac{E_0}{Lp}\sin\left(pt - \frac{\pi}{2}\right)$$

and $$I_C = \frac{E_0}{\frac{1}{Cp}}\sin\left(pt + \frac{\pi}{2}\right).$$

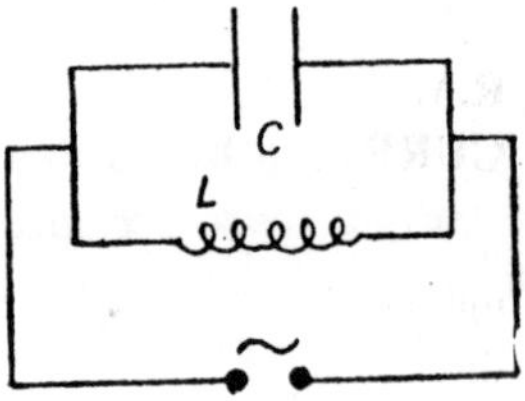

Fig. 2.16

$\therefore$ *total current in the circuit,*

$$I = I_L + I_C = \frac{E_0}{Lp}\sin\left(pt - \frac{\pi}{2}\right)$$

$$+\frac{E_0}{\frac{1}{Cp}} = \sin\left(pt + \frac{\pi}{2}\right)$$

Or, $$I = E_0\left(Cp - \frac{1}{Lp}\right)\cos pt.$$

*The current thus leads or lags ths emf by a phase angle $\pi$.*

For a particular value of p, we have Cp = 1/Lp, when $p^2 = 1/LC$ or $p = 1/\sqrt{LC}$.

The current through the inductance is equal in magnitude to that through the capacitance, the two differing in phase by $\pi$. The net current through the circuit is thus *zero*.

This frequency p is called the *resonant frequency* of the circuit and the circuit itself is referred to as a *parallel resonance circuit.*

As in the series resonant circuit, therefore, we have $p = 2\pi n = 1/\sqrt{LC}$ or $n = 1/2\pi\sqrt{LC}$.

The circuit, under this condition, oscillates with its natural frequency, without requiring any supply of energy from an outside source. In actual practice, however, the inductance coil does possess some resistance and the oscillations of the circuit can be maintained only by some supply of energy from an external source.

It will be seen at once that *whereas in a series circuit, the current at resonance is the maximum and the impedance minimum (or zero), in a parallel circuit, the current at resonance is zero and the impedance maximum.*

The parallel resonance circuit is thus used In wireless transmitting circuits to filter or to cut out the current of resonant frequency, allowing currents of other frequencies to pass through the circuit. It is also, therefore, called a *rejector circuit.*

## R.M.S. OR EFFECTIVE VALUES OF ALTERNATING CURRENT AND EMF

Since an alternating current and emf vary continuously from a maximum in one direction, through zero, to a maximum in the opposite direction, their average values over one complete cycle or time-period are zero each. We, therefore, obtain their mean or average value $I_{(av)}$ and $E_{(av)}$ over half their lime-periods and these can be shown to be equal to $2I_0/\pi$ and $2E_0/\pi$ respectively.

For most of our purposes, however, we are interested in knowing the *effective value* of an alternating current or an alternating emf in terms of its equivalent direct current or emf.

*This effective or virtual value of an alternating current is defined as that steady (or direct) current which would produce the same heating effect as the alternating current in a given resistance in a given time.*

Thus, if $I_v$ be the effective or virtual value of an alternating current $I = I_0 \sin \omega t$, clearly, *the heating effect produced by this steady or direct current in a resistance R in* time $= t = I_v^2 Rt$.

And *the heating effect produced by the alternating current in the same resistance in the same time t* = (*average value of* $I^2 = I_0^2 \sin^2 \omega t$ *over a cycle or time-period* T) $\times R \times t = (I_0^2/2) \times R \times t$, (because the average value of $\sin^2 \omega t$ over a whole cycle or time-period is 1/2).

We, therefore, have $I_v^2\,Rt = \left(\dfrac{I_0^2}{2}\right) Rt.$

Or, $$I_v^2 = \frac{I_0^2}{2},$$

whence $$I_v = \frac{I_0}{\sqrt{2}},$$

*i.e., effective or virtual value of alternating current* $= \dfrac{I_0}{\sqrt{2}}.$

Since this is the square root of the mean square value ($I_0^2/2$) of the current, it is also called the *root mean square* or the r.m.s. *value of* the alternating current.

Similarly, we have effective, virtual or r.m.s. value of alternating emf equal to $E_0/\sqrt{2}$.

## POWER IN AN ALTERNATING CURRENT CIRCUIT

*The power (P) in an electrical circuit is the amount of energy delivered per second by the source to the circuit*, and in a direct current (or D.C.) circuit is measured by the product EI, where E is the emf in the circuit and I, the current.

Since in a D.C. circuit, we have only resistance and no inductance or capacitance, E = IR and, therefore, power $P = I^2R$, appearing in the form of heat.

In an *alternating current* (or A.C.) *circuit*, say in an LCR circuit, the emf and current both vary continuously, *sinusoidally*, and are given by $E = E_0 \sin pt$ and $I = I_0 \sin(pt - \phi)$ respectively. We, therefore have

*power* $P = EI = (E_0 \sin pt)\,[I_0 \sin (pt - f)]$

$$= \frac{E_0 I_0}{2}\,[\cos\phi - \cos(2pt - \phi)]$$

Since the average value of $\cos(2pt - \phi)$ over a whole cycle or time-period is *zero*, we have

$$P = \frac{E_0 I_0}{2}\cos\phi = \frac{E_0}{\sqrt{2}}\frac{I_0}{\sqrt{2}}\cos\phi = E_v I_v \cos\phi$$

$$\left[\because \frac{E_0}{\sqrt{2}} = E_v \text{ and } \frac{I_0}{\sqrt{2}} = I_v\right]$$

where $\cos\phi = \frac{R}{Z} = \frac{R}{\sqrt{R^2 + \left(Lp - \frac{1}{Cp}\right)^2}}$.

The unit of power is the watt = 1 joule/sec.

Obviously, the *maximum power* will be $E_vI_v$ (when cos ϕ = 1) and, therefore,

$$\frac{\text{true power expended}}{\text{maximum power}} = \frac{E_v I_v \cos\phi}{E_v I_v} = \cos\phi.$$

Thus, *true power expended = maximum power* × cos ϕ.

This is why $\cos\phi = \frac{R}{Z}$ is called the **power factor** *of the A.C. circuit.*

As we know, ϕ = 0 and, therefore, cos ϕ = 1 *only in the case of resonance, for reactance is then zero and impedance Z = R.* So that, in *this case, true power = maximum power* = $E_vI_v = I_v^2R$

[∵ $E_v = I_vR$ here,]

It may as well be pointed out that, even otherwise, *the power in an A.C. circuit, like that in a D.C. circuit, appears in the form of heat in the resistance alone, because the power expended in a pure inductance or a pure capacitance is always zero.* This may also be seen from the following:

We have $P = E_vI_v \cos\phi$. Or, since $E_v = I_vZ$, we have

$$P = (I_vZ)\, I_v \cos\phi.$$

Or, $$P = Iv^2 Z\left(\frac{R}{Z}\right) = Iv^2 R,$$

showing that the power appears in the form of heat in the resistance R.

Finally, as in the case of a mechanical oscillator, so also here, the *quality factor of the circuit,*

$$Q = \frac{Lp}{R} = \frac{L\omega_0}{R} \qquad (\because p \approx \omega_0, \text{ at or near resonance})$$

$$= \omega_0\tau = \frac{\omega_0}{2\Delta p}$$ = *frequency at resonance/full width of power-frequency curve at half maximum power, and is a measure of the sharpness of resonance of the LCR circuit.*

## SOLVED EXAMPLES

***Example 1:***

*Write down the equation of motion of the damped harmonic oscillator of example 4 above and calculate the time in which (i) its amplitude (ii) its energy falls to $l/e^{th}$ of its undamped value.*

*If the mass of the oscillator be 1.127 gm, calculate its average rate of loss of energy.*

***Solution:***

The differentia) equation of a damped harmonic oscillator, as we know (See 2.2), is

$$\frac{d^2x}{dt^2} + 2k\frac{dx}{dt} + \omega_0^2 x = 0.$$

which, when the oscillator is underdamped, *i.e.*, when $k << \omega_0$, gives

$$x = a_0 e^{-kt} \sin(\omega t + \phi).$$

So that its amplitude is given by $a = a_0 e^{-kt} = a_0 e^{-t/2\tau}$, where $a_0$ is its amplitude the absence of damping.

(i) Now, as we have seen under example 4 above, the damping constant k for the given oscillator works out to 0.02. If, therefore, its amplitude is reduced to $1/e^{th}$ of its undamped value, we have

$$a = \frac{a_0}{e} = a_0 e^{-1}$$

and, therefore, $a_0 e^{-1} = a_0 e^{-kt}$. Or, $kt = 1$,

whence, $t = \frac{1}{k} = \frac{1}{0.02} = 50$ sec.

Thus, *the amplitude of the oscillator falls to $1/e^{th}$ of its undamped value in 50 sec.*

(ii) Again, as we have seen in 2.3, the energy of a damped harmonic oscillator at a given instant t is given by $E = E_0 e^{-2kt}$. So that, when its energy falls to $1/e^{th}$ of its value $E_0$ in the absence of damping, we have

$$E = \frac{E_0}{e} = E_0 e^{-1}.$$

And, therefore, $E_0 e^{-1} = E_0 e^{-2kt}$. Or, $2kt = 1$,

whence, $t = \frac{1}{2k} = \frac{1}{2} \times 0.02 = 25$ sec.

Thus, *the energy of the oscillator falls to $1/e^{th}$ of its value in the absence of damping in 25 sec.*

The *rate of loss of energy* $= \frac{E}{\tau} = 2kE.$

Now, $E = \frac{1}{2} m a_0^2 \omega_0^2 e^{-2kt},$

where m = 1.127 gm, $a_0$ = 2.011 cm,

$$\omega_0 = \frac{2\pi}{T} = \frac{2\pi}{1.15}, \ k = 0.02$$

and t = 25 sec. So that,

*rate of loss energy*

$$= 2 \times 0.02 \times \frac{1}{2} \times 1.127 \times (2.011)^2 \times (2\pi/1.15)^2 \, e^{-2(0.02 \times 25)}$$

$$\frac{0.02 \times 1.127 \times (2.011)^2 \times 4\pi^2}{(1.15)^2 \times e} = \frac{0.02 \times 1.127 \times (2.011)^2 \times 4\pi^2}{(1.15)^2 \times 2.7183}$$

$$= 1 \text{ erg/sec.}$$

***Example 2:***

*If the amplitude of a seconds pendulum, with a bob of mass 200 gm, is reduced to half its undamped value in 200 seconds, what fa its quality factor Q? What should be the mass of the bob, its size and shape remaining unchanged, in order that the damping of the pendulum may become critical? (Take $\log_e 10 = 2.30$).*

***Solution:***

We know that the amplitude of the pendulum (a damped harmonic oscillator) at a given instant t is given by $a = a_0 e^{-kt} = a_0 e^{-t/2\tau}$.

Here, $a = \frac{a_0}{2}$ and t = 200 sec.

So that, $\frac{a_0}{2} = a_0 e^{-200/2\tau} = a_0 e^{-100/\tau}.$

Or, $\frac{100}{\tau} = \log_e 2 = 2.30 \log_{10} 2 = 2.30 \times 0.3010$

whence, $\tau = \frac{100}{2.30} \times 0.3010 = \frac{100}{0.6923}$.

And, since $\omega_0 = \frac{2\pi}{T} = \frac{2\pi}{2} = \pi$ ($\because$ T = 2 sec), we have

*Quality factor of the pendulum,* $Q = \omega_0\tau = \pi \frac{100}{0.6923} = 453.7$.

For *critical damping*, as we know, the condition is that $k = \omega_0$. So that, $\tau'$ be the value of the *relaxation time* in this case, we have $1/2\tau' = \omega_0$

Or, since $\omega_0 = \pi$, we have $1/2\tau' = \pi$ and, therefore, $\tau' = \frac{1}{2\pi}$.

If m and m' be the masses of the bob in the two cases respectively, we have $\gamma/m = 1/\tau$ and $\gamma/m' = 1/\tau'$ ($\therefore$ $\gamma$ remains the same in either case).

$$\therefore \quad \frac{\tau'}{\tau} = \frac{\left(\frac{\gamma}{m}\right)}{\left(\frac{\gamma}{m'}\right)} = \frac{m'}{m}, \text{ whence, } m' = m \times \frac{\tau'}{\tau}.$$

Substituting the values of m, $\tau'$ and $\tau$, therefore, we have

$$m' = 200 \times \frac{1}{2\pi} \times \frac{0.6923}{100} = \frac{0.6923}{\pi} = 0.2204 \text{ gm.}$$

Thus, *for the damping to be critical, the mass of the bob must be* 0.2204 gm.

***Example 3:***

*If the quality factor of an underdamped harmonic oscillator of frequency 512 be 8 × $10^4$, calculate the time in which its energy is reduced to $1/e^{th}$ of its energy in the absence of damping. How many oscillations does the oscillator make in this time?*

*Calculate the percentage reduction in the frequency of the oscillator due to damping. What inference do you drew from it?*

***Solution:***

The energy of a damped harmonic oscillator at an instant t, as we know, is given by $E = E_0 e^{-2kt} = E_0 e^{-t/\tau}$, where $E_0$ is its energy in the absence of damping.

Since here $\quad E = \frac{E_0}{e} = E_0 e^{-1}$,

we have $E_0 e^{-1} = E_0 e^{-t/\tau}$,

whence, $1 = t/\tau$ or, $t = \tau$.

And since $Q = \omega\eta\tau$, we have

$$\tau = \frac{Q}{\omega_0} = \frac{Q}{2\pi\eta_0} = \frac{8\times10^4}{2\pi\times512} = \frac{10^4}{2\pi\times64} = \frac{10^4}{128\pi}.$$

And therefore, the time in which the energy of the oscillator falls to 1/eth of its energy in the absence of damping, *i.e.*,

$$t = \tau\ 10^4/128\pi = 24.87 \text{ sec.}$$

And, clearly, *number of oscillations made by the oscillator in this time*

$$= nt = \left(\frac{\omega_0}{2\pi}\right)\tau = \frac{Q}{2\pi} = 8\times\frac{10^4}{2\pi} = 12740.$$

Now, in the case of an underdamped harmonic oscillator, we have

$$\omega = \sqrt{\omega_0^2 - k^2},$$

Or, $$\omega^2 = \omega_0^2 - k^2 = \omega_0^2 - \left(\frac{1}{2\tau}\right)^2 = \omega_0^2 - \frac{1}{4}\tau^2.$$

So that, dividing by $\omega_0$ throughout, we have

$$\frac{\omega^2}{\omega_0^2} = 1 - \frac{1}{4\omega_0^2\tau^2} = 1 - \frac{1}{4Q^2} \qquad [\because \omega_0\tau = Q.]$$

And, therefore, $$\frac{\omega}{\omega_0} = \left(1 - \frac{1}{4Q^2}\right)^{1/2} - \left(1 - \frac{1}{2}\times\frac{1}{4Q^2} + \ldots\right) = 1 - \frac{1}{8Q^2}.$$

Or, $$\frac{2\pi n}{2\pi n_0} = \frac{n}{n_0} = 1 - \frac{1}{8Q^2},$$

where $n_0$ is the frequency of the undamped oscillator.

$\therefore$ *percentage reduction in the frequency of the oscillator*

$$= \frac{n_0 - n}{n_0}\times100$$

$$= \left(1 - \frac{n}{n_0}\right)\times100\left[1 - \left(1 - \frac{1}{8Q^2}\right)\right]\times100 = \frac{100}{8Q^2} = \frac{12.5}{Q^2}$$

$$= \frac{12.5}{(8\times10^4)^2} = 1.953\times10^{-9}.$$

The obvious *inference* from the above result is that *for oscillators with large values of the quality factor Q (which is usually the case), the reduction in frequency due to damping is negligibly small.*

***Example 4:***

*If in problem 14 above, the rate of dissipation of energy is desired to be halved, what should be (i) the amplitude of the applied periodic force, its frequency remaining unchanged, (ii) the frequency of the applied periodic force, its amplitude remaining unchanged?*

***Solution:***

(i) We know that the *rate of dissipation of energy* α (*amplitude of the force*)$^2$.

Thus, if $R_1$ and $R_2$ be the rates of dissipation of energy and $f_{01}$ and $f_{02}$ the amplitudes of the force in the two cases respectively, *the frequency of the force remaining the same in either case,* we have

$$\frac{R_1}{R_2} = \left(\frac{f_{01}}{f_{02}}\right)^2$$

We know that the *effective current in an LCR circuit*

$$= \frac{\text{effective emf}}{\text{impedance of}} = \frac{E}{Z}.$$

Here, *effective* emf E = 500 *volts* and *impedance*

$$Z = \sqrt{R^2 + L^2 p^2}$$

$$= \sqrt{R^2 + L^2 (2\pi n)^2},$$

$$\text{Or, } Z = \sqrt{(4)^2 + (0.68)^2 (2\pi \times 120)^2} = \sqrt{16 + 262800} = 512.6 \text{ ohms.}$$

$$\therefore \textit{ effective current in the circuit, } I = \frac{500}{512.6} = 0.9754 \text{ amp.}$$

$$\text{And, } \textit{power factor of the circuit } = \frac{R}{Z} = \frac{4}{512.6} = 0.0078.$$

***Example 5:***

*An electric lamp which runs at 40 volts D.C. and consumes 10 amperes current is connected to A.C. mains at 100 volts, 50 cycles. Calculate the inductance of the choke.*

***Solution:***

Clearly, *resistance of the lamp,* $R = \frac{\text{voltage}}{\text{current}} = \frac{40}{10} = 4$ ohms.

And, *impedance of A.C. circuit,* $Z = \sqrt{R^2 + (Lp)^2}$, where L is the *inductance of the choke.*

Here, R = 4 ohms, π = 2πn × 50 = 100π and I = 10.

$$\therefore \quad I = \frac{E}{Z} = \frac{E}{\sqrt{R^2 + (Lp)^2}} = \frac{100}{\sqrt{4^2 + (L + 100\pi)^2}} = 10.$$

So that, $\frac{10^4}{16 + 10^4 \pi^2 L^2} = 100.$

Or, $10^6\pi^2L^2 = 10^4 - 1600 = 8400.$

So that, $L^2 = \frac{8400}{10^6 \pi^2}.$

Or, $L = \sqrt{\frac{84}{10^4 \pi^2}} = \frac{\sqrt{84}}{100\pi} = 0.02917.$

Thus, the *inductance of the choke* = 0.02917 henry.

***Example 6:***

*An inductance coil of inductance 0.05 henry and resistance 12.4 ohms is connected to an A.C. supply at 200 volts and 50 cycles. Find (a) power factor, and (b) angle of lag. (c) Show that the power consumed in just equal to the rate of production of heat.*

***Solution:***

(a) We know that the power factor,

$$\cos\phi = \frac{R}{Z} = \frac{R}{\sqrt{R^2 + (Lp)^2}}.$$

Here, R = 12.4 ohms, L = 0.05 henry

and π = 2πn = 2π × 50 = 100π.

$$\therefore \textit{ power factor, } \cos\phi = \frac{12.4}{\sqrt{(12.4)^2 + (0.05 \times 100\pi)^2}}$$

$$= \frac{12.4}{\sqrt{153.8 + 246.7}} = \frac{12.4}{20} = 0.62.$$

(b) And, therefore, *angle of lag*, f = $\cos^{-1}$ (0.62) = 51° 41'.

(c) Clearly, *power consumed*, $E_vI_v \cos \phi$.

Now, $E_v$ = 100 volts,

$$I_v = \frac{E_v}{Z} = \frac{100}{\sqrt{(12.4)^2 + (0.05 + 100\pi)^2}}$$

$$= \frac{100}{20} = 5 \text{ ampere and } \cos \phi = 0.62.$$

∴ *power consumed* = 100 × 5 × 0.62 = 310 watts.

And, *rate of production of heat* = $I^2R$ = 25 × 12.4 = 310 watts.

Thus, *power consumed* = *rate of production of heat.*

***Example 7:***

*A massless spring, suspended from a rigid support, carries a mass of 500 gm at its lower end and the system oscillates with a frequency of S/sec, with the amplitude reduced to half its undamped value in 20 sec. Calculate (i) the force constant of the spring. (ii) the relaxation time of the system and (iii) its quality factor.*

*(iv) What would be the quality factor of the system if the suspended mass be reduced to 150 gm?*

***Solution:***

(i) We have the relation $\omega_0 = \sqrt{C/m}$ or $C = m\omega_0^2$, where C is the *force constant* of the spring and m, the *mass* suspended from it.

Here, $\omega_0 = 2\pi n \times 2\pi \times 5 = 10\pi$ radian/sec and m = 500 gm.

So that, *force constant of the spring*, $C = 500 \times (10\pi)^2 = 5\pi^2 \times 10^4$

$= 4.934 \times 10^5$ dynes/cm.

(ii) The amplitude at an instant t is given by $a = a_0e^{-kt}$.

Since $a = a_0/2$ and t = 20 sec, we have $a_0/2 = a_0e^{-20k}$.

Or, $\log_e 2 = 20k$.

Or, $2.30 \log_{10} 2 = 20k$.

Or, $2.30 \times 0.3010 = 20k$,

whence, $k = 2.30 \times 0.3010/20$.

∴ *relaxation time of the system,*

$$\tau = \frac{1}{2k} = \frac{1\times20}{2\times2.30\times0.3010} = 14.44 \text{ sec.}$$

(iii) The *quality factor of the system*, $Q = \omega_0\tau = 10\pi \times 14.44 = 453.7$.

(iv) Since the force constant (C) of the spring remains the same irrespective of the mass suspended from it, and so does the damping factor $\gamma = 2km = 2 \times 2.30 \times 0.010 \times 500/20 = 115 \times 0.3010$, we have

$$\omega'_0 = \sqrt{\frac{C}{m'}} = \sqrt{5\pi^2 \times \frac{10^4}{150}} = 18.26\ \pi, \text{ since } m' = 150 \text{ gm.}$$

∴ $\tau'$, the relaxation time now $= 1/2k = m'/g = \frac{150}{115} \times 0.3010$.

And, therefore, the *quality factor now*, $Q' = \omega_0'\tau'$

$$= 18.26\pi\times\frac{150}{115}\times0.3010 = 248.5.$$

***Example 8:***

*The restoring torque exerted by the suspension fibre of a ballistic galvanometer is 103 dyne-cm/radian. If the period of vibration of the coil be $2\pi$ seconds and its amplitude be reduced to $1/e^{th}$ of its undamped value in 100 seconds, calculate the values of (i) the moment of inertia of the suspended system, (ii) the damping factor $\gamma$ and (iii) the quality factor Q of the system.*

***Solution:***

(i) The oscillating system in a ballistic galvanometer, as we know, is an underdamped one (2.51 (ii)) and, therefore, the angular frequency of the system is given by $\omega = \sqrt{\omega_0^2 - k^2}$, where $\omega_0$ is the undamped angular frequency of the system and k, the damping constant.

∴ *time-period of the suspended system or the coil,*

$$T = \frac{2\pi}{\sqrt{\omega_0^2 - k^2}}.$$

Or, since $T = 2\pi$ sec, we have

$$2\pi = \frac{2\pi}{\sqrt{\omega_0^2 - k^2}},$$

whence, $\sqrt{\omega_0^2 - k^2} = 1.$

Or, $\omega_0^2 - k^2 = 1.$

Now, the amplitude of the coil at an instant t is given by $\theta = \theta_0 e^{-kt}$, where $\theta_0$ is its undamped amplitude. And, since the amplitude is reduced to $\theta_0/e = \theta_0 e^{-1}$ in time t = 100 sec, we have

$$\theta_0 e^{-1} = \theta_0 e^{-100k}$$

or, $100\,k = 1$

or $k = \frac{1}{100} = 0.01.$

$\therefore$ $\omega_0^2 - k^2 = \omega_0^2 - (0.01)^2 = 1.$

Or, $\omega_0^2 \approx 1$ and $\therefore\ \omega_0 \approx 1$ rad/sec.

Hence $C/I = \omega_0^2 = 1$, where C is the torque exerted by the suspension fibre per radian twist of the fibre.

Since C = $10^3$ *dyne-cm/radian* (given), we have $\frac{10^3}{I} = 1.$

Or, $I = 10^3$ dyne-cm$^2$.

Thus, *the moment of Inertia of the suspended system about the suspension fibre, i.e.*

$$I = 10^3 \text{ or } 1000 \text{ dyne-cm}^2.$$

(ii) Now, $(\gamma \times$ ^/R)/I = 2k. Or, $\gamma/I = 2k$ (the electromagnetic damping being obviously ignored in the problem). So that,

*damping factor* $\gamma = I \times 2k = 1000 \times 2 \times 0.1 = 20$ dyne-cm/sec.

(iii) *The quality factor of the system,*

$$Q = \omega_0 \tau = \frac{\omega_0}{2k} = \frac{1}{2} \times 0.01 = 50.$$

### *Example 9:*

*Give the theory of oscillations in a series LCR circuit with small damping. Deduce the frequency and quality factor for a circuit with L = 2mH, C = 5mF and R = 0.2 ohm.*

### *Solution:*

For answer to the first part of the question. The circuit being equivalent to an *underdamped harmonic oscillator*, we have

$$\omega = \sqrt{\omega_0^2 - k^2}$$

and, therefore, *frequency of the oscillations*,

$$n = \frac{\omega}{2\pi} = \sqrt{\omega_0^2 - \frac{k^2}{2\pi}}.$$

Or, $$n = \frac{1}{2\pi}\sqrt{\frac{1}{LC} - \frac{R^2}{4L^2}}.$$

Here, $L = 2\text{mH} = 0.002$ henry,

$C = 5\mu F = 5 \times 10^{-6}$ farad

and $R = 0.02$ ohm. We, therefore, *have frequency of the circuit*,

$$n = \frac{1}{2\pi}\sqrt{\frac{1}{0.002\times5\times10^{-6}} - \frac{(0.02)^2}{4(0.002)^2}} = \frac{1}{2\pi}\sqrt{\frac{4\times10^8 - 10^4}{4}}$$

$$= \frac{10^2}{2\pi}\sqrt{10^4} - 1 = \frac{10^4}{2\pi} = 1.592 \times 10^3 \text{ cps.}$$

Now, the resistance being small, *quality factor*

$$Q = \frac{L\omega_0}{R} = \frac{L\omega}{R},$$

where $\omega = 2\pi n = 2\pi \times 10^4/2\pi = 10^4$ rad/sec.

∴ *quality factor for the circuit*,

$$Q = 0.002 \times \frac{10^4}{0.2} = 0.01 \times 10^4 = 100.$$

### *Example 10:*

*In an experiment on forced oscillations, the frequency of a sinusoidal during force is changed while its amplitude is kept constant. It is found that the amplitude of vibrations is 0.01 mm at very low frequency of the driver and-goes upto a maximum of 5.0 mm at driving frequency 200 sec⁻¹. Calculate (i) Q of the system, (ii) relaxation time T and (iii) half width of resonance curve.*

### *Solution:*

We know that if the driving frequency be negligibly small and the damping be low, the *amplitude of the driven oscillator*, $A = f_0\omega_0^2$ and that at low damping, the maximum amplitude (or amplitude at resonance) is $A_{max} = f_0\tau/\omega_0$, where $\omega_0$ is the *natural frequency of the oscillator* (or the system) and T, its relaxation time.

(i) We, therefore, have $\dfrac{A_{max}}{A} = \dfrac{\dfrac{f_0\tau}{\omega_0}}{\dfrac{f_0}{\omega_0^2}} = \omega_0\tau = Q.$

Since $A_{max}$ ∴ 5.0 m and A = 0.01 mm, we have

*quality factor of the system*, $Q = \dfrac{5}{0.01} = 500.$

(ii) Since Q = $w_0t$, we have $\tau = \dfrac{Q}{\omega_0}.$

Here, Q = 500 and $\omega_0 = 2\pi n = 2\pi \times 200 = 400\pi$

∴ *relaxation time of the system*, $\tau = \dfrac{500}{400\pi} = 0.3971 \approx 0.40$ sec.

(iii) For *low damping, half width of the resonance curve*, $\Delta p = \sqrt{3}$.

Since $k = \dfrac{1}{2\tau} = \dfrac{1}{2} \times 0.40 = 1.25$, we have *half width of the resonance curve*, $\Delta p = \sqrt{3} \times 1.25 = 2.165$ radian/sec

$$= \frac{2.165}{2\pi} = 0.3445 \text{ c.p.s.}$$

***Example 11:***

*A harmonic oscillator of quality factor 10 is subjected to a sinusoidal applied force of frequency one and a half times the natural frequency of the oscillator. If the damping be small, obtain (i) the amplitude of the forced oscillation in tarns of its maximum amplitude and (ii) the angle θ by which it will be out of phase with the driving force.*

***Solution:***

As we know, the *amplitude of the forced oscillation* is given by

$$A = \frac{f_0}{\sqrt{(\omega_0^2 - p^2)^2 + 4k^2p^2}} = \frac{f_0}{\omega_0^2\sqrt{\left(1 - \dfrac{p^2}{\omega_0^2}\right)^2 + 4\dfrac{k^2p^2}{\omega_0^4}}}$$

where $\omega_0$ is the *natural angular frequency of the oscillator and p, the angular frequency of the driving force.*

Now, the quality factor $Q = \omega_0 \tau = \dfrac{\omega_0}{2k} = 10$.

So that, $2k = \dfrac{\omega_0}{10}$ and $\dfrac{p}{\omega_0} = \dfrac{3}{2}$ (given). We, therefore, have

$$A = \frac{f_0}{\omega_0^2\sqrt{\left(1-\dfrac{p^2}{\omega_0^2}\right)^2 + \dfrac{1}{100}\times\dfrac{p^2}{\omega^2}}} = \frac{f_0}{\omega_0^2\left(\dfrac{25}{16}+\dfrac{1}{100}\times\dfrac{9}{4}\right)^{1/2}}$$

$$= \frac{f_0}{\omega_0^2\left(\dfrac{634}{400}\right)^{1/2}} = \frac{20f_0}{\omega_0^2\sqrt{634}}$$

And, *for low damping,* amplitude is maximum when $p = \omega_0$. So that,

$$A_{max} = \frac{f_0}{\omega_0^2\sqrt{(1 \quad 1)^2 + \dfrac{1}{100}\times 1}} = \frac{f_0}{\omega_0^2\left(\dfrac{1}{100}\right)^{1/2}}$$

$$= \frac{f_0}{\omega_0^2\left(\dfrac{1}{10}\right)} = \frac{10f_0}{\omega_0^2}$$

$$\therefore \quad \frac{A}{A_{max}} = \frac{20f_0}{\omega_0^2\sqrt{634}}\times\frac{\omega_0^2}{10f_0} = \frac{2}{\sqrt{634}} = 0.07943 \approx 0.08.$$

Or, $A = 0.08\ A_{max}$.

Thus, *the amplitude of the forced oscillation will be 0.08 of the maximum amplitude.*

Now, $$\tan\theta = -\frac{2kp}{\omega_0^2 - p^2} = \frac{p\omega_0/10}{\omega_0^2 - p^2} = \frac{\left(\dfrac{3\omega_0}{2}\right)\left(\dfrac{\omega_0}{10}\right)}{\omega_0^2 - \dfrac{9\omega_0^2}{4}}$$

$$= \frac{3\omega_0^2}{20}\times\frac{4}{5\omega_0^2} = \frac{3}{25} = 0.12.$$

Or, $\theta = \tan^{-1} 0.12 = 6°51'$

Therefore, *the forced oscillation is* (180° – 6°51') = 173°9' *out of phase with the driving force.*

### Example 12:

*A torsion pendulum of moment of inertia $10^5$ gm-cm$^2$ suffers an angular displacement of 4° under a constant torque of 2000 dynes-cm. What should be the frequency of the periodic torque that would set the pendulum in resonant vibration?*

### Solution:

Here, *angular di placement of the pendulum* $= 4° = 4 \times \frac{\pi}{180}$ radian and the torque applied = 2000 dyne-cm.

$$\therefore \textit{torque per unit angular displacement}\ (C) = \frac{\text{torque applied}}{\text{angular displacement}}$$

$$= \frac{2000}{4\pi/180} = \frac{2000 \times 180}{4\pi} \text{ dyne-cm/rad.}$$

Now, *for setting the pendulum into resonant vibration, the frequency of the applied torque must be the same, or very nearly the same, us the natural frequency of the pendulum,*

$$n = \frac{1}{2\pi}\sqrt{\frac{C}{I}}.$$

Substituting the values of C and I, therefore, we have

*required frequency of the periodic torque,*

$$n = \frac{1}{2\pi}\sqrt{\frac{2000 \times 180}{4\pi \times 10^5}}$$

$$= 0.1704 \text{ or } 0.17 \text{ sec}^{-1}.$$

### Example 13:

*A particle of mass 50 gm, moving with an initial velocity of 100 cm/sec, is acted upon by a damping force which brings it to rest in a distance of. 10 metres. Assuming the damping force to be proportional to velocity, calculate (i) its relaxation time, (ii) the time in which (a) its velocity is halved, (b) its kinetic energy is halved, (iii) the damping force on it when its velocity is 20 cm/sec. (Take $\log_e 10 = 2.30$).*

***Solution:***

We have the relation $x = v_0\tau\,(1 - e^{t/\tau})$ for the distance covered by a particle in time t, where $v_0$ is its *initial velocity* and $\tau$, its relaxation time.

(i) Clearly, x will have its maximum value as $\tau \to \infty$ and, therefore, the quantity $e^{-t}/\tau \to 0$.

So that, $x_{max} = v_0\tau$

and $\therefore$ $\tau = x_{max}/v_0$.

Here, $x_{max} = 10$ m $= 1000$ cm and $v_0$ 100 cm/sec. We, therefore, have *relaxation time* $\tau = \dfrac{1000}{100} = 10$ sec.

(ii) (a) The velocity of the particle after time t from the start is given by the relation $v = v_0 e^{-t/\tau}$,

whence, $\log_e (v/v_0) = -t/\tau$.

Or, $t = \tau \log_e(v_0/v)$.

$\therefore$ the velocity will be reduced to half, *i.e.*, 50 cm/sec in time t given by

$$t = 10 \times 2.30 \log_{10}\left(\frac{100}{50}\right)$$

$$= 10 \times 2.30 \times \log_{10} 2$$

$$= 10 \times 2.30 \times 0.3010 = 6.92 \text{ sec.}$$

(b) We know that if T be the kinetic energy of the particle after time t and $T_0$, its initial kinetic energy, we have $T = T_0\, e^{-2t/\tau}$.

$\therefore$ $$\log_e\left(\frac{T}{T_0}\right) = -\frac{2t}{\tau},$$

whence, $$2t = \tau \log_e\left(\frac{T_0}{T}\right).$$

Or, $$2\tau = 10 \times 2.30 \log_{10}\left(\frac{2}{1}\right).$$

Or, $$t = 5 \times 2.30 \log_{10} 2$$

$$= 5 \times 2.30 \times 0.3010 = 3.46 \text{ sec.}$$

Or, we could have obtained this result directly from (a) above, since, as we know, relaxation time for K.E. is half that for velocity.

Thus, *the K.E. of the particle will be reduced to half its initial value in 3.46 sec.*

(iii) The *damping force* = $\gamma v$, where, $\gamma = \frac{m}{\tau}$.

$\therefore$ *damping force at velocity* 20 cm/sec

$$= mv/\tau = 50 \times \frac{20}{10} = 100 \text{ dynes.}$$

***Example 14:***

*A charged oil drop falls through air under the combined action of an electric field E and gravity g, both acting along the same direction. If m be the mass of the drop and q, the charge on it, show that the terminal velocity attained by it (i.e., when $t \to \infty$) is given by $v_{terminal}$ = $(qE/m)\tau + g\tau$, assuming a solution of the form $v = A + Be^{\alpha\tau}$. (Assume damping force to be proportional to velocity).*

***Solution:***

Since the electric field E and g are both similarly directed, they both act downwards. So that, the force acting on the drop is given by

$$F = m\left(\frac{dv}{dt}\right) = mg + qE - \gamma v,$$

where $\gamma v$ is the damping force.

Or, $$m\left(\frac{dv}{dt}\right) + \gamma v = mg + qE.$$

Or, $$\left(\frac{dv}{dt}\right) + \left(\frac{\gamma}{m}\right)v = g + \left(\frac{q}{m}\right)E.$$

Or, since $\frac{\gamma}{m} = \frac{1}{\tau}$, where $\tau$ is the *relaxation time*, the *equation of motion* of the drop is

$$\frac{dv}{dt} + \frac{v}{\tau} = g + \frac{qE}{m}. \qquad ...(I)$$

Now, let us assume a solution of this equation to be $v = A + Be^{-\alpha t}$ where A and B are *arbitrary constants* to be determined from the initial conditions.

Since at t = 0, *i.e., when the drop Just starts from rest*, $v_0 = 0$, we have 0 = A + B, whence, B = – A. And, therefore,

$$v = A + Ae^{-\alpha t} = A(1 - e^{-\alpha t}). \quad ...(II)$$

Hence, $\dfrac{dv}{dt} = A\alpha e^{-\alpha t}$.

Substituting this value of dv/dt in the equation of motion (I) above, we have

$$A\alpha e^{-\alpha t} + \frac{1}{\tau}A\left(1 - e^{-\alpha t}\right) = g + \frac{qE}{m}.$$

Clearly, at $t \to \infty$, we have $\dfrac{A}{\tau} = g + \dfrac{qE}{m}$.

Or, $$A = \tau\left(g + \frac{qE}{m}\right) \quad ...(III)$$

And, at t = 0, we have $A\alpha = g + \dfrac{qE}{m}$. ...(IV)

From relations III and IV, we thus have $\alpha = \dfrac{1}{\tau}$.

Hence from equation II, we have

$$v = \tau\left(g + \frac{qE}{m}\right)\left(1 - e^{-\alpha t}\right) = \tau\left(g + \frac{qE}{m}\right)\left(1 - e^{-t/\tau}\right) \quad ...(V)$$

Clearly, therefore, when $t \to \infty$, we have

$$v_{terminal} = \tau\left(g + \frac{qE}{m}\right) = \left(\frac{qE}{m}\right)\tau + g\tau.$$

***Example 15:***

*A small ball bearing is falling freely through a viscous medium, of coefficient of viscosity $\eta$, under the action of gravity alone. The viscous drag on it due to the medium is $-6\pi\eta av$, where a is its radius and v, its velocity. Obtain (i) the relaxation time, (ii) the terminal velocity of the ball bearing.*

*(iii) If the ball bearing has an initial horizontal velocity u, what will be the maximum horizontal distance covered by it?*

*(iv) If the radios of the ball bearing be doubled, how will it affect its terminal velocity?*

***Solution:***

(i) Here, obviously, the *damping force* = $6\pi\eta av = \gamma v$, where $\gamma = 6\pi\eta a$ and is thus *proportional to velocity.*

$$\textit{relaxation time } \tau = \frac{m}{\gamma} = \frac{m}{6\pi\eta a}.$$

(ii) Proceeding as in worked example 2 above and remembering that, here, the ball bearing (a tiny sphere) is falling only under the action of gravity, we have terminal velocity of the ball bearing given by

$$v_{terminal} = gt = g\left(\frac{m}{6\pi\eta a}\right) = \frac{mg}{6\pi\eta a}.$$

(iii) The *maximum horizontal distance covered by the ball bearing* is, as we know, equal to its *initial velocity* (u) × relaxation time (τ) = $u(m/6\pi\eta a) = mu/6\pi\eta a$.

(iv) We have seen under (u) above that the *terminal velocity of the ball bearing*

$$= \frac{mg}{6\pi\eta a} = \frac{4}{3}\frac{\pi a^3 \rho g}{6\pi\eta a} = \frac{2a^2 \rho g}{9\rho}$$

where $\frac{4}{3}\pi a^{2}$ is the volume and,

therefore, $\frac{4}{3}\pi a^3 \rho$, the mass (m) of the ball bearing.

Thus, the *terminal velocity* $\alpha\ a^2$, (ρ and g being constants).

Hence, *if the radius of the ball bearing be doubled, its terminal velocity will be four times its previous value, i.e.*, equal to

$$\frac{4mg}{6\pi\eta a} = \frac{2mg}{3\pi\eta a}.$$

***Example 16:***

*An underdamped harmonic oscillator has its amplitude reduced to 1/10th of its initial value after 100 oscillations. If its time-period be 1.15 sec, calculate (i) the damping constant, (ii) the relaxation time.*

*If the observed value of the first amplitude of the oscillator be 2 cm. what would be its value in the absence of damping?*

***Solution:***

(i) The amplitude (a) of an underdamped oscillator is, as we know, given by the relation $a = a_0 e^{-kt}$ where $a_0$ is its amplitude in the absence of damping and k, the *damping constant.*

Now, for the first amplitude ($a_1$), obviously, $t = \frac{T}{4}$ where T is the *time-period* of the oscillator. So that, $a_1 = a_0 e^{-kT/A}$.

Since successive amplitudes occur at intervals of T/2 sec each, the 201th amplitude after 100 oscillations occurs after (200T/2 + T/4) = (100T + T/4) sec. And, therefore,

$$a_{201} = a_0 e^{-k(100T + T/4)}$$

Hence $$\frac{a_{201}}{a_1} = \frac{a_0 e^{-k}(100T + T/4)}{a_0 e^{-kT/4}} = e^{-100kT}$$

But we are given that $\frac{a_{201}}{a_1} = \frac{1}{10}$

or $10^{-1}$ and T = 1.15 sec.

∴ $10^{-1} = e^{-100k(1.15)} = e^{-115k}$.

Or, $\log_e 10 = 115$ k.

Or, $230 \log_{10} 10 = 115k$,

*i.e.*, $2.30 = 115l$,

whence, $k = \frac{2.30}{115} = 0.02$.

Thus, the *damping constant* = 0.02.

(ii) And, therefore, *relaxation time* $\tau = \frac{1}{2k} = \frac{1}{2} \times 0.02 = 25$ sec.

Now, since *the first amplitude* $a_1 = a_0 e^{-kT/4} = 2$ cm, we have

$$2 = e_0 e^{-(0.02 \times 1.15/4)} \text{ or, } a_0 = 2e^{0.023/4}$$

or, $\log_{10} a_0 = \log_{10} 2 + \frac{0.023}{4} \times \frac{1}{2.30}$

Or, $\log_{10} a_0 = 0.3010 + 0.0025 = 0.3025$,

whence, $a_0 = 2.011$ cm.

Thus, *the undamped amplitude of the oscillator would be* 2.011 cm.

***Example 17:***

*If in problem 14 above, the rate of dissipation of energy is desired to be halved, what should be (i) the amplitude of the applied periodic force, its frequency remaining unchanged, (ii) the frequency of the applied periodic force, its amplitude remaining unchanged?*

***Solution:***

(i) We know that the *rate of dissipation of energy* $\alpha$ (*amplitude of the force*)$^2$.

Thus, if $R_1$ and $R_2$ be the rates of dissipation of energy and $f_{01}$ and $f_{02}$ the amplitudes of the force in the two cases respectively, *the frequency of the force remaining the same in either case,* we have

$$\frac{R_1}{R_2} = \left(\frac{f_{01}}{f_{02}}\right)^2.$$

We know that the *effective current in an LCR circuit*

$$= \frac{\text{effective emf}}{\text{impedance of}} = \frac{E}{Z}.$$

Here, *effective* emf E = 500 *volts* and *impedance*

$$Z = \sqrt{R^2 + L^2 p^2}$$

$$= \sqrt{R^2 + L^2 (2\pi n)^2},$$

$$\text{Or, } Z = \sqrt{(4)^2 + (0.68)^2 (2\pi \times 120)^2} = \sqrt{16 + 262800} = 512.6 \text{ ohms.}$$

$$\therefore \textit{ effective current in the circuit, } I = \frac{500}{512.6} = 0.9754 \text{ amp.}$$

$$\text{And, } \textit{power factor of the circuit} = \frac{R}{Z} = \frac{4}{512.6} = 0.0078.$$

***Example 18:***

*An electric lamp which runs at 40 volts D.C. and consumes 10 amperes current is connected to A.C. mains at 100 volts, 50 cycles. Calculate the inductance of the choke.*

***Solution:***

$$\text{Clearly, } \textit{resistance of the lamp, } R = \frac{\text{voltage}}{\text{current}} = \frac{40}{10} = 4 \text{ ohms.}$$

And, *impedance of A.C. circuit,* $Z = \sqrt{R^2 + (Lp)^2}$, where L is the *inductance of the choke.*

Here, R = 4 ohms, $\pi = 2\pi n \times 50 = 100\pi$ and I = 10.

$$\therefore \qquad I = \frac{E}{Z} = \frac{E}{\sqrt{R^2 + (Lp)^2}} = \frac{100}{\sqrt{4^2 + (L + 100\pi)^2}} = 10.$$

So that, $\dfrac{10^4}{16 + 10^4 \pi^2 L^2} = 100.$

Or, $10^6\pi^2L^2 = 10^4 - 1600 = 8400.$

So that, $L^2 = \dfrac{8400}{10^6 \pi^2}.$

Or, $L = \sqrt{\dfrac{84}{10^4 \pi^2}} = \dfrac{\sqrt{84}}{100\pi} = 0.02917.$

Thus, the *inductance of the choke* = 0.02917 henry.

***Example 19:***

*An inductance coil of inductance 0.05 henry and resistance 12.4 ohms is connected to an A.C. supply at 200 volts and 50 cycles. Find (a) power factor, and (b) angle of lag. (c) Show that the power consumed in just equal to the rate of production of heat.*

***Solution:***

(a) We know that the power factor,

$$\cos\phi = \frac{R}{Z} = \frac{R}{\sqrt{R^2 + (Lp)^2}}.$$

Here, $R = 12.4$ ohms, $L = 0.05$ henry

and $\pi = 2\pi n = 2\pi \times 50 = 100\pi.$

$$\therefore \textit{ power factor, } \cos\phi = \frac{12.4}{\sqrt{(12.4)^2 + (0.05 \times 100\pi)^2}}$$

$$= \frac{12.4}{\sqrt{153.8 + 246.7}} = \frac{12.4}{20} = 0.62.$$

(b) And, therefore, *angle of lag*, $f = \cos^{-1}(0.62) = 51° 41'$.

(c) Clearly, *power consumed*, $E_v I_v \cos\phi$.

Now, $E_v = 100$ volts,

$$I_v = \frac{E_v}{Z} = \frac{100}{\sqrt{(12.4)^2 + (0.05 + 100\pi)^2}}$$

$$= \frac{100}{20} = 5 \text{ ampere and } \cos\phi = 0.62.$$

∴ *power consumed* = 100 × 5 × 0.62 = 310 watts.

And, *rate of production of heat* = $I^2R$ = 25 × 12.4 = 310 watts.

Thus, *power consumed* = *rate of production of heat.*

***Example 20:***

*A harmonic oscillator consisting of a 60 gm mass attached to a massless spring has a quality factor 200. If it oscillates with an amplitude of 2 cm in resonance with a periodic force of frequency 20 c.p.s., calculate (i) the average energy stored in it, (ii) the rate of dissipation of energy.*

***Solution:***

In the case of resonant vibration, the average energy stored in the oscillator is equal to its maximum potential energy $= \frac{1}{2}Cx_0^2 = \frac{1}{2}m\omega_0^2 x_0^2$.

$[\because C = m\omega_0^2]$.

Here, $m = 50$ gm, $\omega_0 = 2\pi n = 2\pi \times 20 = 40\pi$

and $x_0 = 2$cm. So that,

*average energy absorbed or stored up in the oscillator*

$$= \frac{1}{2} \times 50 \times (40\pi)^2 \times (2)^2 = 16 \times 10\pi^2 = 1.58 \times 10^6 \text{ ergs.}$$

Now, by its very definition, the quality factor

$$Q = 2\pi \frac{\text{everage energy stored}}{\text{energy dissipated per cycle}}$$

$$\therefore \textit{energy dissipated per cycle} = 2\pi \frac{\text{average energy stored}}{Q}$$

$$= 2\pi \frac{1.58 \times 10^6}{200} = \pi\ 1.58 \times 10^4 \text{ ergs.}$$

Since there are 20 cycles to the second, we have

*energy dissipated per second* or *rate of dissipation of energy*

$$= 20\pi \times 1.58 \times 10^1 = 9.926 \times 10^5 \approx 10^6 \text{ erg/sec.}$$

## EXERCISES

1. Show that in a driven oscillator, the maximum power is absorbed at the frequency of velocity resonance and not at the frequency of amplitude resonance.

2. Assuming the results of forced oscillations, discuss the sharpness of resonance.

   Explain the statement '*the quality factor (Q) is a measure of ths sharpness of resonance in the case of a driven oscillator.*

3. Show that the half width of the *velocity versus driving frequency curve* and that of the *power absorbed versus driving frequency curve* are $\sqrt{3k}$ and k respectively.

4. (a) Explain the terms *reactance, impedance and power factor.*

   (b) Define *virtual current and virtual potential* in an A.C. circuit.

5. Show that a driven oscillator is always behind the driving force in phase and that its displacement is maximum when the driving force is zero and vice versa.

   Illustrate for zero, low and medium damping, the general nature of the phase lag of a driven oscillator as the driving frequency gradually increases and passes through the natural frequency of the oscillator.

6. Give the theory of oscillations in a LCR circuit with small resistance, deducing expressions for (i) frequency of oscillations, (ii) power dissipation and (iii) quality factor of the circuit.

7. Write explanatory notes on the terms: *reactance, impedance, rms value of current, rms value of emf and power factor* in a driven LCR circuit.

   Show that the power in such a circuit like that in a D.C. circuit, appears in the form of heat in the resistance alone.

8. Obtain an expression for the current in a driven LCR circuit and discuss how the current Jags or leads the applied emf in phase (a) when the net reactance in the circuit is (i) capacitive, (ii) inductive and (b) when reactance in the circuit is equal to resistance. Illustrate your answer with the help of suitable graphs.

9. A damped harmonic oscillator is subjected to a sinusoidal driving force whose frequency is altered but amplitude kept constant.

It is found that the amplitude of the oscillator increases from 0.02 mm at very low driving frequency to 8.0 mm at a frequency of 100 cps. Obtain the values of (i) the quality factor, (ii) the relaxation time, (iii) the damping factor and (iv) the half-width of the resonance curve.

**Ans.** (i) Q = 400, (ii) $\tau$ = 0.4245 sec, (iii) k = 1.178, (iv) 2.04 rad/sec.

10. A damped harmonic oscillator of quality factor 20 is subjected to a sinusoidal driving force of frequency twice the natural frequency of the oscillator. If the damping be small, what fraction will the amplitude of the oscillator be of its maximum value and by what angle will it differ in phase from the driving force?

**Ans.** a = 0.017 $a_{max}$; 178.5° *out of phase with the driving force.*

11. Obtain the frequency of the periodic torque which, when applied to a torsion pendulum of moment of inertia 900 gm-cm$^2$, sets it into resonant oscillation. The torsion constant of the suspension fibre may be taken to be 4× $10^4$ dyne-cm/ radian. It is found that a small increase of 0.2% in the frequency of the periodic torque results in a fall of 80% in the amplitude of the pendulum. Calculate the quality factor of the pendulum.

**Ans.** p = 1.06 cps; Q = 187.

12. An alternating sinusoidal emf is applied to a circuit containing an inductance L and a resistance R. Calculate the current at any instant in the circuit.

13. Find the natural frequency of a circuit containing inductance of 50 microhenry and a capacitance of 0.0005 microfarad. Fine the wavelength to which it corresponds.

**Ans.** 1.007 cps; 298.1 metres.

[**Hint:** c = n$\lambda$, where c is the velocity of light (or electromagnetic radiation)]

14. An alternating potential of 110 volts and 50 cycles frequency is applied to a circuit containing a resistance of 200 ohms, inductance of 5 henry and a capacitance of 2 mF. Calculate the maximum current in the circuit.

**Ans.** *peak value of current* = 0.78 amp.

15. To a circuit containing an inductance of 10 millihenry, a capacitance of 1 μF and a resistance of 10 ohms in series, an rms voltage of 100 V is applied. Calculate the frequency at which the circuit will be in resonance with the current of the same frequency and find the value of the current.

    **Ans.** 71 = 1592 and $I_v$ = 10 amp.

16. Find the capacitance of the capacitor which must be placed- in series with a resistance of 5 ohms and an inductance of 200 millihenry to bring the current in phase with the voltage if the frequency be 60 cycles per sec. What current will flow if 100 volts be imposed on the circuit? **Ans.** 35.20 μF; 20 amp.

17. Find the value of the inductance which should be connected in series with a capacitance of 0.5 μF, a resistance of 10 ohms and an A.C. source of frequency 50 cps, so that the power factor of the circuit be unity. **Ans.** 20.29 henry.

18. An alternating voltage of 4 volts is applied at the resonant frequency to a coil of inductance 0.004 H and resistance 20 ohms in series with a 0.001 μF capacitor. Find (i) the resonant frequency (ii) the current flowing, (iii) the voltage across the capacitor. **Ans.** (i) 79600 cps; 0.2 amp; 400 volts.

19. An alternating voltage of 100 rms is applied to a circuit of resistance 0.5 ohm and an inductance of 0.01 henry, the frequency being 50 cycles/sec. What is the current and the time lag?

    **Ans.** 31 amp; 0.0045 sec.

20. (a) Find whether the discharge of a condenser through the following inductive circuit is oscillatory: C = 0.1 μF, L = 10 millihenry. R = 200 ohms. If the circuit is oscillatory, calculate its frequency. **Ans.** Oscillatory; 4772 cps.

    (b) In an oscillatory circuit, L = 0.2 henry, C = 0.0012 μF. What is the maximum value of resistance for the circuit to be oscillatory? **Ans.** Less than $2.582 \times 10^4$ ohms.

21. If the damping in the case of a damped harmonic oscillator be very small obtain expressions for (i) average total energy of the oscillator and (ii) average rate of energy dissipation. (iii) Show that the latter is equal to the work done by the damping force per second.

**Ans.** (i) $E = \frac{1}{2} m\omega^2 a_0^2 e^{-2kt}$, (ii) $\frac{E}{\tau}$, (iii) $-\frac{\gamma dx}{dt} = P$.

22. Show that the relaxation lime of a damped harmonic oscillator has the dimensions of time. Also show that (i) in this time ($\tau$) the amplitude of the oscillator falls to 0.6084 of its undamped value and that (ii) it is reduced to half its undamped value in time 1.3846 $\tau$. ($e = 2.718$ and $\log_e 10 = 2.30$).

23. (a) Show that damping has little or no effect on the frequency of a harmonic oscillator if its quality factor (Q) be large.

    (b) If the quality factor of a- sonometer wire of frequency 300 cps be $2 \times 10^3$, in what time will its energy be reduced to $1/e^{th}$ of its energy in the absence of damping.

    **Ans.** In 1.061 sec.

24. The amplitude of a simple pendulum, with a bob of mass 400 gm and oscillating in air, falls to half its undamped value on completion of 50 complete oscillations. (i) If its time-period be 2 sec, calculate its quality factor, (ii) What should be the mass of the bob, its size and shape remaining unchanged, so that the damping may become, critical? **Ans.** (i) 453.7, (ii) 0.44 gm.

25. A capacitor of capacitance 1 $\mu$F is discharged through a resistance of 2 ohms and an inductance of 1 henry. Is the discharge of the capacitor oscillatory? If so, obtain its frequency and calculate the time in which the amplitude of the oscillations falls to 10% of its undamped value.

    **Ans.** *Discharge oscillatory*; n = 159; In 2.3 sec.

26. (a) In an LCR circuit, L = 0.2 henry and C = 2$\mu$F. What should be the maximum resistance in the circuit for the discharge of the capacitor to be oscillatory in character?

    (b) The quality factor of an oscillatory circuit is 100 and its frequency 8 kilocycles per second. How will its frequency change if the resistance in the circuit be reduced to zero?

    **Ans.** (a) *Less than 20 ohms*, (b) *The frequency will increase by 1 cycle per sec.*

27. Find whether the discharge of a condenser for the following inductive circuit is oscillatory: C = 0.1 $\mu$F, L = 10 mH, R = 200

ohms. If the circuit is oscillatory, calculate its frequency.

**Ans.** *Discharge oscillatory*; n = 4772 cps

28. Write down and solve the differential equation of a damped harmonic oscillator subjected to a sinusoidal force and obtain expressions for its maximum amplitude and quality factor.

29. For a driven harmonic oscillator of natural frequency $\omega/2\pi$, described by mx + rx – kx = F sin pt, deduce an expression for velocity amplitude $x_0$. Show that for the cases $p >> \omega$, $p = \omega$ and $p << \omega$, this amplitude is governed by the inertia factor, the resistance factor and the spring factor respectively.

30. Show that in the case of a system undergoing a forced oscillation, the response is

    (i) independent of its mass if $p << \omega_0$, and

    (ii) independent of the spring constant if $p >> \omega_0$.

31. What is meant by a damping or a dissipative force? What is its effect on a harmonic oscillator? Illustrate your answer by one or two homely examples.

    Obtain an expression for the velocity of a damped harmonic oscillator and deduce from it the definition of *relaxation time* ($\tau$) of the oscillator

32. Show that both velocity and kinetic energy of a damped harmonic oscillator fall exponentially with time, the relaxation time in the latter case being half that in the former.

33. Write down the differential equation for a damped harmonic oscillator (electrical case). Obtain a solution and represent it with a graph, taking a case of small damping.

34. Deduce the differential equation of a damped harmonic oscillator and discuss in detail the case of critical and underdamping.

35. Write the differential equations for the following cases of damped harmonic oscillator: (i) mechanical, (ii) electrical.

31. A coil connected to a sinusoidal alternating supply of emf 240 volts (rms) and of frequency 50 cps takes a current of 5.0 amp (rms). A wattmeter in the circuit reads 1000 watts. Find (a) the power factor of the circuit, (b) the resistance of the coil, (c) the reactance of the coil, (d) the maximum stored energy.

37. What is meant by the terms (i) *logarithmic decrement* and (ii) *quality factor* (Q) of a damped harmonic oscillator? Obtain the usual expressions for them.

38. Obtain the differential equation for the motion of the coil of a moving coil galvanometer and discuss the conditions under which its motion is (i) *dead beat* (ii) *critical* and (iii) *ballistic.* Enumerate the essential requirements of a ballistic galvanometer.

39. Discuss an LCR circuit as an example of a damped harmonic oscillator and show that it is the resistance alone which is responsible for the damping of the oscillations. Discuss the conditions under which the discharge of the capacitor is aperiodic, critical and oscillatory. What is the quality factor or such a circuit?

40. A body is propelled through a viscous medium with a speed of 300 cm/sec On the propelling force being suddenly removed, it comes to rest after covering a distance of 45 metres. What is its relaxation time? How long would it be before its speed falls to 30 cm/sec? **Ans.** 15 sec, 34.5 sec.

41. Show that whereas the amplitude of a pendulum oscillating in a viscous medium like air is affected, its time-period remains practically unaffected.

[**Hint.** Damping force due to a viscous medium = $-6\pi\eta rv$ = $-6\pi\eta rl\, d\theta/dt$, where r is the radius of the bob, v, its velocity and $l$, the length of the pendulum The equation of motion of the pendulum is thus $ml d^2\theta/dt^2 + 6\pi\eta rl\, d\theta/dt +$ where m is the mass of the bob. This reduces to $d^2\theta/dt^2 + (6\pi\eta r/m)\, d\theta/dt + (g/l)\theta = 0$ which gives $k = 3\pi\eta r/m = 3\pi\eta r/4/3\,\pi r^3\rho = 9\eta/4r^2\rho$, material of the bob) and $w_0^2 = g/l$.

Now, *time-period* $T = 2\pi/\omega = 2\pi/\sqrt{\omega_0^2 - k^2}$ / Since k is small compared with $\omega_0$, we have $T = 2\pi/\sqrt{\omega_0^2} = 2\pi/\omega_0 = 2\pi/\sqrt{g/l} = 2\pi\sqrt{l/g}$, the same as in the absence of damping.

And the *amplitude* is of course given by $a = a_0 e^{-kt}$, where $a_0$ is the amplitude in the absence of damping.]

3

# WAVE MOTION

## INTRODUCTION

A wave motion is a means of translatory energy from one point to another without there being any transfer of matter between the points. We are all familier with the ideas of waves. The most common waves perhaps are water waves produced say by throwing a stone on the water surface. We also have invisible waves like sound waves propagating through air. As mentioned earlier, in these examples there is energy transfer from on location to another by means of a periodic disturbance between two point.

We now try to understand the basic ideas of waves by considering a few examples. It is a matter of common experience that when we drop a stone into a quiet ponds a circular pattern spreads out from the point of throwing. Such a disturbance is called a wave. Suppose we watch the water as such a wave moves across the surface. We will find that although the water is disturbed locally, it does not move forward with the waves. This is quite clear if we look at a piece of cork that has been thrown on the water surface. Then cork moves up and down as the wave pass, it does not travel along with the waves. Thus a wave can travel over distances but once the disturbance has passed every drop of the water is left where it was before.

We may also consider the case of a flag as it flirts in the breeze top of the flag post. The dipples of the wave travel out along the cloth. Individual spots on the cloth of the flag, however stay at their positions as the waves passed by the Charka in the middle always remains in the middle and its distance from the four edges of the flag remains unchanged. Just as the water does not travel with the water waves so the cloth of the flag remains in place when the waves have passed through it.

It is clear than that the following are the general features of wave motion:

(1) The disturbance travels through some medium—through the water, through the cloth of the flag etc.

(2) The medium does not trave along with the disturbances.

Disturbances which travel through the media are what we call waves. The observed motion of the wave is that of the state of matter and not a movement of matter itself.

## TYPES OF WAVES

One of the commonest modes of transference of energy from one point to another is through the agency of waves with which we are all familiar in a general way. Thus, for example, we know that *water waves, sound waves, light waves and radio waves* all carry energy of one form or the other from one place to another, without any bulk motion of the interventing medium in the case of the former two and in the absence of any medium at all in that of the latter. This mode of transference of energy is called *wave motion* and may be divided into two broad categories:

(i) *Mechanical wave* motion, and

(ii) *Non-mechanical* or *electromagnetic wave motion.*

(i) *Mechanical wave motion* is possible only in material media (solids, liquids and gases) which possess inertia as well as elasticity; so that, water waves and sound waves are both examples of this type of wave motion and are, therefore referred to a *mechanical waves.*

(ii) *Non mechanical or electromagnetic wave motion,* on the other hand, requires no material medium for its propagation; so that, light and radio waves, which can travel through empty space, belong to this Category and are, therefore referred to as *non mechanical* or *electromagnetic waves.*

We are concerned here with only the mechanical type of wave motion and shall, for simplicity, refer to it merely as wave motion.

Since all material media possess elasticity as well as inertia, it is easy to see how a wave motion is produced and propagated through them. For, no particle of an elastic medium can be disturbed without affecting its immediate neighbour and, tending to recover its original position, it first stores up potential energy and then coverts it back into kinetic energy. The neighbouring particle which has thus been disturbed then performs a similar motion, so that each successive particle repeats, in turn, the movements of its predecessor a little later than it and hands the same on to its successor, resulting in the transference of energy from particle to particle all along the line. One complete oscillation of a particle of the medium obviously produces *one single wave* or a *pulse* and its repeated periodic motion, a succession of waves or a *wave train*.

We may thus define a *wave motion* as a *disturbance (or a condition) that travels onwards through a medium due to the repeated periodic motion of its particles about their mean or equilibrium position, each particle repeating the movements of its predecessor a little later than it and handing it on to its successor, so that there is a regular phase difference between one particle and the next.*

The simplest type of periodic motion performed by a particle is, of course, the simple harmonic motion and the corresponding wave motion is therefore called a *simple harmonic or a sinusoidal wave motion*, which alone is the most general type of wave motion and the one we shall deal with in the following discussion.

It may be emphasised again that, but for the properties of elasticity and intertia, no wave motion could be produced in, or propagated through, a medium. These two properties in fact determine the velocity of propagation of the wave motion through the medium, as we shall presently see. And, in order that a wave may travel through a medium over fairly large distances without attenuation (*i.e.*, without any decrease in its amplitude), a third property is also necessary, *viz.*, that the medium should offer the least frictional resistance so as not to unduly damp the periodic motion of the particles.

A wave motion which thus progresses onwards through a medium, with energy transferred across every section of it, is called *a travelling or a progressive wave motion to distinguish it from what a called a standing or a stationary wave motion*, which we shall study a little later and in which there is no onward movement of the wave motion through the medium and hence no transference of energy across any section of it.

## TRANSVERSE AND LONGITUDINAL WAVE MOTION

There are two distinct classes of wave motion:

(i) *Transverse,* and

(ii) *Longitudinal.*

(i) In a *transfers wave motion*, the particles of the medium oscillate up and down about their mean or equilibrium position, at right angles *to the direction of propagation of the waves motion itself.* This form of wave motion therefore travels in the form of *crests* and *troughs* (Fig. 3.1), as, for example, waves travelling along a stretched string. *A crest and an adjoining trough constitute one wave or pulse* and a succession of them, a *wave train.*

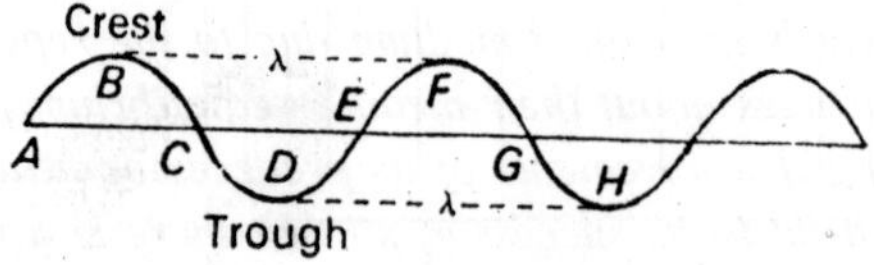


**Fig. 3.1**

This type of wave motion is possible in media which possess elasticity of shape or rigidity, *i.e.*, in *solids*. But, as we know, they are also possible in water and liquids, in general, even though they do not possess the property of rigidity. This is because they possess another equally effective property of resisting any vertical displacement of their particles (or keeping their level). Gases, however, possess neither rigidity nor do they resist any vertical displacement of their particles (or keep their level). *A transverse wave motion is, therefore, not possible in a gaseous medium.*

**Note :** It may just be mentioned for the information of the student that cm electromagnetic wave is necessarily a transverse wave because of the electric and magnetic fields being perpendicular to its direction of propagation.

(ii) In a *longitudinal wave motion*, the particles of the medium oscillate to and fro about their mean or equilibrium position, *along the direction of propagation of the wave motion itself.* This type of wave motion, therefore, travels in the form of compressions (or *condensations*) and *rarefactions*, *i.e.*, in the particles of the medium getting closer together and further apart alternately, (Fig. 3.2), and is possible in media possessing elasticity of volume, *i.e.*, in solids, liquids as well as gases. As examples may be mentioned sound waves in air [Fig. (a)] and waves in a spring or helix when one end of it is suddenly compressed or pulled out and then released [Fig. (b)].

Here, *one compression and the adjoining rarefaction constitute one wave or pulse* and, as in case (i), a succession of them, a wave train.

In some cases, the waves are neither purely transverse nor purely longitudinal as, for example, *ripples* or *surface waves* on water (produced by dropping a stone in water), in which the particles of the medium (here, water) oscillate across as well as along the direction of propagation of the wave motion, describing elliptical paths. We need hot, however, bother ourselves with any such types of waves here.

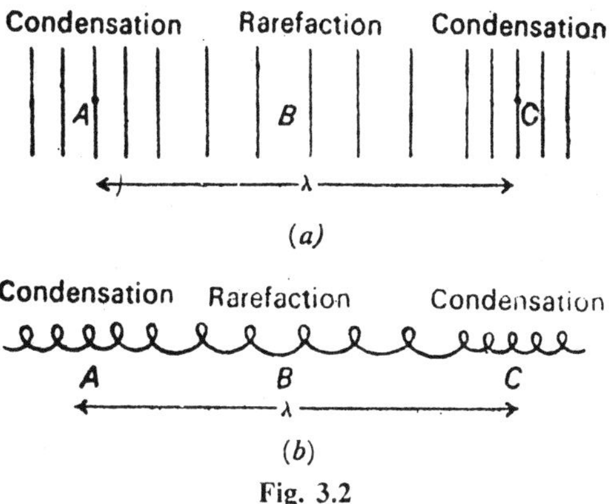


**Fig. 3.2**

Again, waves may be *one-dimensional, two-dimensional or three dimensional* according as they propagate energy in just one, two or three dimensions. Thus, transverse waves along a string or longitudinal waves along a spring are one-dimensional, surface waves or ripples on water are two-dimensional and sound waves proceeding radially from a point-source are three dimensional.

Before proceeding further, we had better summarise here the important characteristics of a wave motion, in general, whether transverse or longitudinal, *viz*,, that

(i) It is simply a disturbance, a condition or a state of motion that travels through the medium and nothing material, *i.e., there is no transference of any part of the medium (or matter) from one point to another*. For, as we have seen, the particles of the medium simply oscillate up and down or to and fro about their mean position and do not move onwards with the wave motion itself.

(ii) Each particle of the medium receives the disturbance a little later than its predecessor, repeats its movements and hands it on to the next succeeding particle, *i.e., there is a regular phase lag between one particle and the next.*

(iii) *The velocity of the particles of the medium or the particle velocity, as it is referred to, is entirely different from the velocity of the wave motion or the wave velocity.*

## WAVE FRONT

A plane or a surface on which all particles of the medium are in an identical state of motion at a given instant, *i.e.*, in the same phase, is called a *wave front.*

In a homogeneous and isotropic medium, the wave front is always perpendicular to the direction of propagation of the wave motion and the position of a given wave front shifts onwards with the wave. In other words, it moves onwards through the medium with the velocity of the wave motion itself. A line normal to the wave front thus gives the direction of propagation of the wave and is called a *ray.*

Although wave fronts may have various shapes, the two most important types are plane and the spherical wave fronts. In case the waves are one-dimensional, such as those travelling along a stretched string or a spring or produced by the motion of a plane surface back and forth in a gaseous medium, we obtain a *plane wave front* and such waves are, therefore, also referred to as *plane wires.*

On the other hand, in the case of *spherical wives*, or *waves proceeding from a point-source*, we obtain a *spherical wave front.* Since the radius of a spherical wave front goes on increasing as the wave moves onwards, its curvature goes on progressively decreasing until at an infinite or a large distance from the source it becomes almost a plane surface or a plane wave front, implying thereby that *at large distances from the source, spherical waves—and, infact, waves of all types—may be treated as plane waves.*

We shall therefore, restrict our discussion here to only plane or one-dimensional waves. For, this, without in any way detracting from the applicability of our result to waves in general, will greatly facilitate our task by doing away with undue mathematical complexities.

## WAVE LENGTH, FREQUENCY AND WAVE NUMBER

It may be recalled that in above, a crest and an adjoining trough, in the case of a transverse wave, and a compression and an adjoining rarefaction, in the case of a longitudinal wave, have been said to constitute one wave. The distance between the particles at the two ends of the wave or the pulse, therefore, constitutes a *wavelength,* usually denoted by the symbol $\lambda$. Since these two panicles are in an identical state of motion or *in the same phase*, a more general definition of wavelength would be the *distance between any two nearest particles of the medium in the same phase*. Thus, in the case of a transverse wave (Fig. 3.1) the *distance between the centres of two nearest crests are two nearest troughs* (*i.e.*, distance BF or DH) also constitutes one wave length. Similarly, in the case of a longitudinal wave (Fig. 3.2) the *distance between the centres of two nearest compressions or two nearest rarefactions* (*e.g.*, the distance AC) also constitutes one wavelength.

Again, since a wave or a pulse is produced in the time taken by a particle of the medium to complete its one full oscillation about its mean position, *a wavelength may also be defined as the distance covered by the disturbance during one full time-period (T) of a particle of the medium.*

*The number of waves produced per second* (which is obviously the same as the number of complete oscillations made by a particle of the medium per second) is called the *frequency* (n) of the wave motion or the wave, and is obviously the *reciprocal of the time-period T*, *i.e.*, $n = 1/T$.

*The time rate of change of phase* (*i.e.*, its rate of change with time) is called the *angular frequency* ($\omega$) of the wave motion and is equal to $2\pi/T$ because a complete cycle of phase change $2\pi$ occurs in one full time-period T of a particle of the medium.

On the same analogy, we have a quantity, called the *wave number*, k, representing the *space rate of change of phase* (*i.e.*, its rate of change with distance) and since one complete cycle of phase change $2\pi$ occurs in a distance $\lambda$ (the wave length), we have $k = 2\pi/\lambda$.

Quite often, however, the wave number is defined as the *number of waves in unit distance*, *i.e.*, equal to $1/\lambda$ and is denoted by the symbol $\nu$. So that, $k = 2\pi/\lambda$ is, in this case, referred to as the *propagation constant*.

It will also be seen from the above that if the number of waves produced per second, *i.e.*, the frequency of the wave motion, be n and the length of each wave (or wavelength) be $\lambda$, we have

$$v = n\lambda \text{ or, since } n = \frac{1}{T}$$

we also have $v = \dfrac{\lambda}{T}$,

where v is the *phase velocity*, usually referred to simply as the *velocity of the wave motion or of the wave.*

## DISPLACEMENT, VELOCITY AND PRESSURE CURVES

(i) **Displacement Curve:** In the case of a transverse wave, as we know, the particles of the medium oscillate up and down about their mean position, at right angles to the direction of propagation of the wave. So that, if we plot the displacements of the various particles at a given instant against their respective distances from the origin or, in fact any fixed point, we obtain the *displacement curve* of the wave, which, for a sinusoidal wave, will be a series of sine curves, (Fig. 3.1), with one complete sine curve representing one full wave, as shown in Fig. 3.3. The displacement curve thus gives the form of the wave or the wave profile, as it were.

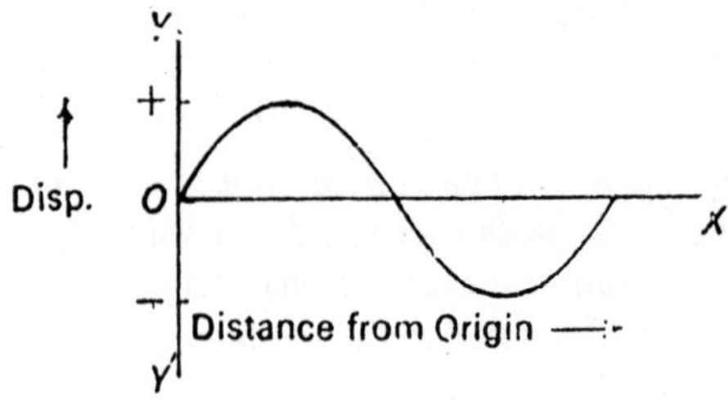


**Fig. 3.3**

In the case of a longitudinal wave, on the other hand the particles oscillate to and fro about their mean position *along* the direction of the wave motion, so that the displacement curve would, in this case, be a straight line along the direction of wave propagation itself. For our convenience, however, we *conventionally* represent the displacement curve of a longitudinal wave motion too by a sine curve of the type shown in Fig. 3.1, the curve here being referred to as the *associated displacement* curve and it is clearly understood that the displcements of the particles shown perpendicular to the direction of wave propagation are in fact along it.

(ii) **Velocity Curve:** The graph between the velocity (U) of an oscillating particle of the medium and its distance from the origin or a fixed point is called the velocity curve of the wave motion. Using the same convention as in case (i) above, the velocity curve in the case of both a transverse and a longitudinal wave is of the form shown in

Fig. 3.4, which is obviously *also a sine curve, a phase angle $\pi/2$ or a quarter period $T/4$ ahead of the displacement curve.*

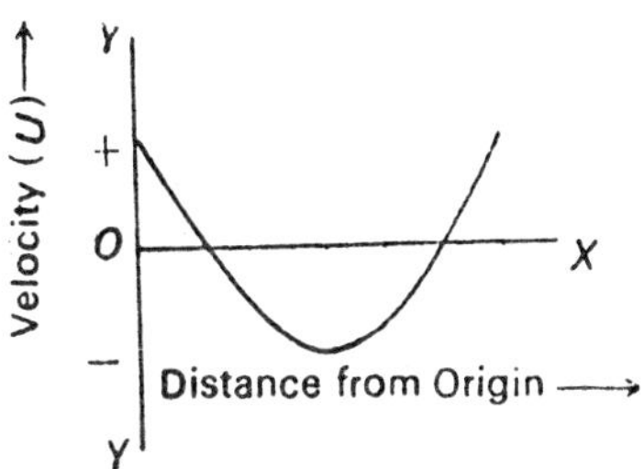


Fig. 3.4

(iii) **Pressure Curve:** Let ABCDE (Fig. 3.5.) be the associated displacement curve of a longitudinal wave travelling towards the right along the axis ACE of a cylinder of unit area of cross section (shown dotted).

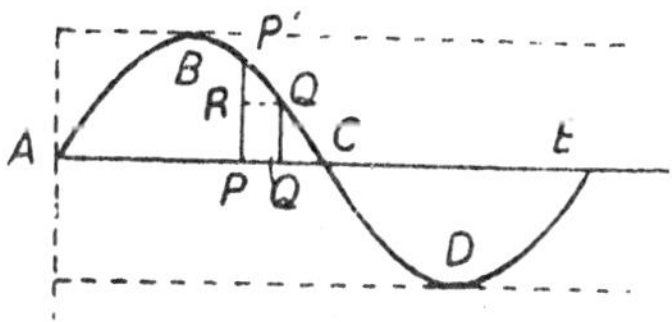


Fig. 3.5

Consider two particles P and Q of the medium (say, *air*), a small distance $\delta x$ apart. Before the wave is set up in the cylinder, *i.e.*, when the air inside it is yet undisturbed, the *volume of the air enclosed between P and Q* is obviously $PQ \times 1 = \delta x \times 1 = \delta x$ (the cross section of the cylinder being unity). After the wave is set up and takes the form shown in the figure, the displacement at P is PP' and that at Q is QQ' and the change in displacement is, therefore, $PP' - QQ' = P'R = \delta y$, say, where Q'R is the perpendicular from Q' on to PP'. So that, remembering that the displacements are in fact longitudinal, along the direction of wave propagation, the *change in*

*volume between P and Q* $= (PP' - QQ') \times 1$

$$= P'R \times 1 = \delta y \times 1 = \delta y.$$

∴ *volume strain set up in the medium = change in volume/original volume* = $\delta y/\delta x$ = $dy/dx$ If PQ be really small

But $dy/dx$ is the *slope of the displacement curve* at P and represents a compression or a rarefaction according as it has a negative or a positive sign. If, therefore, we plot $dy/dx$ for the particles along the axis of y and their diplacements from the origin along the axis of x, we obtain the *pressure curve* of the wave (Fig. 3.6) which is of the same form as the velocity curve (Fig 3.4). Thus, *both the velocity and the pressure curves of a wave are $\pi/2$ or T/4 ahead of the displacement curve in phase.*

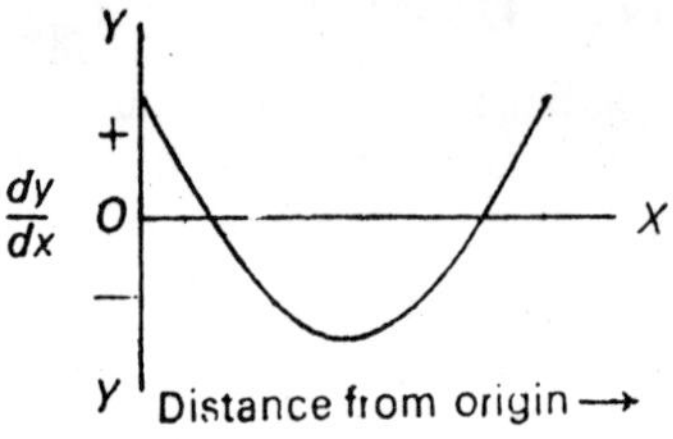


**Fig. 3.6**

Incidentally, we also see that *volume strain at a point = slope of the displacement curve at that point.*

## EXPRESSION FOR A PLANE PROGRESSIVE HARMONIC WAVE

We know that a plane progressive wave is one which travels onwards through the medium *in a given direction without attenuation, i.e.,* with its amplitude constant.

Now, a progressive wave may be transverse or longitudinal. In either case, however, there exists a regular phase difference between any two successive particles of the medium.

Thus, suppose a wave, originating at O, travels to the right along the axis of x (Fig. 3.7). Then, if we start counting time when the particle at O just passes through its mean position in the positive direction (*i.e.*, upwards in the case of a transverse wave and forwards in the case of a longitudinal wave) the equation of motion of this particle (at O) is obviously $y = a \sin wt$, where y is its *displacement* at time t, a, its *amplitude* and ω, its *angular velocity.*

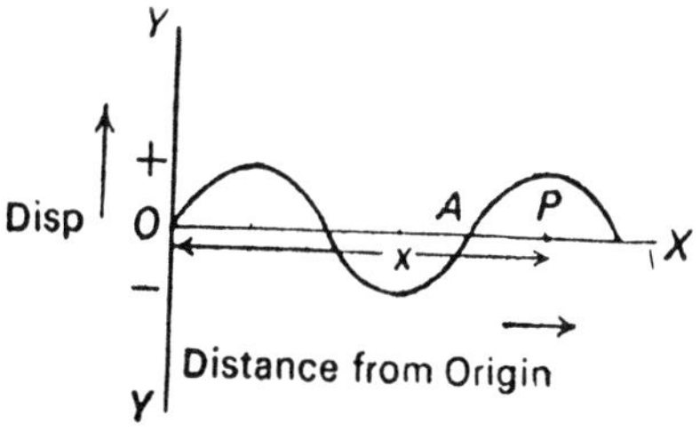


Fig. 3.7

Since the successive particles to the right of 0 receive and repeat its movements after definite intervals of time, the phase lag goes on increasing as we proceed away from O towards the right. So that, for a particle at P, *distant x from* O, it is $\phi$, say. The equation of motion of the particle at P is thus $y = a \sin(\omega t - \phi)$.

Now, as we know, in a distance $\lambda$ the phase lag increases by $2\pi$. In a distance x, therefore, the phase lag increases by $2\pi/\lambda$, *i.e.*, $\phi = 2\pi x/\lambda$. Substituting this value of $\phi$ in the expression for y above, we have

$$y = a \sin\left(\omega t - \frac{2\pi x}{\lambda}\right).$$

Or, $$y = a \sin(\omega t - kx) \qquad \text{...(i)}$$

where $\dfrac{2\pi}{\lambda} = k$, the *propagation constant.*

Again, $\omega = \dfrac{2\pi}{T}$, where T is the *time-period* of each particle of the medium. And, since $v = n\lambda = \dfrac{\lambda}{T}$, where v is the *phase velocity* or the *wave velocity* and n, the frequency of the oscillating particles or of the wave, we have $\dfrac{1}{T} = \dfrac{v}{\lambda}$ and, therefore, $\omega = \dfrac{2\pi v}{\lambda}$. We thus have from relation (i) above,

$$y = a \sin\left(\frac{2\pi vt}{\lambda} - \frac{2\pi x}{\lambda}\right).$$

Or, $$y = a \sin\frac{2\pi}{\lambda}(vt - x).$$

Or, $$y = a \sin k(vt - x). \qquad \text{...(ii)}$$

Any one of these expressions or any one of their variations, such as

$$y = a\sin\frac{2\pi v}{\lambda}\left(t - \frac{x}{v}\right). \qquad ...(iii)$$

$$y = a\sin 2\pi n\left(t - \frac{x}{v}\right) \qquad \left[\because \frac{v}{\lambda} = n\right] \qquad ...(iv)$$

Or, $$y = a\sin\frac{2\pi}{T}\left(t - \frac{x}{v}\right) \qquad ...(v)\ [\text{because } n = 1/T.]$$

Or, $$y = a\sin 2\pi\left(\frac{t}{T} - \frac{x}{\lambda}\right) \qquad ...(vi)\ [\text{because } v = n\lambda = \lambda/T.]$$

is referred to as the *equation of a plane, progressive harmonic or sinusoidal wave*—the one most commonly used being equation (ii).

Since $$\lambda = \frac{v}{T},\ \omega = \frac{2\pi}{T}$$

and $k = \dfrac{2\pi}{\lambda}$, it will be easily seen that

$$v = \frac{\lambda}{T} = \frac{\lambda}{\frac{2\pi}{\omega}} = \frac{\omega}{\frac{2\pi}{\lambda}} = \frac{\omega}{k}.$$

For a wave travelling *towards the left*, since x will be negative, we shall have

$$y = a\sin\frac{2\pi}{\lambda}(vt + x)$$

or, $$y = a\sin(\omega t + kx).$$

**Note:** In deducing these equations, we have assumed that the particle at 0 just passes through its mean position, in the positive direction, at t = 0, *i.e.*, y = 0 at t = 0. If this be not so, the particle is said to have an *initial phase* θ, say, and the equation of the wave then becomes

$$y = a\sin\frac{2\pi}{\lambda}(\omega t - x + \theta).$$

Or, $$y = a\sin(\omega t - kx + \theta),$$

where θ is referred to as the *phase constant.*

Thus, if θ = 90°, we have y = a at x = 0 and t = 0.

The following points that emerge from the above discussion may be carefully noted.

(i) *At a given point* x (*i.e.*, for a constant value of x) the displacement (y) varies simple harmonically with time, completing one full cycle in time λ/v which, therefore, gives the *periodic time* T of the wave and hence also its frequency n = 1/T = v/λ.

(ii) *At a given instant* t (*i.e.*, for a constant value of t), the displacement (y) varies simple harmonically with distance from the origin and for x = λ, y is restored to its original value, so that the wave lengths

(iii) The phase lag for a distance x is 2πx/λ and hence for distance λ it is 2πλ/λ = 2π which is, in effect, the same thing as *zero*. So that, *particles separated by a distance λ or an integral multiple of λ are in the same phase.*

(iv) If we increase time t by δt, and distance x by vδt, the value of y remains the same, showing that a disturbance at one point is repeated an interval δt later at a point vδt further away. This means, in other words, that the *disturbance or the wave travels, without attenuation, with velocity v.*

## PARTICLE VELOCITY

We have the equation of plane progressive wave motion,

$$y = a \sin \frac{2\pi}{\lambda}(vt - x),$$

where y is the *displacement* of a particle of the medium at distance x from the origin at instant t, a, its *amplitude* and v, the *wave velocity* (*i.e.*, the *phase velocity*).

∴ differentiating this expression for y with respect to t, we have

$$\textit{particle velocity } U = \frac{dy}{dt} = \frac{2\pi v}{\lambda} a \cos \frac{2\pi}{\lambda}(vt - x). \qquad ...(i)$$

And, differentiating the same expression for y with respect to x, we have *slope of the displacement curve (or strain or compression),*

$$\frac{dy}{dx} = -\frac{2\pi}{\lambda} a \cos \frac{2\pi}{\lambda}(vt - x) \qquad ...(ii)$$

From relations (i) and (ii) we thus obtain $U = \frac{dy}{dt} = -v\frac{dy}{dx}$, ...(iii)

*i.e.*, particle velocity at a point = – (wave velocity) × (slope of displacement curve at that point,).

## DIFFERENTIAL EQUATION OF A WAVE MOTION

Differentiating equation (i), under 3.7 above, for the particle velocity again with respect to t, we have

*Wave Motion of the particle,* $\frac{d^2y}{dt^2} = -\frac{4\pi^2v^2}{\lambda^2} a \sin\frac{2\pi}{\lambda}(vt - x)$

$$= -\frac{4\pi^2v^2}{\lambda}y. \qquad ...(iv)$$

And, similarly, differentiating equation (ii), under the same article, for strain or compression, again with respect to x, we have

*rate of change of compression with distance,*

$$\frac{d^2y}{dx^2} = -\frac{4\pi^2}{\lambda^2} a \sin\frac{2\pi}{\lambda}(vt - x) = -\frac{4\pi^2}{\lambda^2}y \qquad ...(v)$$

From relations (iv) and (v) we thus obtain $\frac{d^2y}{dt^2} = v^2\frac{d^2y}{dx^2}$. ...(vi)

This is referred to as the *differential equation* of a plane or one-dimensional progressive wave, in which *the coefficient of $d^2y/dx^2$ straight away gives the square of the wave velocity.*

Any equation of this form can unphesitatingly be declared to represent a plane progressive harmonic wave, the velocity of which, is given by the square root of the coefficient of $d^2y/dx^2$.

Since $d^2y/dx^2$ also gives the curvature of the displacement curve, we may interpret the differential equation to mean that

*particle Wave Motion at a point = (wave velocity)$^2$ × (curvature of displacement curve at that point)*

It may once again be pointed out here that *the essential characteristic of a wave motion, in general, is the onward progression of a disturbance* with a velocity depending upon factors pertaining to the particular situation in question. It does not by any means imply that y in the wave equation must necessarily represent a physical displacement of particles as we have assumed here. In fact, for a wave, in general, we replace the symbol y in the differential equation of the wave by $\phi$ (psi), referred to as the *wave function.* So that, for a one-dimensional wave the differential equation takes the form

$$\frac{d^2\phi}{dt^2} = v^2\frac{d^2\phi}{dx^2},$$

where $\phi$ may represent physical quantities as disparate as, for example, (i) pressure in a medium in the case of a sound wave travelling through it, (ii) the magnitude of the electric vector of a plane-polarised electromagnetic wave or (iii) even the probability amplitude of what is called a matter wave (from which the probability of finding a particle within a given volume of space may be determined).

## DIFFERENTIAL EQUATION OF A THREE-DIMENSIONAL WAVE

A three-dimensional wave, as we have seen, is one that travels out in all directions. For such a wave, the differential equation takes the form

$$\frac{d^2\psi}{dt^2} = v^2\nabla^2\psi,$$

where $$\nabla^2\psi = \frac{\partial^2\psi}{\partial x^2} + \frac{\partial^2\psi}{\partial y^2} + \frac{\partial^2\psi}{\partial z^2}.$$

One possible solution of the equation is $\psi = a \sin(\omega t - kr + \theta)$, where $\psi$ is the *displacement* at a point r at an instant t and k, the *propagation vector* whose magnitude is equal to $k = 2\pi/\lambda$ (the propagation constant) and direction the same as that of the wave propagation.

## ENERGY DENSITY OF A PLANE PROGRESSIVE WAVE

We have already seen how in a progressive wave motion, the energy derived from the source is passed on from particle to particle, so that *there is a regular transmission of energy across every section of the medium.*

By the *energy density* of a plane progressive wave we mean the *total energy (kinetic + potential) per unit volume of the medium through which the wavers passing.* It is usually denoted by the letter E.

Let us proceed to obtain an expression for it.

As we know, the displacement of a particle of the medium distant x from the origin or the source, at an instant t, is given by

$$y = a\sin\frac{2\pi}{\lambda}(vt - x),$$

where the symbols have their usual meanings.

∴ *the velocity of th particle,* $U = \frac{dy}{dt} = \frac{2\pi v}{\lambda} a \cos \frac{2\pi}{\lambda}(vt - x)$

and *Wave Motion of the particle,* $\frac{dU}{dt} = \frac{d^2y}{dt^2}$

$$= -\frac{4\pi^2 v^2}{\lambda^2} a \sin \frac{2\pi}{\lambda}(vt - x) = -\frac{4\pi^2 y^2}{\lambda^2} y.$$

Now, if we consider *unit volume* of the medium in the form of an extremely thin element of the medium parallel to the wave front, we have

*mass of the element* = ρ, its density (because density is mass per unit volume).

And since the layer is very thin, the velocity of all the particles in it may be assumed to be the same. We therefore have

*K.E. per unit volume of the medium* $= \frac{1}{2}$ (mass) (velocity)$^2$ $= \frac{1}{2}\rho U^2$.

$$= \frac{1}{2}\rho \frac{4\pi^2 v^2}{\lambda^2} a^2 \cos^2 \frac{2\pi}{\lambda}(vt - x) = \frac{2\pi^2 v^2 \rho}{\lambda^2} a^2 \cos^2 \frac{2\pi}{\lambda}(vt - x) \quad ...(i)$$

Clearly, *force, acting on unit volume × mass × Wave Motion*

$$= \rho \frac{d^2y}{dt^2} = \frac{4\pi^2 v^2 \rho}{\lambda^2} y,$$

ignoring the –ve sign which merely indicates that the Wave Motion is directed oppositely to the displacement.

∴ *work done in a small displacement dy of the layer*

$$= \text{force} \times \text{displacement} = \frac{4\pi v^2 v^2 \rho}{\lambda^2} y dy.$$

Hence *work done during the whole displacement from 0 to y*

$$= \int_0^y \frac{4\pi^2 v^2}{\lambda^2} y dy = \frac{4\pi^2 v^2 \rho}{\lambda^2} \int_0^y y dy.$$

$$= \frac{4\pi^2 v^2 \rho}{\lambda^2} \cdot \frac{y^2}{2} = \frac{2\pi^2 v^2 \rho}{\lambda^2} y^2 = \frac{2\pi^2 v^2 \rho}{\lambda^2} a^2 \sin^2 \frac{2\pi}{\lambda}(vt - x)$$

This work done must obviously be stored up in the medium in the form of potential energy. So that,

*P.E. per unit volume of the medium*

$$= \frac{2\pi^2 v^2 \rho}{\lambda^2} a^2 \sin^2 \frac{2\pi}{\lambda}(vt - x).$$

$\therefore$ *total energy per unit volume of the medium or energy density of the plane progressive wave, i.e.,* E = K.E. + P.E.

$$\text{Or, } E = \frac{2\pi^2 v^2 \rho}{\lambda^2} a^2 \cos^2 \frac{2\pi}{\lambda}(vt - x) + \frac{2\pi^2 v^2 \rho}{\lambda^2} a^2 \sin^2 \frac{2\pi}{\lambda}(vt - x)$$

$$= \frac{2\pi^2 v^2 \rho}{\lambda^2} a^2 \qquad \text{....(ii)}$$

$$= 2\pi^2 \left(\frac{v}{\lambda}\right)^2 \rho a^2 = 2\pi^2 n^2 a^2 \rho, \qquad \text{...(iii)} \left[\because \frac{v}{\lambda} = n\right]$$

where a is the *frequency* of the wave.

In case the cross section of the beam be unity, expressions (ii) and (iii) give the *total energy of the beam or the wave per unit length.*

A point of interest that must clearly be noted here is that *although both kinetic and potential energies of the wave depend upon the values of x and t, its total energy or the energy density is quite independent of either.*

## DISTRIBUTION OF ENERGY IN A PLANE PROGRESSIVE WAVE

Having calculated the total energy of a plane progressive wave in 3.10 above, let us see how it is distributed over a complete wave length.

Clearly, *average K.E. over a complete wavelength*

$$= \frac{1}{\lambda}\int_0^\lambda \frac{2\pi^2 v^2 \rho}{\lambda^2} \cos^2 \frac{2\pi}{\lambda}(vt - x)\,dx$$

$$= \frac{2\pi^2 n^2 a^2 \rho}{\lambda} \int_0^\lambda \frac{1}{2}\left[1 + \cos\frac{4\pi}{\lambda}(vt - x)\right]dx \qquad \left[\because \frac{v^2}{\lambda^2} = n^2\right].$$

$$\text{Now, } \int_0^\lambda \cos\frac{4\pi}{\lambda}(vt - x)\,dx = -\frac{\lambda}{4\pi}\left[\sin\frac{4\pi}{\lambda}(vt - x)\right]_0^\lambda$$

$$= -\frac{\lambda}{4\pi}\sin\frac{4\pi}{\lambda}\left[(vt - x) - \sin\frac{4\pi}{\lambda}vt\right]$$

$$= -\frac{\lambda}{4\pi}\left[\sin\left(\frac{4\pi}{\lambda}vt - 4\pi\right) - \sin\frac{4\pi}{\lambda}vt\right] = 0$$

So that, *average K.E. of the wave over a complete wavelength*

$$= 2\pi^2 n^2 a^2 \rho\ \lambda/2\lambda$$

$$= \pi^2 n^2 a^2 \rho = \frac{1}{2}\ \textit{total energy.}$$

Similarly, average P.E, of the wave over a complete wavelength

$$= \frac{1}{\lambda}\int_0^{\lambda} \frac{2\pi^2 v^2 a^2 \rho}{\lambda^2} \sin^2 \frac{2\pi}{\lambda}(vt - x)dx = \pi^2 n^2 a^2 \rho = \frac{1}{2}\ \textit{total energy.}$$

Thus, *at any given instant, the energy of a plane progressive harmonic wave is, on an average, half kinetic and half potential in form.*

## ENERGY CURRENT—INTENSITY OF A WAVE

It has been pointed out earlier under 3.10 above that if the cross section of tne beam or the wave be taken to be unity, we may regard the total energy per unit volume or the energy density $E = 2\pi^2 n^2 a^2 \rho$ as the *total energy per unit length of the wave.* And, since in a progressive wave train, a new length v of the medium is set into motion every second, the energy transferred per second must be the energy contained in a length v. *This rate of flow of energy per unit area of cross section of the wave front along the direction of wave propagation is called the* **energy current** (C) *or the* **energy flux** *of the wave and is obviously equal to* $E \times v$.

Thus, *energy current or energy flux of a plane progressive wave,*

$$C = 2\pi^2 n^2 a^2 \rho v \text{ erg/sec. cm}^2.$$

Now, we define the *intensity of the we* (I) as the *quantity of incident energy per wit area of the wave front per unit time.* It is thus the same thing as the energy current or the energy flux of the wave. So that, we have

$$I = 2\pi^2 n^2 a^2 \rho v,$$

indicating that in any given case it is *proportional to the square of the amplitude of the wave, i.e.,* $I \propto a^2$.

Since in a medium with little or no frictional resistance, a plane or *one-dimensional wave* travels without attenuation, *i.e.*, with its amplitude undiminished, *the intensity of the wave remains the same throughout* in the case of a *three-dimensional or a spherical wave*, on the other hand, such as the one emanating from a point-source, the wave fronts are spherical shells of successively increasing radii and the intensity obeys the well known *inverse square law, viz.*, $I \propto 1/r^2$ and thus goes on falling as the square of the distance (r) from the source.

And, since $I \propto a^2$, we have $a^2 \propto 1/r^2$ or $a \propto 1/r$, *i.e.*,

*the amplitude in the case of a spherical wave goes on decreasing progressively with distance from the source.*

Thus, whereas *for a plane progressive wave, we have*

$$y = a \sin \frac{2\pi}{\lambda}(vt - x),$$

*for a spherical wave*, $y = \frac{A}{r} \sin \frac{2\pi}{\lambda}(vt - x)$, where A is a constant.

## SUPERPOSITION OF WAVES

All types Of waves, mechanical as well as electromagnetic, are, in general, subject to the *principle of superposition,* first enunciated by *Huyghens*, according to which *if two or more independent waves are propagated through a medium or space, all at the same time, the resultant physical quantity (i.e., displacement etc) at any point is the vector sum of the quantities due to each individual wave, i.e.,*

$$\psi = \psi_1 + \psi_2 + \psi_3 + \ldots$$

Thus, *in the case of one-dimensional or plane waves, the resultant displacement at any point is the linear sum of the displacements due to the individual waves.*

Or, $$y = y_1 + y_2 + y_3 + \ldots$$

This means, in other words, that two or more waves can traverse the same medium or space independently of one another. This is amply borne out by the fact that we can hear the individual notes of the various instruments constituting an orchestra, even though the resultant sound wave reaching our ears is a complex one. Similarly, we can tune our radio receiver to a particular frequency or wave-length even though waves Of several frequencies or wavelengths may be simultaneously crossing the aerial.

The *only one condition* for the principle of superposition to apply is that *the equation of the waves in question must be a linear one*. Thus, it holds good for waves in an elastic medium because the relation between the deformation produced and the recovering force is linear; and for electromagnetic waves because the mathematical relationship between the electric and magnetic fields is also linear.

The principle fails in the case of waves for which the equations governing their propagation are not linear, as, for example, *shock- waves* produced by large explosions. Even though these waves too are longitudinal elastic waves in air, the equation governing their propagation is quadratic and not linear.

The importance of the principle of superposition lies in the fact that in the case where it holds good, it is possible to analyse a complicated wave motion into a set of simple waves. We shall revert back to it and discuss it more fully later in 3.24.

Another important consequence of this principle is that it gives rise to the phenomenon referred to as *interference of waves.*

## INTERFERENCE—BEATS

Let two *identical* waves (*i.e.*, waves having the *same amplitude and wavelength*) with a *path-difference* δ between them, travelling along the same, or very nearly the same path and in the same direction, be represented respectively by

$$y_1 = a\sin\frac{2\pi}{\lambda}(vt - x)$$

and $$y_2 = a\sin\frac{2\pi}{\lambda}[vt(x+\delta)].$$

Then, the resultant wave will, in accordance with the principle of superposition, be clearly represented by

$$y = y_1 + y_2 = a\sin\frac{2\pi}{\lambda}(vt - x) + a\sin\frac{2\pi}{\lambda}[vt - (x+\delta)]$$

$$= 2a\sin\frac{2\pi}{\lambda}\left[vt - \left(x + \frac{\delta}{2}\right)\right]\cos\frac{\pi\delta}{\lambda}$$

Or, $$y = 2a\cos\frac{\pi\delta}{\lambda}\sin\frac{2\pi}{\lambda}\left[vt - \left(x + \frac{\delta}{2}\right)\right],$$

*i.e.*, the resultant wave is also a simple harmonic wave, of the *same wavelength, frequency and velocity* as the two constituent waves but has an amplitude 2a cos $p\delta/\lambda$, indicating that the amplitude changes with the path difference $\delta$ (or the phase difference $2\pi\ \delta/\lambda$) between the two waves.

Clearly, the amplitude of the resultant wave will be the maximum (+ 2a or – 2a) when cos $\pi\delta/\lambda = +1$ or $-1$, *i.e.* when $\pi\delta/\lambda = 0, \pi, 2\pi$, $3\pi$ etc. and therefore, $\delta = 0, \lambda, 2\lambda, 3\lambda$ or $n\lambda$, where n = 0, 1, 2, 3 etc. or $2\pi\ \delta/\lambda = 0, 2\pi, 4\pi$ or $n\ (2\pi)$ where n = 0, 1, 2, 3 etc.

In other words, the amplitude, and hence also the intensity of the resultant wave (which is proportional to the square of the amplitude) will be the *maximum*, and equal to twice and four times respectively of the amplitude and intensity of one single wave when the two wave arrive at a point *in phase* (*i.e.*, with a path difference equal to an integral multiple of $\lambda$ or a phase difference equal to an even multiple of $2\pi$); for, then, the crest or the condensation of one wave falls on the crest or the condensation of the other and so also the trough or the rarefaction of one wave over the trough or rarefaction of the other.

The two waves, in this case, thus *reinforce* each other and are said to interfere constructively.

On the other hand, the amplitude, and hence the intensity, of the resultant wave will be the minimum or zero when cos $\pi\delta/\lambda = 0$, *i.e.*, when $\pi\delta/\lambda = \pi/2, 3\pi/2, 5\pi/2$ etc. and, therefore $\delta = \lambda/2, 3\lambda/2, 5\lambda/2$ etc. or $(2n + 1)\lambda/2$ or when $3\pi\delta/\lambda = \pi, 3\pi, 5\pi$ or $(2n + 1)\ \pi$, where n = 1, 2, 3 etc.

Thus, the amplitude, and hence the intensity, will be the *minimum* or *zero* when the two waves arrive at a point *out of phase* (*i.e.*, with a path difference equal to an odd multiple of $\lambda/2$ or a phase difference equal to an odd multiple of $\pi$); for, then, the crest or the condensation of one wave falls on the trough or the rarefaction of the other. The two; waves here thus *nullify* each other and are said to *interfere destructively.*

Because the energy of the resultant wave at the former points (of constructive interference) is *four times* that due to one single wave and at the latter points (of destructive interference), *zero*, we must not jump to the conclusion that energy has been created at the former and destroyed at the latter points. *It has dimply been redistributed in the medium or space, i.e.,* transferred from the points of minimum or zero intensity to those of maximum intensity, the total energy of the resultant wave being the same as the sum of the total energies of the two constituent waves.

This may be easily seen from the fact that energy being proportional to $a^2$, the total energy of the two waves before interference is proportional to $2a^2$ throughout the medium (or space). After interference, the energy of the resultant wave is proportional to $4a^2$ at some points and zero at others. The average energy of the wave is thus again proportional to $2a^2$.

The phenomenon of interference thus does not by any means violate the law of conservation of energy. On the other hand, it is a direct consequence of this law, for we cannot possibly have points of increased intensity without there being also corresponding points of decreased intensity. In fact, *interference may more aptly be defined as the redistribution of energy due to superposition of waves.*

In case the two identical waves travel along the same path in *opposite directions*, (as, for example, an incident and a reflected wave), we have a new type of waves produced, called *standing or stationary waves* for the simple reason that they do not move onward into the medium or space but appear and disappear within the limited region in which they are produced. This is, therefore, referred to as *interference in space.*

The maxima and minima of displacement do not move forwards into the medium or space but throughout remain fixed in their respective positions alternating with each other and there is thus no transference of energy across any section of the medium or space. We shall deal fully with these in 3.21.

**Beats.** If the two waves, travelling along the same path and in (he same direction, have *slightly different frequencies or wavelengths* and *also different amplitudes*, they arrive *in phase* at some points and out *of phase* at others, thus producing maximum displacements, and hence also maximum intensity, at the former, and minimum (but not *zero*) displacements and hence minimum intensity at the latter points as they proceed along.

In view, however, of the frequencies of the two waves being slightly different, their relative phase at various points keeps on changing progressively. For, if at a particular instant the two waves arrive to *phase* at a given point, we have the maximum displacement or amplitude there and hence also the maximum intensity. But, as soon as the wave with the higher frequency has gained half a period or half a wavelength over the other, there is the minimum displacement and hence the minimum intensity at the point instead of the previous maximum which has shifted further on to the point where half a period before there was the minimum displacement and hence also minimum intensity. And, as the higher frequency wave gains one full time-period or wavelength over the other, the two again arrive *in phase* at the point in question, producing the maximum displacement and intensity there, with the condition of minimum displacement and intensity shifting further up to the point where half a period before there had been the maximum of displacement and intensity.

Thus, the condition of maximum and minimum displacement and intensity travels onwards through successive points as the two waves proceed forwards in the medium or space and we have, as it were, *an interference pattern travelling onwards with respect to the observer.* This is referred to as *interference in time*, and *these alternations of maxima and minima of intensity are called* **beats**, *one maximum and one succeeding minimum* (or *vice versa*) *constituting one beat.* The number of beats produced per second is equal to the *difference between the frequencies of the two waves.*

If the two waves be sound waves in air, the maxima and minima of displacement and intensity manifest themselves as louder and fainter sounds alternating with each other as the two waves travel through the air, *one loudness and one succeeding faintness constituting one beat.* Due to the limitations of our ear, we cannot, however, hear more than 16 beats per second. In point of fact, we can hardly ever hear more than 7 or 8 beats per second.

## VIBRATIONS OF STRINGS

A string may be made to execute longitudinal, transverse as well as *torsional* vibrations. We shall, however, concern ourselves here only with its transverse mode of vibration, this being the chief source of most musical sounds and hence the basis of a host of musical instruments.

Now, a transverse vibration in a string may be excited by *plucking, hammering* or *bowing* it, as in the cases of the *guitar,* the *pianoforte* and the *violin* respectively, all these different modes of excitation having their own subtle differences. Again, therefore, we shall restrict ourselves hereto the transverse vibration in a string excited mainly by plucking it.

Before proceeding further, we may as well understand clearly that a string, for our purpose, may be defined as *a wire or a cord, homogeneous in composition, and having a uniform area of cross section, so that its mass per unit length or its line density is the same all along it.* Further, *it should be perfectly flexible so as to be able to bend without giving rise to any viscous forces in its material, i.e.*, it should have no *stiffness.* And, finally, *it should not yield when under tension*, so that there is no increase in its length.

No actual wire or cord can, of course, satisfy these ideal conditions fully, particularly that of flexibility (or non-stiffness), but if its length be very much greater than its thickness or diameter, it approximates more or less to our definition of a string. *For all practical purposes, therefore, a long, infinitely thin and flexible wire or cord of a uniform composition and area of cross section may be considered to be a string.*

## PROPAGATION OF TRANSVERSE VIBRATION (OR A TRANSVERSE WAVE) ALONG A STRING

Consider a long string with its one end fixed and with its normal or equilibrium position along the axis of x, *i.e.*, with the tension in it sufficiently large so as to make the effects of gravity quite negligible. It will be found that on giving its free end a smart and quick up and down jerk, the hump thus produced travels along it towards the fixed end. Or, if the string under tension be fixed at both ends and plucked at any point, a hump travels towards either fixed end.

In either case, the string is thrown into transverse vibration, with each successive portion of it executing an up and down motion perpendicular to its length, *i.e.*, a transverse vibration perpendicular to the direction of propagation of the wave (constituted by these vibrations) which obviously travels along the length of the string.

Now, as we know, the-two essential conditions for the production and propagation of a wave in a medium are that (i) *it must possess inertia and (ii) a restoring force must be called into play, tending to restore the displaced particles back to their normal positions.* The former is here supplied by the *mass of the particles of the string* and the latter, by the *transverse component of the tension along the string.*

It is thus clear that *in the absence of any tension, no transverse waves can possibly travel along a string.*

## VELOCITY OF A TRANSVERSE WAVE ALONG A STRING

Suppose we have a string of *mass per unit length or line density* $\delta$, with its normal position along the axis of x, *i.e.*, with the tension large enough to offset the effect of gravity.

Consider a portion PQ of the string of length $\delta x$ displaced slightly in the vertical plane into the position P'O from its equilibrium position along the axis of x, (Fig. 3.8). The displacement being small (though shown greatly exaggerated in the figure), we have PQ = P'Q' = $\delta x$. And because the string is supposed to be perfectly flexible, the tension T is the same at all points on it both in its displaced and undisplaced positions. Since, however, it acts *tangentially* to the string at the point considered, its inclination to the axis of x (or the normal position of the string) is different at different points.

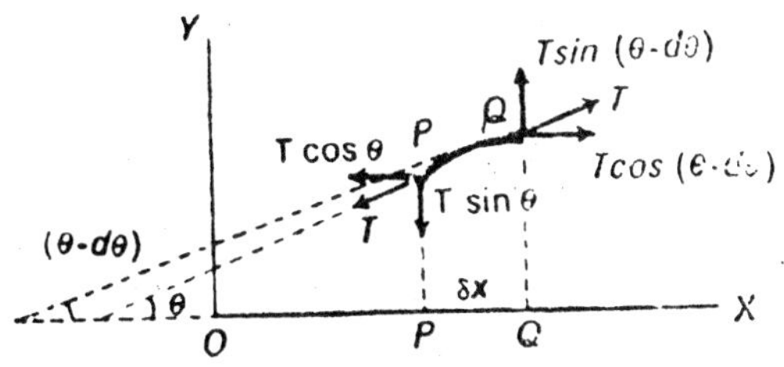


Fig. 3.8

So that, acting tangentially at P' and Q' to the portion P'Q' of the string, as shown, it is inclined at angles θ and (θ – dθ) respectively to the axis of x.

Clearly, the horizontal and vertical components of the tension T at P are F cos θ and F sin θ and those at Q', T cos (θ – dθ) and T sin (θ – dθ) respectively.

The element $\delta x$ of the siring and its displacement being infinitesimally small, the horizontal and vertical components of T at P' and Q' may be taken to act along the same horizontal and vertical lines respectively, though of course in opposite directions to each other. So that, we have

*resultant horizontal force on element $\delta x$ of the string*

$= T \cos(\theta - d\theta) - T \cos\theta = 0$, since $d\theta$ is much too small.

And, *resultant vertical force on element $\delta x$ of the string*

$= T \sin\theta - T \sin(\theta - d\theta) = T \tan\theta - T \tan(\theta - d\theta)$

[$\because$ θ and (θ – dθ) are small.]

Clearly, tan θ is the *slope of the curve at* P' and tan (θ – dθ), its *slope at* Q'.

∴ *resultant downward force on element* $\delta x$ *of the string* T (*slope of the curve at* P' – *slope of the curve at* Q')

Now, *slope of the curve at* P' $= \frac{dy}{dx}$ and since the rate of change of slope with distance is $\frac{d}{dx}\left(\frac{dy}{dx}\right)$, the *change in slope for distance* $\delta x$

$$= \frac{d}{dx}\left(\frac{dy}{dx}\right)\delta x = \frac{d^2 y}{dx^2}\delta x.$$

And, therefore, *slope of the curve at* $Q' = \frac{dy}{dx} - \frac{d^2 y}{dx^2}\delta x.$

Hence *resultant downward force on element* $\delta x$ *of the string*

$$= T\left[\frac{dy}{dx} - \left(\frac{dy}{dx} - \frac{d^2 y}{dx^2}\delta x\right)\right] = T\frac{d^2 y}{dx^2}\delta x.$$

Now, if the Wave Motion of the element $\delta x$ downwards (*i.e.*, along the axis of y) be $\frac{d^2 y}{dt^2}$, the *downward force acting on it is also equal to mass* × Wave Motion $= \sigma\delta x\left(\frac{d^2 y}{dt^2}\right).$

Equating the two values of the downward force on element $\delta x$, we, therefore, have

$$\sigma\delta x\frac{d^2 y}{dt^2} = T\frac{d^2 y}{dx^2}\delta x.$$

Or, $$\frac{d^2 y}{dt^2} = \frac{T}{\sigma}\cdot\frac{d^2 y}{dx^2}.$$

This is clearly, an equation of the same form as the differential equation of a wave motion (Equation (vi), 3.8, and thus represents a transverse wave along the string, with its velocity given by $v^2 = T/\sigma$.

Or, $$v = \sqrt{\frac{T}{\sigma}} = \sqrt{\frac{\text{tension in the string}}{\text{mass per unit length of the string}}}$$

It is thus clear that the velocity of the transverse wave (v) along the string depends only upon (i) the *tension* (T) *applied to the string* and

(ii) the *mass per unit length* (or *line density* $\sigma$) *of the string* and is quite independent of the shape and the amplitude of the hump or the displacement initially produced in it, provided it be small, as it invariably is. This means, in other words, that *the velocity of a transverse wave along a string is quite independent of the actual waveform.*

It also follows at once that the solution of the expression for v is

$$y = a\sin\frac{2\pi}{\lambda}(vt - x) = a\sin\frac{2\pi}{\lambda}\left(\sqrt{\frac{T}{\sigma}}t - x\right).$$

## LONGITUDINAL WAVES (OR SOUND WAVES) IN A GASEOUS (OR A FLUID) MEDIUM

In a gaseous medium, as we know, only a longitudinal wave motion is possible, with its particles executing a simple harmonic motion along the direction of propagation of the wave and that the wave thus travels in the form of *condensations* (*i.e.*, crowding together of the particles) and *rarefactions* (*i.e.*, the particles getting further apart); so that, *there is a continuous variation of pressure all along its direction of propagation.*

To obtain an expression for the velocity of such a wave in a gaseous (or a fluid) medium, let us imagine the wave to be travelling from left to right along the axis of x through a uniform cylindrical tube of area of cross section a and with its axis coinciding with the axis of x, (Fig. 3.9).

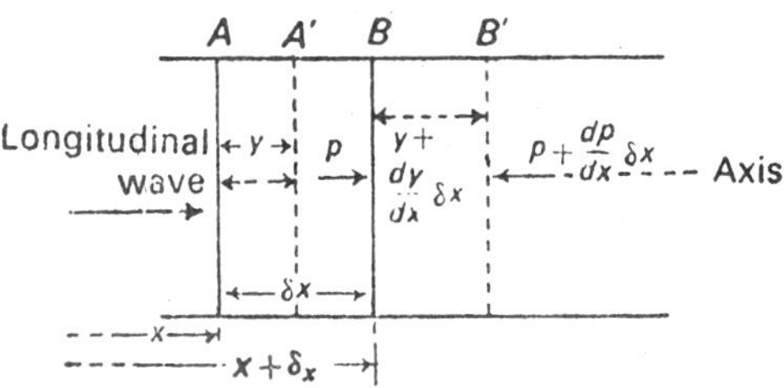


Fig. 3.9

Let A and B be two right plane sections of the tube (*i.e.*, planes perpendicular to the axis) whose positions before the passage of the wave are at distances x and x + $\delta$x respectively from an arbitrary origin, so that they lie $\delta$x apart and the volume of the cylindrical slab or slice of the gas (or the fluid) enclosed in between them is $\delta x.\alpha$.

On the passage of the wave, let the two plane sections get displaced to positions A' and B' respectively. Then, if the displacement of section A be y, *i.e.*, if AA' = y, the displacement of section B, *i.e.*, BB' = y + (dy/dx) δx, where dy/dx is the *rate of change of displacement with distance.*

The two right sections thus get displaced further apart and the *increase in the distance between them*

$$= y + (dy/dx)\,\delta x - y = (dy/dx)\,\delta x.$$

So that, *increase in volume of the cylindrical slab or slice of the gas (or fluid) in between them* $= \left(\frac{dy}{dx}\right)\delta x.\alpha$.

$$\therefore \text{ volume strain } = \frac{\text{increase in volume}}{\text{original volume}} = \frac{\left(\frac{dy}{dx}\right)\delta x.\alpha}{\delta x.\alpha} = \frac{dy}{dx}.$$

If, therefore, K be the *volume elasticity* of the gas (or the fluid) and p, the *excess pressure* (*i.e.*, pressure over and above that of the atmosphere) at section A, now at A', we have

$$K = -\frac{p}{\frac{dy}{dx}},$$

the –ve sign indicating that the excess pressure and the change in volume bear opposite signs, *i.e.*, if p be –ve, the change in volume is +ve or *an increase* and vice versa.

Thus, *excess pressure at section A, now at A'*, is given by

$$p = -K\left(\frac{dy}{dx}\right).$$

If the *pressure gradient*, or the rate of change of pressure .with distance, along the axis of x be dp/dx, we have *excess pressure at section* B, *now at*

$$B' = p + \left(\frac{dp}{dx}\right)\delta x$$

$$= p + \frac{d}{dx}\left(-K\frac{dy}{dx}\right)\delta x = p - K\frac{d^2y}{dx^2}\delta x.$$

∴ *resultant pressure on the slice of gas (or fluid) enclosed between*

$$\text{A' and B'} = p - K\left(\frac{d^2y}{dx^2}\right)\delta x - p = -K\left(\frac{d^2y}{dx^2}\right)\delta x,$$

the – ve sign indicating that it is directed from B' to A', opposite to the direction of propagation of the wave.

∴ force acting on this slice of gas (or fluid), say,

$$F = -K\left(\frac{d^2y}{dx^2}\right)\delta x.\alpha.$$

Now, force is also equal to *mass* × *Wave Motion*. So that, if $-\frac{d^2y}{dt^2}$ be the *Wave Motion* of the slice of gas (or fluid) enclosed by the two sections (in the direction B' to A'), the *force acting on it is also* = *mass of the slice* × $\left(-\frac{d^2y}{dt^2}\right)$.

If ρ be the density of the gas (or the fluid), we have *mass of the slice* = δx.α.ρ. And, therefore, *force acting on it, i.e.,*

$$F = -\delta x.\alpha.\rho\left(\frac{d^2y}{dt^2}\right)$$

Equating the two values of F, we have

$$-\delta x.\alpha.\rho\left(\frac{d^2y}{dt^2}\right) = -k\left(\frac{d^2y}{dx^2}\right).\delta x.\alpha.$$

Or,
$$\frac{d^2y}{dt^2} = \left(\frac{K}{\rho}\right)\frac{d^2y}{dx^2},$$

which is of the same form as the standard differential equation of a wave motion. So that, *velocity of the longitudinal wave through the gas (or the fluid) is given by*

$$v = \sqrt{\frac{K}{\rho}},$$

*i.e., velocity of the wave* $= \sqrt{\dfrac{\text{volume elasticity of the gas (or fluid)}}{\text{density of the gas (or fluid)}}}$

Now, Newton, who first deduced this relation for v assumed that during the passage of a sound wave through a gas (or air) the temperature of the gas (or air) remains unaffected, the heat passing from the hotter regions of condensations to the adjacent colder regions of rarefactions, thereby equalising the temperature. In other words, he assumed that a sound wave passes through a gas (or air) under *isothermal* conditions, and hence took K to be the *isothermal elasticity* of the gas (or air), which, as we know, is equal to its pressure (P). So that, *Newtons's formula for the velocity of a sound wave (or a longitudinal wave) in a gaseous medium* becomes

$$v = \sqrt{\frac{P}{\rho}}.$$

If, however, we calculate the velocity of sound in-air at N.T.P. with the help of this formula, we have $P = 76 \times 13.6 \times 981$ dynes/cm$^2$ and $\rho = 0.001293$ gm/c.c. and, therefore,

$$v = \sqrt{\frac{76\times13.6\times981}{0.001293}} = 280 \text{ metres/sec, very nearly.}$$

Actually, the velocity of sound in air at N.T.P., as measured by *Newton* himself, is found to be 332 metres/sec.

*Newton* could offer no satisfactory explanation for this large discrepancy between his theoretical and experimental results. Strangely enough, the error in Newton's formula remained undetected for almost a century and a half (or 140 years, to be exact) until La' *place*, a French mathematician, pointed it out in the year 1816.

## LA' PLACE'S CORRECTION OF NEWTON'S FORMULA—EFFECT OF PRESSURE AND TEMPERATURE ON THE VELOCITY OF SOUND IN A GAS (OR AIR)

*La' place* correctly argued that as a sound wave passes through a gas (or air) the condensations and rarefactions succeed each other much too rapidly. This coupled with the fact that a gas (or air) is a poor conductor of heat, does not make for any equalisation of temperature due to passage of heat from the hotter regions of condensation to the colder ones of rarefaction. So that, although the total quantity of heat in the medium (gas or air) remains constant, *there are temperature variations throughout the medium and that, therefore, a sound wave passes through a gas (or air) under adiabatic, and not isothermal, conditions. Newton should thus have used the adiabatic, and not the isothermal*, elasticity of the gas (or air) in his formula.

Now, *adiabatic elasticity* = $\gamma$ × *isothermal elasticity* = $\gamma$P, where $\gamma$ is the ratio between the two principal specific heats of the gas (or air) *viz.*, the specific heats at constant pressure and at constant volume.

Thus, Newton's formula, *as corrected by La' place*, becomes

$$v = \sqrt{\frac{\gamma P}{\rho}}.$$

The value of $\gamma$ is the highest, 1.67, for a monatomic gas, like helium or mercury vapour, 1.4 for a diatomic gas like hydrogen, oxygen etc. and 1.33 for a triatomic gas like ozone ($0_3$) or water vapour ($H_2O$), *i.e.*, its value goes on decreasing with increasing atomicity of the gas or vapour but is always greater than 1.

For air, which is a mixture of gases, $\gamma$ = 1.41. So that, in air,

$$v = \sqrt{\frac{1.41P}{\rho}},$$

which gives 331.6 metres/sec as the velocity of sound (in air) at

N.T.P., in excellent agreement with Newton's own experimental result, *thus fully justifying La' places' correction of Newton's formula.*

*Effect of pressure and temperature on the velocity of sound in a gas (or air).* Considering a gram-molecule of a gas (or air), we have PV = RT, where V is the *volume of a gram-molecule of the gas (or air), T, its absolute temperature and R, the gas constant.*

If M be the molecular weight of the gas (or air), we have $V = \frac{M}{\rho}$.

So that $P\left(\frac{M}{\rho}\right) = RT$, whence, $\rho = \frac{PM}{RT}$. And, therefore,

$$v = \sqrt{\frac{\gamma P}{\rho}} = \sqrt{\frac{\gamma P}{\frac{PM}{RT}}} = \sqrt{\gamma \frac{RT}{M}},$$

showing that the *velocity of sound in a gas (or air} is quite independent of its pressure and directly proportional to the square root of its absolute temperature.*

Thus, if the velocity of sound in a gas (or air) at t°C or (t + 273) or T° abs. be v and its velocity at 0°C or (0 + 273) or $T_0^\circ$ abs. be $v_0$, we get,

$$\frac{v}{v_0} = \sqrt{\frac{T}{T_0}}.$$

Or, in general, $\dfrac{v_1}{v_2} = \sqrt{\dfrac{T_1}{T_2}}$,

where $v_1$ and $v_2$ are the velocities of sound in a gas (or air) at absolute temperatures $T_1$ and $T_2$ respectively.

## LONGITUDINAL WAVES IN RODS

Newton's formula for the velocity of a longitudinal wave in a fluid (*viz.*, $v = \sqrt{K/\rho}$) may also be extended to the case of an isotropic solid if it be in the form of a *thin rod* or *wire*, so that lateral contraction and expansion (in the regions of condensations and rarefactions) are permissible. Obviously, the elasticity involved here will be the *linear elasticity* or *Young's modulus* Y. So that, if a *thin rod* or *wire* (*i.e.*, one whose diameter is very much smaller than its length) be set into longitudinal vibration by rubbing it (in one direction only) with a chamois leather or a resined cloth, the velocity of the longitudinal wave, thus set up in it and travelling along its 'length, is given by $v = \sqrt{Y/\rho}$, where $\rho$ is the *density* of the material of the rod or the wire.

This relation may, however, also be deduced directly in the same manner as used in the case of a gas (or a fluid).

Thus, consider two right plane sections A and B of a thin rod to be a distance $\delta x$ apart before the passage of a longitudinal wave (or a sound wave) through it, (Fig. 3.10).

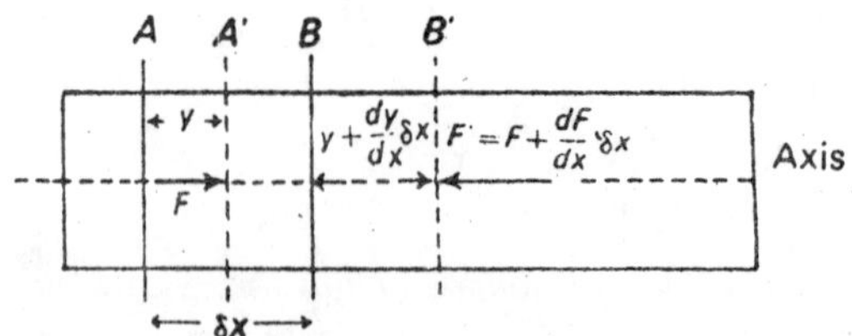


**Fig. 3.10**

On the passage of the wave, let the two sections be displaced to positions A' and B' respectively, such that the displacement of section

A, *i.e.*, AA' = y and that of section B, *i.e.*, BB' = $y+\left(\frac{dy}{dx}\right)\delta x$, where $\frac{dy}{dx}$ is the rate of change-of displacement with distance.

Then, obviously, *increase in length of element δx of the rod*

$$= y+\left(\frac{dy}{dx}\right)\delta x - y = \left(\frac{dy}{dx}\right)\delta x.$$

∴ *linear or tensile strain set up m the rod* = $\frac{\text{increase in length}}{\text{original length}}$

$$= \frac{\left(\frac{dy}{dx}\right)\delta x}{\delta x} = \frac{dy}{dx}.$$

This increase in the length of the element gives rise to restoring forces in the material of the rod, tending to bring it back to its original length. If F be force acting at section A, now at A', clearly, *tensile stress* $\frac{F}{\alpha}$, where α is the *area of cross section* of the rod.

∴ *Young's modulus for the material of the rod*, $Y = \frac{\text{tensile stress}}{\text{tensile strain}}$

Or, $$Y = \frac{\frac{F}{\alpha}}{\frac{dy}{dx}},$$

whence $$F = Y.\alpha\left(\frac{dy}{dx}\right)$$ in the direction A' to B'.

∴ *force acting at section B, now at B'*, say, $F' = F+\left(\frac{dF}{dx}\right)\delta x$, in *the direction* B' to A', where $\frac{dF}{dx}$ is the rate of change of force with distance.

Hence *resultant force acting on element δx of the rod*

$$= F+\left(\frac{dF}{dx}\right)\delta x - F = \left(\frac{dF}{dx}\right)\delta x$$

$$= \frac{d}{dx}\left(Y\alpha\frac{dy}{dx}\right)\delta x = Y\alpha\frac{d^2y}{dx^2}\delta x$$ *in the direction B' to A'.*

Also if $\frac{d^2y}{dt^2}$ be the *instantaneous Wave Motion* produced in the element δx of the rod by this force, in the direction B' to A', we *force acting on element*

$$\delta x = \textit{mass of element} \times \frac{d^2y}{dt^2}.$$

$$= \delta x.\alpha.\rho.\frac{d^2y}{dt^2}, \text{ where } \rho \text{ is the } \textit{density} \text{ of the material of the rod.}$$

Equating the two values of the force, therefore, we have

$$\delta x.\alpha.\rho.\left(\frac{d^2y}{dt^2}\right) = Y.\alpha.\left(\frac{d^2y}{dx^2}\right).\delta x,$$

whence,
$$\left(\frac{d^2y}{dt^2}\right) = \left(\frac{Y}{\rho}\right)\left(\frac{d^2y}{dx^2}\right).$$

This equation being of the same form as the standard differential equation of a wave motion, we have

= *velocity of the longitudinal wave in the rod, i.e.* $v = \sqrt{\frac{Y}{\rho}}$.

**Note :** The relation, it may be emphasised once against is valid only if the solid be *isotropic* and in the form of a *thin rod* or *wire.*

In the case of an extended solid, where lateral contractions and extensions are not possible, as for example, in the case of the earth's crust (*i.e.*, the upper layers of the earth), the elasticity involved is what is called *elongational elasticity* given by $\left(K + \frac{4}{3}n\right)$, where K is the *volume elasticity* or *bulk modulus* and *n*, the *modulus of rigidity* of the material of the solid. So that, in such a case,

$$v = \sqrt{\frac{\left(K + \frac{4}{3}n\right)}{\rho}}.$$

The value of v here it always about 1.1 times that if the material be in the form of a thin rod or wire.

For different modes of vibration of rods.

## STATIONARY OR STANDING WAVES IN A LINEAR BOUNDED MEDIUM

A *linear medium* is one in which a wave is constrained to travel along a fixed linear path. Since *plane waves*, as we know, *being one-dimensional*, also travel along a fixed linear path, they too *may just as well be treated as waves in a linear medium*, as, for example, transverse waves along a string or a longitudinal wave along a thin rod or a pipe.

If the medium be of an infinite or unlimited length, the waves just continue to travel along or through it for an infinite time.

In case, however, the length of the medium be a limited or a finite one, it is spoken of as a *linear bounded medium* and if the waves travelling along it suffer normal reflection at the boundary, we have two identical waves (*i.e.*, having the *same wavelength, frequency and amplitude*) travelling along the same linear path in opposite directions. The superposition of such waves gives rise to a very special case of interference in which the positions of maxima and minima of displacement remain fixed throughout, so that the interference pattern or the resulting waves appear to remain stationary in space, with no onward or progressive movement. This is the reason why they are called stationary or standing waves.

Now, obviously, two main cases arise: (i) *when the boundary of the linear medium is fixed or rigid and (ii) when the boundary is free.*

A possible third case may be that of a boundary separating two media of different densities and hence of different velocities of the wave in them, so that the incident wave at the boundary is partly reflected and partly transmitted (or absorbed) into the second medium. We need, however, concern ourselves here with only the former two, with which alone, therefore, we shall deal in proper detail

It is pretty obvious that in case (i), the particle at *the fixed or rigid boundary, like a wall, the fixed end of a string or the closed end of a pipe, for example, is not free to move and hence the displacement (y) of the particle there will be zero at all times. But equally obviously, dy/dx must not simultaneously be zero or else it would mean zero displacement at all points of the medium and hence no wave at all. Thus, although the displacement at the fixed or rigid boundary is zero throughout, the strain (dy/dx) and the pressure variation there are not simultaneously zero (since* $p = -K dy/dx$*).*

This means, in other words, that whereas the displacement of the particle at the fixed or rigid boundary is exactly cancelled by an equal and opposite displacement due to reflection, *i.e., the displacement, and hence also the particle velocity, undergoes a phase change of w, the strain and the pressure variation do not undergo any such phase change and are not, therefore, zero*. This if expressed by saying that *at a fixed or rigid boundary, a wave is reflected without change of type, i.e., a crest (or a condensation) as a crest (or a condensation) and a trough (or a rarefaction), as a trough (or a rarefaction), but with change of sign, i.e., with the direction of displacement of the particles reversed.*

On the other hand, in case (ii), there being no insurmountable opposing force or resistance encountered, the strain (dy/dx) and *the excess pressure p(= – Kdy/dx) become zero, i.e., the phase of pressure variation gets reversed or undergoes a phase change of π, but, obviously, y is not zero or else the displacement will be zero everywhere and there will be no wave at all. This means in other words, that the particles of the medium continue to have their displacements, and hence their velocities, (i.e., particle velocities) in the same direction as before, with no phase change, and only the direction of strain and pressure variation gets reversed. So that, a crest (or a condensation) is reflected back as a trough (or a rarefaction) and vice versa, with the direction of displacement of the particles remaining the same.* This is expressed by

saying that *at a free boundary, a wave is reflected with change of type but without change of sign.*

The difference between reflection at a fixed or rigid boundary and that at a free boundary may, for ready reference, he summarised as follows:

| Reflection at a fixed or rigid boundary | Reflection at a free boundary |
|---|---|
| 1. Displacement, and hence also particle velocity, undergoes a phase change of w, *i.e.*, gets reversed. | 1. Displacement and particle velocity undergo no phase change and do not, therefore, get reversed. |
| 2. Strain and pressure .variation undergo no phase change and, therefore, reflection occurs without change of type but with change of sign. | 2. Strain and pressure variation undergo a phase change of π and thus get reversed, so that reflection occurs with change of type but without change of sign. |

We are now in a position to study the formation of stationary or standing-waves in the two cases and, therefore, proceed on to it.

**Case I :** *When reflection occurs at a fixed or rigid boundary:* Let the equation of a simple harmonic wave of amplitude a and wavelength λ, travelling along the positive direction of the x-axis be

$$y_1 = a\sin\frac{2\pi}{\lambda}(vt - x),$$

where v is the velocity of the wave ir the medium.

Then, if it be incident *normally* on, and reflected from, a *fixed* or *rigid* boundary, the equation of the reflected wave will be

$$y_2 = -a\sin\frac{2\pi}{\lambda}(vt + x),$$

since both, the direction of displacement of the particles and the direction of travel of the wave itself, get reversed.

Now, both the waves, travelling along the same linear path, get superposed and the equation of the resultant stationary or standing wave is, therefore,

$$y = y_1 + y_2 = a\sin\frac{2\pi}{\lambda}(vt - x) - a\sin\frac{2\pi}{\lambda}(vt + x)$$

$$\text{Or, } y = a\left[\sin\frac{2\pi}{\lambda}(vt - x) - \sin\frac{2\pi}{\lambda}(vt + x)\right] = -2a\cos\frac{2\pi vt}{\lambda}\sin\frac{2\pi v}{\lambda}$$

$$\text{Or, } y = -2a\sin\frac{2\pi x}{\lambda}\cos\frac{2\pi vt}{\lambda} = -2a\sin\frac{2\pi x}{\lambda}\cos\omega t, \qquad \text{...(I)}$$

$$\left[\because \frac{v}{\lambda} = n \text{ and } 2\pi n = \omega\right]$$

showing that *the resulting wave is also a simple harmonic wave of the same time-period and wavelength as the two constituent waves but with an amplitude*

$$= -\frac{2a\sin 2\pi x}{\lambda}.$$

$$\text{Clearly, particle velocity, } U = \frac{dy}{dt} = \frac{4\pi v}{\lambda}a\sin\frac{2\pi x}{\lambda}\sin\omega t, \qquad \text{...(II)}$$

$$\text{strain}\frac{dy}{dx} = -\frac{4\pi}{\lambda}a\cos\frac{2\pi x}{\lambda}\cos\omega t \qquad \text{...(III)}$$

$$\text{and } \textit{excess pressure } p = -K\frac{dy}{dx} = K\frac{4\pi}{\lambda}a\cos\frac{2\pi x}{\lambda}\cos\omega t$$

$$= v^2 \rho \frac{4\pi}{\lambda} a \cos \frac{2\pi x}{\lambda} \cos\omega t, \qquad ...(IV)$$

because $v = \sqrt{\frac{K}{\rho}}$ and, therefore, $K = v^2\rho$.

Let us now consider the changes (hat occur (a) *with respect to position of* a particle, (b) *with respect to time.*

(a) *Changes with respect to position*

(i) **Displacement:** The displacement of a particle of the medium in the resulting wave is given by relation I above and, as can be seen at once, *varies simple harmonically with time.*

It is, however, zero at all times for points for which sin $2\pi x/\lambda = 0$ or cos $2\pi x/\lambda = \pm 1$, or $2\pi x/\lambda = m\pi$, where m is an integer, 0, 1, 2, 3 etc. and, therefore,

$$x = 0, \frac{\lambda}{2}, \frac{3\lambda}{2}, \text{ etc.}$$

Thus, *the displacement is zero at points distant 0, λ/2, 3λ/2 etc from the fixed or the rigid boundary, irrespective of time. These points of permanent zero displacement are called* **nodal points or nodes**, or, *more correctly,* **displacement nodes**, *and obviously lie λ/2 apart..*

On the other hand, the displacement will be the maximum (positive or negative) or the amplitude ± 2a, irrespective of the value of at and hence of time, for points for which sin $2\pi x/\lambda = \pm 1$

or $\qquad \cos 2\pi x/\lambda = 0$

or $\qquad 2\pi x/\lambda = (2m + 1)\pi/2$,

where $\qquad m = 0, 1, 2, 3$ etc.,

and therefore, $\quad x = \lambda/4, 3\lambda/4, 5\lambda/4$ etc.

*Thus, the displacement is the maximum at points distant λ/4, 3λ/4, 5λ/4 etc. from the fixed or the rigid boundary, irrespective of time. These points of maximum displacement (positive or negative) are called* **antinodal points** *or* **antinodes**, *or, more correctly,* **displacement antinodes**, *and also lie λ/2 apart.*

It may be noted that *the displacement at the antinodal points also varies simple harmonically with time but is always the maximum there it any given instant relative to that at all other points, where the displacement lies between the two extremes, decreasing from the maximum it an antinodal point to zero at the preceding or succeeding nodal point.*

Further, it will be easily seen that *no two nodes can exist without an antinode in between and vice versa, so that the distance between a node and a succeeding or preceding antinode* is $\lambda/4$.

(ii) **Particle velocity:** *At the nodal points or displacement nodes,* as we have seen under (i) above, $\sin 2\pi x/\lambda = 0$. So that, *the particle velocity* U(given by relation II) *is zero threat all times. These points, therefore, permanently remain at rest throughout the passage of the wave.*

And, *at the antinodal points or displacement antinodes, since* $\sin 2\pi x/\lambda = \pm 1$, *the particle velocity (U) is the maximum (positive or negative) irrespective of time.* It may be noted that the particle velocity here too varies simple harmonically with time but is, at any given instant, maximum relative to that at all other points, where it lies between these two extremes, decreasing from the maximum at an antinodal point to zero at the preceding or succeeding nodal point, its amplitude being obviously

$$\frac{4\pi v}{\lambda} a \sin\frac{2\pi x}{\lambda}.$$

(iii) **Strain:** Since $\sin 2\pi x/\lambda = 0$ and, therefore, $\cos 2\pi x/\lambda = \pm 1$ at *the nodal points, the strain (dy/dx) there,* (given by relation III above), *is the maximum at all times,* due, obviously, to the condensations or the rarefactions of the oppositely travelling waves coming across each other, as shown in Fig. 3.11 (a).

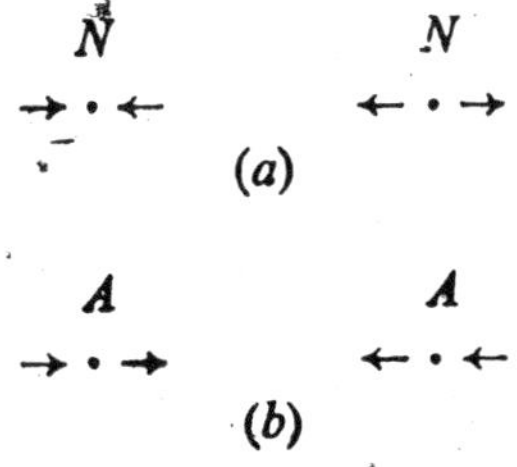


**Fig. 3.11**

At the antinodal points, on the other band, $\sin 2\pi x/\lambda = \pm 1$ and, therefore, $\cos 2\pi x/\lambda = 0$. So that, *the strain at these points is zero at all times*, due, no doubt, to the condensation of one wave coming across the rarefaction of the other, as shown in Fig. 3.11 (b).

(iv) **Pressure variation:** In equation IV above for excess pressures p, if we put K.4πa/λ or $\rho v^2$ 4πa/λ = $P_m$, the maximum value of pressure, we have

$$p = P_m \cos\frac{2\pi x}{\lambda}\cos\omega t,$$

where, clearly, $P_m \cos\frac{2\pi x}{\lambda}$

is the *amplitude of pressure variation.* Denoting it by $P_x$, therefore, we have $p = P_x \cos \omega t$, indicating that *the pressure at all points of the medium varies simple harmonically with time, except, of course, at the antinodal points or displacement antinodes*, for which cos 2πx/λ = 0 and, therefore, p = 0.

Thus, there being no excess pressure at the antinodal points, *there is no change of pressure and hence also of density at these points, which thus throughout remain normal.* For this reason these points are also referred to as *pressure nodes.*

*At the nodal points or displacement nodes*, cos 2πx/λ = ± 1 and hence *the excess pressure* (p) *is the maximum there (positive or negative). The pressure and density changes are, therefore, the maximum at these points—higher . and lower than the normal, alternately.* These points are, therefore, also referred to as pressure antinodes.

(v) **Phase:** Considering two points at distances $x_1$ and $x_2$ we have

$$y_1 = -2a\sin\frac{2\pi x_1}{\lambda}\cos\omega t$$

and $$y_2 = -2a\sin\frac{2\pi x_2}{\lambda}\cos\omega t.$$

Or, putting – 2a sin $2\pi x_1/\lambda = A_1$ and – 2a sin $2\pi x_2/\lambda = A_2$, where $A_1$ and $A_2$ are the amplitudes at distances $x_1$ and $x_2$ respectively, we have

$$y_1 = A_1 \cos \omega t \text{ and } y_2 = A_2 \cos \omega t.$$

Thus, although the *amplitudes* of the two particles are *different*, then *phases* are the *same* indicating *that the phase of a particle is quite independent of its position.*

This means, in other words, that *all particles attain their respective maximum displacements (positive or negative) simultaneously* (though, of course, their maxima are all different, those of the antinodal points being the greatest).

Similarly, *all particles pass through their mean position simultaneously, though with their different maximum velocities.*

Further, since the displacement of the particles varies simple harmonically with time, their phases get reversed every half a period. It follows, therefore, that with the nodal points lying $\lambda/2$ apart, *the particles between one pair of nodal points are in opposite phase to those between the next succeeding (or preceding) pair.*

(b) *Changes with respect to time:* From equations I, II and III above, it is clear that when $\cos 2\pi vt/\lambda = 0$ and, therefore, $\sin 2\pi vt = \pm 1$ we have displacement $y = 0$; *particle velocity* $U = dy/dt =$ *maximum* (positive or negative), though this maximum is different for different points; strain $dy/dx = 0$ and hence also *excess pressure* $p = - K\, dy/dx = 0$. (From equation IV)

This will obviously happen when $2\pi\ vt/\lambda = (2m + 1)\ \pi/2$ or $(2\pi/\lambda)\ (\lambda/T)t = (2m + 1)\ \pi/2$ ($\because\ v = n\lambda = \lambda/T$), *i.e.*, when $t = (2m + 1)\ T/4$ or $t = T/4$, $3T/4$, $5T/4$ etc., half a period apart from each other.

Thus, *at instants T/14, 3T/4, 5T/4, half a period apart, all particles pass through their mean positions (y = 0) with their maximum velocities (U = dy/dt = maximum), though these maxima are different for different particles. Also, the strain everywhere at these instants is zero (dv/dx =0).*

Further, since at these *particular instants, there is no change of pressure and density anywhere, they remain normal all through the medium.*

On the other band, when $\cos \omega t$ or $\cos 2\pi vt/\lambda = \pm 1$ and, therefore, $\sin 2\pi vt/\lambda = 0$, we have **displacement** $y =$ *maximum* (*positive or negative*); **particle velocity** $U = dy/dt = 0$; **strain** $dy/dx =$ *maximum* (*positive or negative*) and hence also **excess pressure** $p = -K dy/dx =$ *maximum (positive or negative).*

This will obviously happen when $2\pi vt/\lambda = m\pi$ or when $(2\pi/\lambda)(\lambda/T)t = m\pi$, or, $t = mT/2$, *i.e.*, when $t = 0$, $T/2$, $3T/2$ etc., again half a period apart from each other.

Thus, *at instants 0, T/2, 3T/2 etc., half a period apart, all particles attain their maximum displacements (positive or negative) simultaneously, though their maxima are all different. Also, the strain on the particles is now the maximum compared with that at other instants.*

Further, since at these instants, p is maximum (positive or negative), *the changes of pressure and density are also the maximum.*

It follows at once from the above that (i) *all particles pass through their mean positions and attain their maximum displacements (positive or negative) twice in one time-period, (ii) the particles attain their maximum displacements (positive or negative) an interval T/4 later than they pass through their mean positions and (iii) the pressure changes occur in accordance with the Bernoulli principle, being the maximum where the velocity is zero and vice versa.*

*Case II : When reflection occurs at a free boundary:* As before, let the equation of a simple harmonic wave, of *amplitude* a and *wavelength* λ, travelling along the positive direction of the x-axis be $y_1 = a\sin\frac{2\pi}{\lambda}(vt - x)$. Then, the equation of the wave reflected at the free boundary will be $y_2 = a\sin\frac{2\pi}{\lambda}(vt + x)$. Since only the direction of travel of the wave is reversed and not the direction of displacement of the particles of the medium. The equation of the resulting stationary or standing wave is, therefore,

$$y = y_1 + y_2 = a\sin\frac{2\pi}{\lambda}(vt - x) + a\sin\frac{2\pi}{\lambda}(vt + x) = 2a\sin\frac{2\pi vt}{\lambda}\cos\frac{2\pi x}{\lambda}.$$

$$\text{Or, } y = 2a\cos\frac{2\pi x}{\lambda}\sin\frac{2\pi vt}{\lambda} = 2a\cos\frac{2\pi x}{\lambda}\sin\omega t, \quad \text{...(V)}$$

$$\left[\because \frac{v}{\lambda} = n \text{ and } 2\pi n = \omega\right]$$

showing that *the resulting wave is also a simple harmonic wave, of the same time-period and wavelength as each of the two constituent waves but with an amplitude = 2a cos 2px/λ.*

Clearly, therefore, *particle velocity*

$$U = \frac{dy}{dt} = \frac{4\pi v}{\lambda}a\cos\frac{2\pi x}{\lambda}\cos\omega t \quad \text{...(VI)}$$

$$\text{strain } \frac{dy}{dx} = -\frac{4\pi}{\lambda}a\sin\frac{2\pi x}{\lambda}\sin\omega t \quad \text{...(VII)}$$

*and excess pressure p* $= -K\frac{dy}{dx} = K\frac{4\pi}{\lambda}a\sin\frac{2\pi x}{\lambda}\sin\omega t$

$$= \rho v^2 \frac{4\pi}{\lambda} a \sin\frac{2\pi x}{\lambda} \sin \omega t. \qquad ...(VIII)$$

$$\left[\because v = \sqrt{\frac{K}{\rho}} \text{ and } \therefore K = \rho v^2.\right]$$

Let us now consider the changes (a) *with respect to position of a particle* and (b) *with respect to time.*

(a) *Changes with respect to position*

(i) **Displacement:** From equation V above, for displacement y, it is clear that *the displacement at all points varies simple harmonically with time* but is *zero at points for which* cos $2\pi x/\lambda = 0$ or $2\pi x/\lambda = (2m + 1)\pi/2$, (where m is an integer 0, 1, 2, 3, etc) or, when $x = (2m + 1)\lambda/4$, *i.e.*, equal to $\lambda/4$, $3\lambda/4$, $5\lambda/4$ etc., *irrespective of time.*

*These points of zero displacement thus occur at distances* $\lambda/4$ $3\lambda/4$, $5\lambda/4$ etc. *from the free boundary, $\lambda/2$ apart from each other and are, as we know, called* **nodal points** *or* **nodes**, *or, more correctly* **displacement nodes.**

On the other hand, the *displacement is the maximum (positive or negative) or the amplitude ± 2a at points for which* cos $2\pi x/\lambda = \pm 1$, *i.e.*, when $2\pi x/\lambda = m\pi$ or $x = m\lambda/2 = 0, \lambda/2, 3\lambda/2, 5\lambda/2$ etc. *irrespective of time.*

These points of maximum displacement (positive or negative) thus occur at distances 0, $\lambda/2$, $3\lambda/2$, $5\lambda/2$ etc from the free boundary. *They are the* **antinodal points** *or* **antinodes**, or more correctly, **displacement antinodes**, *and also lie $\lambda/2$ apart but alternating with the nodal points or displacement nodes.*

The displacements at all other points lie between these two extremes, decreasing from the maximum at an antinodal point to zero it the preceding or succeeding nodal point.

(ii) **Particle velocity:** From equation VI for particle velocity U, it can be easily seen that *at the nodal points, since* cos $2\pi x/\lambda = 0$, *the particle velocity is zero, irrespective of time.* In other words, *the particles of the medium throughout remain at rest at these points.*

On the other hand, *at the antinodal points, since* cos $2\pi x/\lambda = \pm 1$, *the particle velocity is the maximum (positive or negative) and this is again so, irrespective of time.*

(iii) **Strain:** *At the nodal points,* cos $2\pi x/\lambda = 0$ and, therefore, sin $2\pi x/\lambda = \pm 1$. So that the strain (dy/dx), given by equation VII above, *is the maximum (positive or negative) at all times.*

And, *at the antinodal points, since* cos $2\pi x/\lambda = \pm 1$ *and, therefore,* sin $2\pi x/\lambda = 0$, *the strain is the minimum or zero at all times.*

**Pressure variation:** As in case I, so also here, putting $\rho v^2 4\pi\ a/\lambda = P_m$, the maximum value of pressure, we have (from equation VIII), $p = P_m \sin\dfrac{2\pi x}{\lambda}\sin\omega t$, so that $P_m \sin\dfrac{2\pi x}{\lambda}$ is the *amplitude of pressure variation.* Putting it equal to $P_x$, we, therefore, have

$$p = P_x \sin \omega t,$$

indicating that *pressure too at all points varies simple harmonically with time.*

*At displacement nodes or nodal points, where* cos $2\pi x/\lambda = 0$ *and, therefore,* sin $2\pi x/\lambda = \pm 1$, *the pressure and density changes are the maximum at all times*, the pressure (and density) being alternately higher and lower than the normal. These points are thus also referred to as the pressure antinodes.

*At displacement antinodes or antinodal points*, on the other hand, cos $2\pi x/A = \pm 1$ and, therefore, sin $2\pi x/\lambda = 0$, *so that* $p = 0$, *i.e., there are no pressure and density changes which, therefore, remain normal throughout.* For this reason, these points are referred to as **pressure nodes.**

(v) **Phase:** As before (case I), we can put displacements at two points $x_1$ and $x_2$ as $y_1 = 2a \cos\dfrac{2\pi x_1}{\lambda} \sin\omega t = A_1 \sin\omega t$ and $y_2 = 2a \cos\dfrac{2\pi x_2}{\lambda} \sin\omega t = A_2 \sin\omega t$, where $2a\cos\left(\dfrac{2\pi x_1}{\lambda}\right) = A$ and $2a\cos\left(\dfrac{2\pi x_2}{\lambda}\right) = A_2$.

Thus, *although the amplitudes of the two particles are different, their phases are the same, showing that the phase of a particle is independent of its position. This means, in other words, that all particles pass through their mean positions or attain their maximum displacements (positive or negative) simultaneously.*

And, since toe displacement of the particles varies cyclically with time, their phases get reversed every half a period; so that the nodal points lying $\lambda/2$ apart, *all particles between one pair of nodal points are in a phase opposite to that of the particles between the preceding or succeeding pair.*

(b) *Changes with respect to time:* At instants t for which sin $\omega t$ = sin $2\pi vt/\lambda = 0$ and, therefore, cos $2\pi vt/\lambda = \pm 1$, we have **displacement** y = 0, **particle velocity** U = dy/dt, *the maximum* (positive or negative) though the value of the maximum velocity is different at different points; **strain**, dy/dx = 0

$$\text{and hence also } p = -\frac{K\,dy}{dx} = 0.$$

This obviously occurs when $2\pi vt/\lambda = (2\pi/\lambda)(\lambda/T)t = m\pi$ or t = mT/2 = 0, T/2, T, 3T/2 etc. Thus, *at instants 0, T/2, T, 3T/2 etc., half a period apart, all particles pass through their mean positions (y = 0), with their maximum velocities (positive or negative), since dy/dt is maximum, but the maxima of velocity are different for different particles.*

Also *strain and hence changes of pressure and density are zero at all points at these instants.*

And, at instants t for which sin $2\pi vt/\lambda = \pm 1$ and, therefore, cos $2\pi vt/\lambda = 0$, we have **displacement** y = *maximum (positive or negative)*; **particle** velocity U = dy/dt = 0; **strain** = dy/dx = *maximum (positive or negative)* and hence also **excess pressure** p = – K dy/dx = *maximum (positive or negative).*

This clearly occurs when $2\pi vt/\lambda = (2\pi/\lambda)(\lambda/T)t = (2m + 1)\pi/2$ or when t = (2m + 1) T/4, *i.e.*, equal to T/4, 3T/4, 5T/4 etc., *again at intervals* of T/2.

Thus, *at instants* T/4, 3T/4, 5T/4, *an interval T/2 apart, all particles attain their maximum displacements (different for different particles), with the strain and hence also changes of pressure and density maximum, relative to those at all other instants.* Again, therefore, the *Bernoulli principle* holds good, *viz.*, that the pressure is the maximum when the velocity is zero and vice versa.

**Flow of energy in a stationary or standing wave:** Considering the case of stationary waves formed by reflection at a free boundary (*i.e.*, case II), we have *excess pressure* $p = P_x \sin \omega t = P_x \sin 2\pi vt/\lambda$. And, since *particle velocity*

$$U = \frac{dy}{dt} = \frac{4\pi v}{\lambda} a \cos\frac{2\pi x}{\lambda} \cos\frac{2\pi vt}{\lambda},$$

we may put it as

$$U = U_x \cos \frac{2\pi vt}{\lambda},$$

where $$U_x = \frac{4\pi v}{\lambda} a \cos \frac{2\pi x}{\lambda}.$$

So that, *energy transferred (equal to the work done) per unit area in an interval of time dt is,* say, dI = p.U.dt.

∴ *energy transferred during the whole time-period* T is given by

$$I = \int_0^T p.U\,dt = \int_0^T P_X \sin \frac{2\pi vt}{\lambda} U_X \cos \frac{2\pi vt}{\lambda} dt$$

And, therefore, *rate of energy transfer or average transferred per second,* say,

$$I_{av} = \frac{P_X U_X}{T} \int_0^T \sin \frac{2\pi vt}{\lambda} \cos \frac{2\pi vt}{\lambda}$$

$$= \frac{P_X U_X}{2T} \int_0^T \sin \frac{4\pi vt}{\lambda} dt = 0.$$

The same result may be obtained by considering case I.

Thus, *there is no transference of energy across any section of the medium in the case of a stationary or standing wave.*

This also follows from the fact that an equal amount of energy is transferred by the two constituent waves in *opposite* directions.

This result may also be expressed differently by saying that no *energy is transferred across any section of a linear bounded medium.*

**Note :** It will be readily seen that the excess pressure (p) and the particle velocity (U) differ in phase by π/2. They are, therefore, said to be in quadrature. The case is, in fact, similar to that of wattless current in an A.C. circuit, where too the voltage and current differ in phase by π/2 (*i.e.*, are in quadrature) and the wattage (or the power) is zero, *i.e.*, no work is done by the source or no energy is drawn from it. That is why the current in this case is dubbed as '*wattless*' current.

*Distinction between progressive and stationary waves:* The main differences between progressive and stationary (or standing) waves may, for ready reference, be summarised in parallel columns as shown below.

| Progressive waves | Stationary waves |
|---|---|
| 1. The *vibration characteristic* of each particle of the medium is the *same* and is *handed on from particle to particle*, so that there is an onward propagation of the wave through the medium. | 1. The *vibration characteristic* of each particle of the medium is its own which it does not pass on to others, so that there is no onward propagation of the wave through the medium, *i.e.*, it remains confined to the limited space in which it is produced. |
| 2. All panicles of the medium attain the same maximum displacement (positive or negative) but one after the other. | 2. All particles attain their maximum displacements simultaneously but their maxima are different, decreasing progressively from an antinode to the adjoining node where it is reduced to zero. |
| 3. All particles pass through their mean positions with the same maximum velocity but one after the other. | 3. All particles pass through their mean positions Simultaneously but with different maximum velocities. |
| 4. No particles of the medium are permomently at rest (except momentarily at the extremities of their vibrations). | 4. Certain particles of the medium (*viz.*, the displacement nodes) throughout remain at rest. |
| 5. All particles of the medium undergo the *same changes of pressure and density but one after the other*. | 5. The changes of pressure and density are the *maximum (positive or negative) at the nodal points and zero at the antinodal points but occur simultaneously* at all points. |
| 6. There is a *regular transference of energy across every section of the medium*. | 6. *There is no transference of energy across any section of the medium*. |

## DIFFERENT MODES OF VIBRATION OF STRINGS, RODS AND AIR COLUMNS

(A) *Different modes of vibration of a stretched string:* Suppose we have a string under tension, fixed at both ends and lying along the axis of x in its *normal, undisturbed* position. If it be plucked at any point, two *identical* waves proceed from the point in opposite directions. Each of these two waves then suffers reflection at the fixed end or the rigid boundary, giving rise to stationary or standing waves.

By plucking the string at different suitable points, it can be set into different modes of vibration, *i.e.*, stationary waves of different wavelengths and frequencies may be produced, though the two fixed points, in every case, remain nodal points or displacement nodes.

Thus, if the string be plucked in the middle, it *vibrates as a whole, in one single segment*, as shown in Fig. 3.12 (a), *giving its lowest or fundamental note* (*i.e.*, of the lowest frequency).

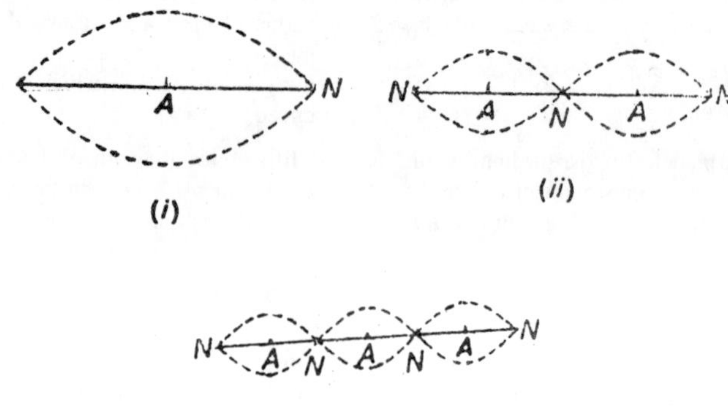


Fig. 3.12

If pressed in the middle and plucked at the mid-point of either half, it *vibrates in two segments*, as shown in Fig. 3.12 (b), *producing a note of twice the frequency of the fundamental.*

And, if pressed at *one-third* of its length and plucked at the mid-point of the smaller length, it *vibrates in three segments*, as shown in Fig. 3.12 (c), *giving a note of thrice the frequency of the fundamental* and so on.

Let us obtain these results analytically.

Let the equation of the simple harmonic wave travelling along the positive direction of the x-axis be $y_1 = a\sin\frac{2\pi}{\lambda}(vt - x)$. It will naturally get reflected at the fixed end of the string at x = 0 and the equation of the reflected wave will, therefore, be $y_2 = -a\sin\frac{2\pi}{\lambda}(vt + x)$. So that, me equation of the resulting stationary wave thus formed will be

$$y = y_1 + y_2 = a\sin\frac{2\pi}{\lambda}(vt - x) - a\sin\frac{2\pi}{\lambda}(vt + x)$$

Or, $$y = -2a\sin\frac{2\pi x}{\lambda}\cos\frac{2\pi vt}{\lambda}.$$

Now, the displacement y = 0 at the other fixed point also, *i.e.*, at x = = – *l*, where *l* is the *length* of the string. So that, we have

$$0 = 2a\sin\frac{2\pi l}{\lambda}\cos\frac{2\pi vt}{\lambda}.$$

Since the displacement at the fixed end is always zero, the relation above must be true for all values of t and this is obviously possible only if $\sin 2\pi l/\lambda = 0$, or, $2\pi l/\lambda = p\pi$, *i.e.*, $\lambda = 2l/p$, where p is an integer, 0, 1, 2, 3 etc.

Now, the *frequency* of vibration of the string, or that of the note emitted by it, is given by $n = \frac{v}{\lambda}$. So that, we have

$$n = \frac{v}{\lambda} = \frac{v}{2l/p} = \frac{pv}{2l},$$

where v is, of course, the velocity of the wave.

Since $v = \sqrt{\frac{T}{\sigma}}$, where T is the *tension* in the string and σ, the *line density* or *mass per unit length of the string*, we have

$$n = \frac{p}{2l}\sqrt{\frac{T}{\sigma}}.$$

If p = 1, we obtain the *lowest* or the *fundamental tone* of frequency $n_1 = \frac{1}{2l}\sqrt{\frac{T}{\sigma}}$ and *l* = 2*l*, with the string vibrating as a whole in *one single segment*, as shown in Fig. 3.12 (a) having *two displacement nodes* at *either fixed end* and a *displacement antinode* at its *mid-point.*

If p = 2, we have $n_2 = \frac{2}{2l}\sqrt{\frac{T}{\sigma}} = \frac{1}{l}\sqrt{\frac{T}{\sigma}}$ = *twice the frequency of the fundamental tone*, and λ = *l*, with the string vibrating in *two equal segments*, as shown in Fig. 3.12 (b), having nodes and antinodes in the positions indicated.

If p = 3, we have $n_3 = \frac{3}{2l}\sqrt{\frac{T}{\sigma}}$ = *thrice the frequency of the fundamental tone*, and λ = 2*l*/3, with the string vibrating in *three equal segments* as shown in Fig 3.12 (c) and having nodes and antinodes indicated in the figure.

In general, therefore, if the string vibrates in p equal segments, we have a note of frequency p times that of the fundamental, *i.e., the frequencies of the notes emitted by the string are proportional to the number of segments or loops into which the string is thrown into vibration, or,*

$$n_1 : n_2 : n_3 \ldots n_p :: 1, 2, 3, \ldots p,$$

Whereas the *lowest tone* is, as already pointed out, referred to as the fundamental tone or, usually simply the *fundamental*, all the rest (*i.e.*, any integral multiples of the fundamental) are called *overtones*, with the special name *octave* reserved for the first overtone *having twice the frequency of the. fundamental.*

Now, if, as is the case here, the fundamental and the overtones have their frequencies in the ratio 1 : 2 : 3 ..., thus forming a *harmonic series*, they are called *harmonics*, the *fundamental* itself being the *first harmonic*, the first overtone, the second harmonic and so on.

N.B. We have, in the above discussion, facility assumed that the string vibrates only in 1, 2, 3 etc. segments into which it has been deliberately set vibrating, giving a pure tone of the corresponding frequency. In actual practice, however, it executes two or more of these vibrations *simultaneously*, with, of course, the one into which it has actually been set vibrating predominating. The note emitted is thus a mixture of the pure tone sounded and some of its overtones. In fact, it is the presence of these different overtones with the given note sounded on different musical stringed instruments that gives it its characteristic or distinctive quality.

(B) *Different modes of longitudinal vibration of a rod.* In the case of a rod, set into longitudinal vibration, the following different case arise:

(i) The rod is *not clamped anywhere*, so that the *boundary at both ends of the rod is free*. It is referred to as a **free-free rod**.

(ii) The rod is *clamped at one end, with the other end free*, so that it has a *fixed or a rigid boundary at the former and a free boundary at the latter end.* It is called a **fixed-free rod**.

(ii) The rod is *clamped at ifs mid-point*, so that the boundary at either end is free.

(iv) The rod is *clamped at both ends*, so that there is a *fixed or a rigid boundary at either end.* It is referred to as the fixed-fixed rod.

It will be readily seen that in the *first three cases*, the longitudinal wave set up in the rod, or the incident wave is reflected at a free boundary. So that, if the equation of the simple harmonic wave (or the incident wave) set up in the rod, travelling along the positive direction of the x-axis along which the rod lies, be $y_1 = a\sin\frac{2\pi}{\lambda}(vt - x)$, that of the reflected wave will be $y_2 = a\sin\frac{2\pi}{\lambda}(vt + x)$. And the equation of the resulting stationary wave thus formed will be

$$y = y_1 + y_2 = a\sin\frac{2\pi}{\lambda}(vt - x) + a\sin\frac{2\pi}{\lambda}(vt + x).$$

Or,
$$y = 2a\cos\frac{2\pi x}{\lambda}\sin\frac{2\pi vt}{\lambda} = 2a\cos\frac{2\pi x}{\lambda}\sin\omega t$$

$$\left[\because \frac{v}{\lambda} = n \text{ and } 2\pi n = \omega.\right]$$

Let us now consider each of these cases individually.

**Case (i)**—*Free-free rod.* Here, we shall obviously have *an antinodal point or a displacement antinode at either free end irrespective of time* and, therefore, the strain at these points will always be zero, *i.e.*, $dy/dx = 0$ at $x = 0$ as well as at $x = -l$, whatever the value of $\omega t$.

Since strain $= \frac{dy}{dx} = -\frac{4\pi}{\lambda}a\sin\frac{2\pi x}{\lambda}\sin\omega t,$

we have putting $x = -l$,

$\because$ at an antinodal point $\cos 2\pi x/\lambda = \pm 1$ and

$\therefore \sin 2\pi x\ x/\lambda = 0.$ *i.e.*, $\frac{2\pi l}{\lambda} = p\pi,$

where p is an integer, 0, 1, 2, 3 etc.

Or,
$$\lambda = \frac{2l}{p}.$$

And, therefore, *frequency of the tone produced, i.e.*,

$$n = \frac{v}{\lambda} = \frac{pv}{2l} = \left(\frac{p}{2l}\right)\sqrt{\frac{Y}{\rho}}. \qquad \left[\because v = \sqrt{\frac{Y}{\rho}}\right]$$

Hence, if p = 1, we have a *tone of frequency* $n_1 = \frac{1}{2l}\sqrt{\frac{Y}{\rho}}$, which is the *lowest* or the *fundamental tone.*

If p = 2, we have a *tone of frequency* $n_2 = \frac{2}{2l}\sqrt{\frac{Y}{\rho}} = \frac{1}{l}\sqrt{\frac{Y}{\rho}}$, the *first overtone* (or the *octave of the fundamental tone*),

If p = 3, we have a *tone of frequency* $n_3 = \frac{3}{2l}\sqrt{\frac{Y}{\rho}}$, *the second overtone*, having *three times* the *frequency of the fundamental tone.*

Thus, $n_1 : n_2 : n_3 :: 1 : 2 : 3$, *i.e., the tones all form a harmonic series.*

In other words, *all possible harmonics or overtones (odd as well as even) can be produced in this mode of vibration of the rod.*

**Case (ii)** *Fixed-free rod.* In this case, we must necessarily have *a nodal point or a displacement node at the fixed end, and an antinodal point or a displacement antinode at the free end, of the rod at all times,*

*i.e.,* $y = 0$ at $x = -l$.

Since $y = 2a\cos\frac{2\pi x}{\lambda}\sin\omega t$,

we have, putting $y = 0$ and $x = -l$,

$$0 = 2a\cos\frac{2\pi l}{\lambda}\sin\omega t.$$

Or, $$\cos\frac{2\pi l}{\lambda} = 0.$$

Or, $$\frac{2\pi l}{\lambda} = (2p+1)\frac{\pi}{2},$$

where, p = 0, 1, 2, 3 etc.

Or, $\lambda = \frac{4l}{(2p+1)}$. So that, $n = \frac{v}{\lambda} = \frac{(2p+1)}{4l}v$, *i.e.,*

$$n = \frac{2p+1}{4l}\sqrt{\frac{Y}{\rho}}. \qquad \left[\because v = \sqrt{\frac{Y}{\rho}}\right]$$

Thus, the *frequency of the lowest or the fundamental tone,* corresponding to p = 0, is

$$n_1 = \frac{1}{4l}\sqrt{\frac{Y}{\rho}}.$$

This, it will be seen, is half the frequency of the fundamental tone in the case of the free-free rod. In other words, *the fundamental tone emitted by a rod not clamped anywhere is the octave (i.e., has twice the frequency) of the fundamental tone emitted by the same rod when clamped at one end.*

The frequency of the second tone emitted by the fixed-free rod, corresponding to p = 1 will be

$$n_2 = \frac{3}{4l}\sqrt{\frac{Y}{\rho}}.$$

*i.e., three times that of the fundamental.*

Similarly, the third tone emitted by the rod, corresponding to p = 2, will be

$$n_3 = \frac{5}{4l}\sqrt{\frac{Y}{\rho}}, \text{ i.e., five times that of the fundamental.}$$

Thus, $n_1 : n_2 : n_3 :: 1 : 3 : 5.$

In other words, *we can have only odd harmonics (or overtones) excited in the rod in this case.*

**Case (iii)** *Rod clamped at its mid-point.* Here, clearly, *there must always be an antinodal point (or displacement antinode) at each free end of the rod and a nodal point (or displacement node) at its mid-point where it is clamped.*

So that, y = 0 at x = – *l*/2 *for all values of t and hence of ωt.*

∴ substituting these values in the expression tor y, we have

$$0 = -2a\cos\frac{2\pi}{\lambda}\cdot\frac{l}{2}\sin\omega t = -2a\cos\frac{\pi l}{\lambda}\sin\omega t.$$

Or, $\cos\frac{\pi l}{\lambda} = 0$, *i.e.*, $\frac{\pi l}{\lambda} = (2p+1)\frac{\pi}{2}$,

where $\lambda = \frac{2l}{(2p+1)}$ with p = 0, 1, 2, 3 etc.

∴ *frequency of the tone emitted by the rod, i.e.,*

$$n = \frac{v}{\lambda} = \frac{2p+l}{2l}v.$$

Or, $$n = \frac{2p+1}{2l}\sqrt{\frac{Y}{\rho}}. \qquad \left[\because v = \sqrt{\frac{Y}{\rho}}\right]$$

Hence, *frequency of the lowest or the fundamental tone*, corresponding to p = 0, will obviously be

$$n_1 = \frac{1}{2l}\sqrt{\frac{Y}{\rho}},$$

*i.e., the same as in the case of the free-free rod, and the octave of that in the case of the fixed-free rod.*

The *frequency of the second tone* emitted by the rod, corresponding to p = 1, will be

$$n_2 = \frac{3}{2l}\sqrt{\frac{Y}{\rho}},$$

*i.e., three times that of the fundamental.*

Similarly, *frequency of the third tone* corresponding to p = 2, will be

$$n_3 = \frac{5}{2l}\sqrt{\frac{Y}{\rho}}$$ *i.e., five times that of the fundamental.*

Here, again, therefore, $n_1 : n_2 : n_3 :: 1 : 3 : 5$,

*i.e., only odd harmonics or overtones can be excited in the rod in this case.*

**Case (iv)**—*Fixed-fixed rod.* Here, the longitudinal wave set up in the rod, *i.e.*, the incident wave, will be reflected at a rigid boundary, so that the equations of the incident and reflected waves will respectively be

$$y_1 = a\sin\frac{2\pi}{\lambda}(vt - x)$$

and $$y_2 = -a\sin\frac{2\pi}{\lambda}(vt + x).$$

∴ the equation of the resultant stationary wave set up in the rod will be

$$y = y_1 + y_2 = a\sin\frac{2\pi}{\lambda}(vt - x) - a\sin\frac{2\pi}{\lambda}(vt + x).$$

Or, $$y = -2a\sin\frac{2\pi x}{\lambda}\cos\frac{2\pi vt}{\lambda} = -2a\sin\frac{2\pi x}{\lambda}\cos\omega t$$
$$\left[\because \frac{v}{\lambda} = n \text{ and } 2\pi n = \omega\right]$$

Since there must always be a nodal point or a displacement node at either fixed or clamped end of the rod, we have y = 0 at x = 0 and at x = – *l*.

Substituting y = 0 and x = a – *l* in the expression for y, we therefore have

$$0 = 2a\sin\frac{2\pi l}{\lambda}\cos\omega t.$$

Or, $$\sin\frac{2\pi l}{\lambda} = 0.$$

Or, $$\frac{2\pi l}{\lambda} = p\pi,$$

whence $$\lambda = \frac{2l}{p}, \text{ where } p = 1, 2, 3, \text{ etc.}$$

Therefore, *frequency of the note emitted, i.e.,* $n = \frac{v}{\lambda} = \frac{p}{2l}v$

$$= \frac{p}{2l}\sqrt{\frac{Y}{\rho}}. \qquad \left[\because v = \sqrt{\frac{Y}{\rho}}\right]$$

So that, *frequency of the lowest or the fundamental tone*, corresponding to p = 1, is given by

$$n_1 = \frac{1}{2l}\sqrt{\frac{Y}{\rho}}.$$

Similarly, *frequency of the second tone*, corresponding to p = 2, is given by

$$n_2 = \frac{2}{2l}\sqrt{\frac{Y}{\rho}} = \frac{1}{l}\sqrt{\frac{Y}{\rho}}$$

and *frequency of the third tone*, corresponding to p = 3 is given by

$$n_3 = \frac{3}{2l}\sqrt{\frac{Y}{\rho}}, \text{ and so on.}$$

Thus, $n_1 : n_2 : n_3 :: 1 : 2 : 3$,

*i.e.*, all possible harmonics (or overtones), odd as well as even, can be excited in the rod in this case.

As will be readily seen, this is thus a case essentially similar to that of the *free-free* rod.

(C) *Different modes of vibration of air columns.* An air column can be set into vibration in what is called an *organ pipe*, which is just a long tube or pipe of wood or metal, rectangular or circular in cross section, fitted with a mouth-piece at one end and either closed or open at the other end. In the former case it is call a *closed pipe* and in the latter, an *open pipe*.

On blowing into the mouth piece, the air column inside the pipe is set into vibration and the longitudinal waves thus produced get reflected at the open or the closed end of the pipe, resulting in the formation of stationary waves.

Since the mouth-piece end of the pipe also serves as an open end, the open and closed pipes, it will be easily seen, correspond to the *free-free* and the *fixed-free* rod respectively dealt with under B-(i) and (ii) *Above.* So that the equations of the incident and reflected waves here too are respectively

$$y_1 = a\sin\frac{2\pi}{\lambda}(vt - x)$$

and $$y_2 = a\sin\frac{2\pi}{\lambda}(vt + x).$$

And, therefore, the equation of the resultant stationary wave is

$$y = y_1 + y_2 = a\sin\frac{2\pi}{\lambda}(vt - x) + a\sin\frac{2\pi}{\lambda}(vt + x).$$

Or, $$y = 2a\cos\frac{2\pi x}{\lambda}\sin\frac{2\pi vt}{\lambda} = 2a\cos\frac{2\pi x}{\lambda}\sin\omega t.$$

Now, in the case of an open pipe, we have an *antinodal point at either end*, irrespective of the value of t and hence of ωt. So that, *excess pressure p at either end is zero, i.e., p = 0 at x = 0 and at x = – l at all times.*

Now, $$p = -K\frac{dy}{dx} = -K\frac{4\pi}{\lambda}a\sin\frac{2\pi x}{\lambda}\sin\omega t.$$

Substituting p = 0 and x = – *l* in the expression for p, therefore, we have

$$0 = K\frac{4\pi}{\lambda}a\sin\frac{2\pi l}{\lambda}\sin\omega t.$$

Or, $$\sin\frac{2\pi l}{\lambda} = 0.$$

Or, $$\frac{2\pi l}{\lambda} = m\pi, \text{ where } m = 1, 2, 3 \text{ etc.}$$

Or, $$\lambda = \frac{2l}{m}$$

And, therefore, *frequency of the tone emitted, i.e.,*

$$n = \frac{v}{\lambda} = \frac{mv}{2l} = \frac{m}{2l}\sqrt{\frac{K}{\rho}}. \qquad \left[\because v = \sqrt{\frac{K}{\rho}}\right]$$

So that, *frequency of the lowest or the fundamental tone,* corresponding to m = 1 is given by

$$n_1 = \frac{1}{2l}\sqrt{\frac{K}{\rho}},$$

and those of the *second* and the *third* tones by

$$n_2 = \frac{2}{2l}\sqrt{\frac{K}{\rho}} = \frac{1}{l}\sqrt{\frac{K}{\rho}}$$

and $$n_3 = \frac{3}{2l}\sqrt{\frac{K}{\rho}} \text{ respectively.}$$

Thus, $n_1 : n_2 : n_3 :: 1 : 2 : 3,$

*i.e., all possible harmonics or overtones (odd as well as even) can be produced in the case of an open pip .*

In the *case of a closed pipe,* since *there must necessarily be a nodal point at the closed end,* irrespective of the value of t and hence of cut of ωt, we have, *as in the case of a fixed free rod,* y = 0 at x = – *l at all times.*

Substituting these values in the expression for y therefore, we have

$$0 = -2a\cos\frac{2\pi l}{\lambda}\sin\omega t.$$

Or, $$\cos\frac{2\pi l}{\lambda} = 0.$$

Or, $$\frac{2\pi l}{\lambda} = (2m+1)\frac{\pi}{2},$$

and, therefore, $\lambda = \dfrac{4l}{(2m+1)}$, where m = 0, 1, 2, 3 etc.

∴ *frequency of the tone emitted, i.e.,*

$$n = \frac{v}{\lambda} = \frac{(2m+1)}{4l}v = \frac{2m+1}{4l}\sqrt{\frac{K}{\rho}}$$

So that, *frequency of the lowest or the fundamental tone*, corresponding to m = 0, is given by

$$n_1 = \frac{1}{4l}\sqrt{\frac{K}{\rho}},$$

which is clearly *half* that of the fundamental tone given by an open pipe of the same length. In other words, *the fundamental tone given by an open pipe is the octave of the fundamental lone given by a closed pipe of the same length.*

The *frequencies of the second and third tones* (corresponding to m = 1 and m = 2) are respectively

$$n_2 = \frac{3}{4l}\sqrt{\frac{K}{\rho}} \text{ and } n_3 = \frac{5}{4l}\sqrt{\frac{K}{\rho}}.$$

Thus $$n_1 : n_2 : n_3 :: 1 : 3 : 5,$$

*i.e., only odd harmonics or overtones can be produced in the case of a closed pipe.*

As in the case of strings, so also here, the note emitted by an air column is usually a mixture of the tone sounded with some of its harmonics or overtones. Obviously, both odd and even harmonics are present if a note is sounded on an open pipe and only odd harmonics when it is sounded on a closed pipe of the same length. This is why the former is always sweeter and richer than the latter.

**Note:** In our discussion above, we have assumed that reflection of the wave at the open end occurs exactly at the level of the open end itself. In point of fact, it occurs just a wee-bit beyond the open end, so that the actual length of the air column involved is slightly greater than the length of the pipe.

*Rayleigh* has shown that the *effective length* of the air column in the case of a closed pipe is 0.3 D more than the actual length of the pipe, where D is the *internal diameter* of the pipe. And, obviously, in the case of an open pipe, it would be $2 \times 0.3$ D = 0.6D more than the actual length of the pipe. This is referred to as the end-correction of the pipe.

## FOURIER'S THEOREM

If we examine the wave pattern, *i.e.*, the *time-displacement curve* (or the pressure variation curve) of music from a musical instrument or a song rendered by an artiste, we find that although it is a *periodic curve* in the sense that it repeats itself periodically, (*i.e.*, after fixed and regular intervals of time) *it is by no means a simple harmonic wave* which may be represented by a sine or a cosine curve. *Waves of this type, having an irregular profile,* are called *complex waves* and the vibrations which give rise to them *complex vibrations.*

The French mathematician, **J.B.J. Fourier** (1768-1830) showed, in the year 1822, that such waves may be analysed into a set of simple harmonic waves and that, in fact, a periodic motion may, in general, be represented as a combination of sinusoidal or simple harmonic motions, whose frequencies are an integral multiple of that of the periodic motion in question.

The limitations under which a periodic function may be so analysed are that (i) *it must he continuous* or may just have a finite number of discontinuities of slope or magnitude within its time-interval of one oscillation but must not be discontinuous as shown at Q in Fig. 3.13. This means, in other words, that it must have only finite values and should never assume an infinite value at any instant. For, an infinite value for a displacement etc, at any given instant, is simply inconceivable; (ii) *it must be a single valued function, i.e.*, it must have only one value at a given instant t, so that the curve representing it may not overhang at any point, as shown at P in the figure, or else it would have two values of y at the same instant t. This is a sheer impossibility in air or in any elastic medium.

And, naturally enough, the frequencies of the component vibrations should not be incommensurable (like, for instance, 500 and $500\sqrt{3}$). For, then, the resultant vibration is no longer periodic and the theorem, as we know, applies only to periodic vibrations.

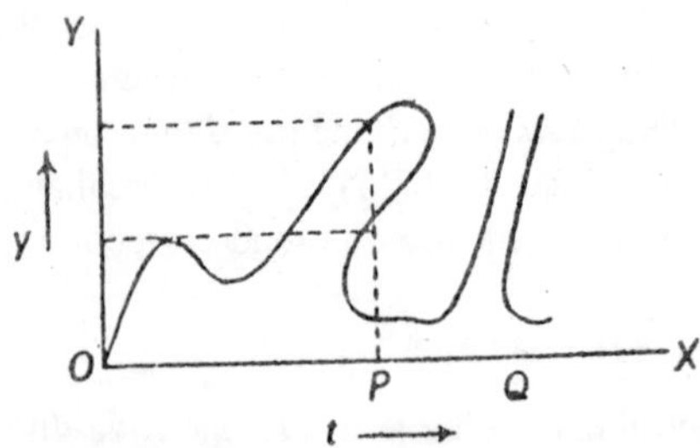


Fig. 3.13

All these conditions are satisfied by mechanical waves, in general, and by sound waves, in particular.

We may, therefore, state *Fourier's theorem* thus:

*Any finite, single-valued periodic function, which is either continuous or which possesses only a finite number of discontinuities of slope or magnitude (all within the interval of one time-period), may be regarded as a combination of simple harmonic vibrations whose frequencies are integral multiples of that of the given function.*

Analytically, then, if we have a *finite, single-valued, continuous function* y = f(t) of t, which is, say, a displacement (or a pressure variation) such that y = f(t) = f(t + T), it is clearly a periodic, function of time-period Tor of frequency $1/T = \omega/2\pi$. This finite, single-valued, continuous periodic function f(t) may, in accordance with *Fourier's* theorem be expanded into a summation or a series, called *Fourier series*. Thus,

$$y = f(t) = A_0 + A_1 \sin \omega t + A_2 \sin 2\omega t + \ldots + A_r \sin r\omega t$$
$$+ B_1 \cos \omega t + B_2 \cos 2\omega t + \ldots + B_r \cos \omega t,$$

*i.e.*, $$y = A_0 + \Sigma A_r \sin r\omega t + \Sigma B_r \cos r\omega t. \quad \ldots(I)$$

Here, $A_0$ represents the *zero frequency term* (in most cases, equal to *zero*) and is, in fact, a measure of the mean displacement of the time-axis. The sine and cosine terms represent the component sinusoidal vibrations of frequencies which are integral multiples of frequency $\omega/2\pi$ of the periodic function and $A_1, A_2 \ldots A_r$ and $B_1\ B_2 \ldots B_r$ are amplitudes of these vibrations, all having definite values for the given periodic function (displacement, in our case) and are referred to as *Fourier coefficients* of the given series.

It is by no means necessary that all the terms of the series should be present in any given case.

**Evaluation of Fourier Coefficients**

The values of the Fourier coefficients may be easily obtained, as we shall presently see—the process of evaluation being termed Fourier (or harmonic) analysis.

(i) *Evaluation of the constant $A_0$.* If we integrate both sides of equation I between the limits t = 0 and t = T, we have

$$\int_0^T y\,dt = A_0\int_0^T dt + A_1\int_0^T \sin\omega t\,dt + ... + A_r\int_0^T \sin r\omega t\,dt$$

$$+B_1\int_0^T \cos\omega t\,dt + ... + B_r\int_0^T \cos r\omega t\,dt = A_0 T$$

the other integrals all vanishing to *zero* under the limits chosen.

We, therefore, have $A_0 = \frac{1}{T}\int_0^T y\,dt,$

where, clearly, $A_0$ *represents the mean value of the function (in our case, displacement) during a full time-period.*

So that; in the event of $A_0$ being equal to zero, the axis of the curve representing the function (the displacement curve, in the present case) would lie along the axis of time.

(ii) *Evaluation of constants* $A_1$, $A_2$ ...$A_r$. Multiplying equation I by sin rωt, if we again integrate between the limits t = 0 and t = T, we have

$$\int_0^T y\sin r\omega t\,dt = A_0\int_0^T \sin r\omega t\,dt + A_1\int_0^T \sin\omega t\sin r\omega t\,dt ...$$

$$+A_r\int_0^T \sin^2 r\omega t\,dt + B_1\int_0^T \sin r\omega t\cos\omega t\,dt + ... B_r\int_0^T \sin r\omega t\cos r\omega t\,dt.$$

Since all other integrals on the right hand side except $A_r\int_0^T \sin^2 r\omega t\,dt$ vanish to *zero* under the chosen limits, we have

$$\int_0^T y\sin r\omega t\,dt = A_r\int_0^T \sin^2 r\omega t\,dt = \frac{A_r}{2}\int_0^T (1-\cos 2r\omega t)\,dt = A_r \times \frac{T}{2}$$

$$\text{whence,}\quad A_r = \frac{2}{T}\int_0^T y\sin r\omega t\,dt = \frac{2}{T}\int_0^T y\sin\frac{2\pi rt}{T}dt \qquad \left[\because \omega = \frac{2\pi}{T}\right]$$

And, therefore, putting r = 1, 2, 3, etc., we can easily obtain the values of the coefficients $A_1$, $A_2$, $A_3$ etc.

(iii) *Evaluation of constants* $B_1$, $B_2$ ...$B_r$. Multiplying both sides of equation I by cos rωt and integrating between the limits t = 0 and t = T, we have

$$\int_0^T y\cos r\omega t\, dt = A_0\int_0^T \cos r\omega t\, dt + A_1\int_0^T \cos r\omega t \sin \omega t\, dt + ...$$
$$+A_r\int_0^T \cos r\omega t \sin r\omega t\, dt + B_1\int_0^T \cos r\omega t \cos \omega t\, dt + ...$$
$$+B_r\int_0^T \cos^2 r\omega t\, dt.$$

Here also, since all the integrals on the right hand side, except $B_r\int_0^T \cos^2 r\omega t\, dt$ vanish to *zero* under the chosen limits, we have

$$\int_0^T y\cos r\omega t\, dt = B_r\int_0^T \cos^2 r\omega t\, dt = \frac{B_r}{2}\int_0^T (1+\cos 2r\omega t)dt = \frac{B_r}{2}\times T,$$

whence, $B_r = \dfrac{2}{T}\displaystyle\int_0^T y\cos r\omega t\, dt = \dfrac{2}{T}\int_0^T y\cos\dfrac{2\pi rt}{T}dt.$ $\left[\because \omega = \dfrac{2\pi}{T}\right]$

Again, putting r = 1, 2, 3, etc., we can easily obtain the values of the constants $B_1$, $B_2$, $B_3$ etc.

Incidentally, the very fact that these constants have particular fixed values for a given function f(t) constitutes a *formal proof of Fourier's theorem.*

N.B. The above process of determining the *Fourier constants* is obviously applicable if the function f(t) can be put into an algebraic form, amenable to integration. In case this is not so, other methods, including a graphical one, are available, for, after all an integration is essentially the same thing as a summation or determination of the area under a curve.

Then, again, a Fourier series may also adopt other forms, *e.g.*, the form of a *series of sine terms or of cosine terms or even an exponential form.*

Thus, if we put $A_r = a_r \sin\phi$ and $B_r = a_r \cos\phi_r$. equation I takes the following form, consisting of a *constant or zero frequency term* $A_0$ *and the rest all sine terms:*

$$y = f(t) = A_0 + \sum_{r=1}^{\infty} a_r \sin(r\omega t + \phi r)$$

$$= A_0 + a_1 \sin(\omega t + \phi_1) + a_2 \sin(2\omega t + \phi_2) + \dots a_r \sin(r\omega t + \phi_r),$$

where $a_1, a_2, \dots a_r$ are the *amplitudes* and $\phi_1, \phi_2, \dots \phi_r$, the *phase, constants* of the constituent harmonic vibrations.

Clearly, $a_r = \sqrt{A_r^2 + B_r^2}$ and $f_r = \dfrac{A_r}{B_r}$. So that, if $A_r = 0$, we have $a_r = B_r$ and $\phi = 0$; and if $B_r = 0$, we have $a_r = A_r$ and $\phi_r = \dfrac{\pi}{2}$.

In the same manner, the Fourier series may also be expressed *wholly as a series of cosine terms.*

These forms of the Fourier series lend themselves to a much easier physical interpretation than the others.

The *exponential form of Fourier series* may be obtained by putting 2 sin ωt

$= (e^{i\omega t} - e^{-i\omega t})$ and $2\cos\omega t = (e^{i\omega t} + e^{-i\omega t})$. So that, equation I becomes

$$y = f(t) = A_0 + \sum_{r=1}^{\infty} A_r\left(e^{ir\omega t} - e^{-ir\omega t}\right)/2i + \sum_{r=1}^{\infty} B_r\left(e^{ir\omega t} + e^{-ir\omega t}\right)/2$$

$$= A_0 + \sum_{r=-1}^{-\infty} \frac{(B_r - iA_r)}{2} e^{ir\omega t} + \sum_{r=1}^{\infty} \frac{(B_r - iA_r)}{2} e^{ir\omega t}$$

If we put $A_0 = C_0$ and $\dfrac{(B_r - iA_r)}{2} = C_r$, we have

$$y = f(t) = C_0 + \sum_{-\infty}^{\infty} C_r e^{ir\omega t},$$

which is the *exponential form of the Fourier series.*

## SOME ILLUSTRATIVE EXAMPLES OF THE APPLICATION OF FOURIER'S THEOREM

Let us now take up a few examples of Fourier analysis of some typical types of curves.

(i) *The square wave.* Consider a *square wave* of the type shown in Fig. 3.14, (perhaps more appropriately called a *top-hat wave* because of its resemblance with the shape of a top hat).

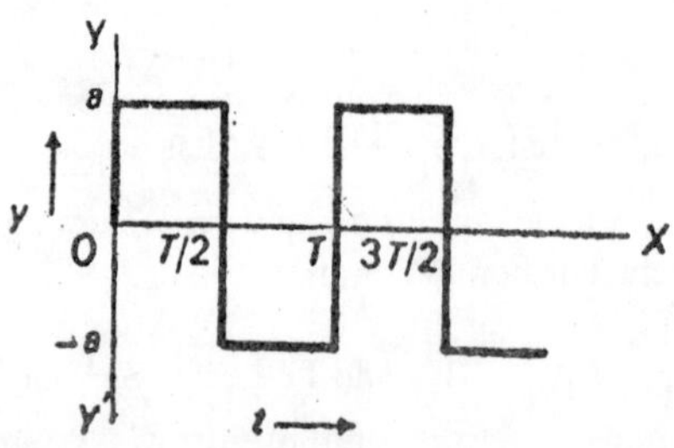


Fig. 3.14

Here, we have y = f(t) = a from t = 0 to r = T/2

and y = f(t) = – a from T = T/2 to t = T.

Now, as we know the *Fourier series* gives $y = f(t) = A_0 + A_1 \sin \omega t + ... + A_r \sin r\omega t + B_1 \cos \omega t + ... + B_r \cos r\omega t$.

Since $$A_0 = \frac{1}{T}\int_0^T y\,dt, \text{ we have}$$

$$A_0 = \frac{1}{T}\left[\int_0^{T/2} a\,dt - \int_{T/2}^{T} a\,dt\right] = 0,$$

indicating that *the mean value of the displacement over a time-period, is zero* or that *the axis of the displacement curve coincides with the time-axis about which the wave is symmetrical.*

And, since $A_r = \frac{2}{T}\int_0^T y \sin r\omega t\,dt$, we have

$$A_r = \frac{2}{T}\left[\int_0^{T/2} a \sin r\omega t\,dt - \int_{T/2}^{T} a \sin r\omega t\,dt\right]$$

$$= \frac{2a}{T}\left[\left(\frac{-\cos r\omega t}{r\omega}\right)_0^{T/2} + \left(\frac{\cos r\omega t}{r\omega}\right)_{T/2}^{T}\right]$$

$$= \frac{2a}{r\omega T}\left[\left(-\cos\frac{r\omega T}{2} + 1\right) + \left(\cos r\omega T - \cos\frac{r\omega T}{2}\right)\right]$$

$$= \frac{2a}{r\omega T}\left[\left(-\cos\frac{r\omega T}{2} + 1\right) + \left(\cos r\omega T - \cos\frac{r\omega T}{2}\right)\right]$$

Since $\omega = \frac{2\pi}{T}$ or $\omega T = 2\pi$, we have

$$A_r = \frac{a}{\pi r}(2 - 2\cos\pi r) = \frac{2a}{\pi r}(1 - \cos\pi r)$$

*For even values of r,* therefore, $A_r = 0$ and *for odd values of r,*

$$A_r = \frac{4a}{\pi r}.$$

Again, since $B_r = \frac{2}{T}\int_0^T y \cos r\omega t \, dt$, we have

$$B_r = \frac{2}{T}\left[\int_0^T a \cos r\omega t \, dt - \int_{T/2}^T a \cos r\omega t \, dt\right]$$

$$= \frac{2a}{T}\left[\left(\frac{\sin r\omega t}{r\omega}\right)_0^{T/2} - \left(\frac{\sin r\omega t}{r\omega}\right)_{T/2}^{T}\right]$$

$$= \frac{2a}{r\omega T}\left[\left(\sin\frac{r\omega T}{2} - 0\right) - \left(\sin r\omega T - \sin\frac{r\omega T}{2}\right)\right]$$

Since $\omega = \frac{2\pi}{T}$ or $\omega T = 2\pi$, we have

$$B_r = \frac{a}{\pi r} \times 0 = 0.$$

Thus, we have $A_0 = 0$, $A_1 = 4a/\pi$, $A_2 = 0$, $A_3 = 4a/3\pi$, $A_4 = 0$, $A_5 = 4a/5\pi$ and so on, and $B_1$, $B_2$, $B_3$ ... $B_r$, *all equal to zero, i.e., the cosine terms are all absent.* The full Fourier series in this case, therefore becomes

$$y = \frac{4a}{\pi}\left(\sin\omega t + \frac{1}{3}\sin 3\omega t + \frac{1}{5}\sin 5\omega t + \ldots\right).$$

Or, $$y = \frac{4a}{\pi}\left(\sin\frac{2\pi t}{T} + \frac{1}{3}\sin\frac{6\pi t}{T} + \frac{1}{5}\sin\frac{10\pi t}{T} + \ldots\right),$$

clearly showing that *the combination here is that of the fundamental and its odd harmonics. In other words, the ratios of the frequencies of the constituents are proportional to the odd values of r and their amplitudes to their reciprocals, i.e., their frequencies are as 1: 3 : 5 ... and their amplitudes as* $1 : \frac{1}{3} : \frac{1}{5}$.

A superposition of the displacement curves (or harmonic waves) representing these constituents should thus give the square wave in question.

Thus, in Fig. 3.15, *curve* (1) represents the simple harmonic wave corresponding to $y_1 = \frac{4a}{\pi}\sin\omega t$, the *dotted curve* (2) represents the simple harmonic wave corresponding to $y_3 = \frac{4a}{3\pi}\sin\omega t$, having $\frac{1}{3}$*rd the amplitude* and *thrice the frequency of the first*, and curve (3) represents the simple harmonic wave, corresponding to $y_5 = \frac{4a}{5\pi}\sin 5\omega t$, with $\frac{1}{5}$th *the amplitude and five times the frequency of the first*, those corresponding to $y_2$, $y_4$ etc. being absent. Their superposition then gives the thick full line curve, which bears only a partial resemblance with the given square wave.

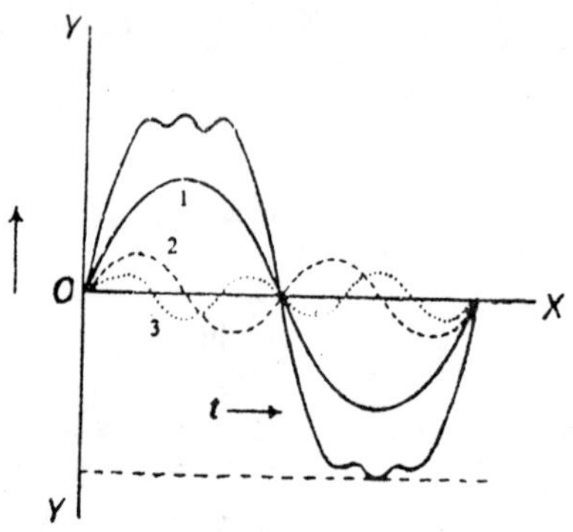


**Fig. 3.15**

If, however, we add up or superimpose more waves corresponding to $y_7$, $y_8$, $y_{11}$ etc, the resultant wave assumes a shape more and more resembling the given, square wave. So that, with a still larger number of them, thus superposed, we shall obtain the square wave we have analysed.

(ii) *The saw tooth wave:* This wave, (Fig. 3.16) is obviously so called because of ifs shape and is commonly met with in electronics.

As will be readily seen, here, y = a at t = 0, then falls linearly to *zero* at t = T. So that, y = 0 at t = T.

Let the displacement at time t be y. Then, clearly,

$$\frac{y}{a} = \frac{T-t}{T} = \left(1 - \frac{t}{T}\right).$$

So the equation of the wave is

$$v = f(t) = a\left(1 - \frac{t}{T}\right).$$

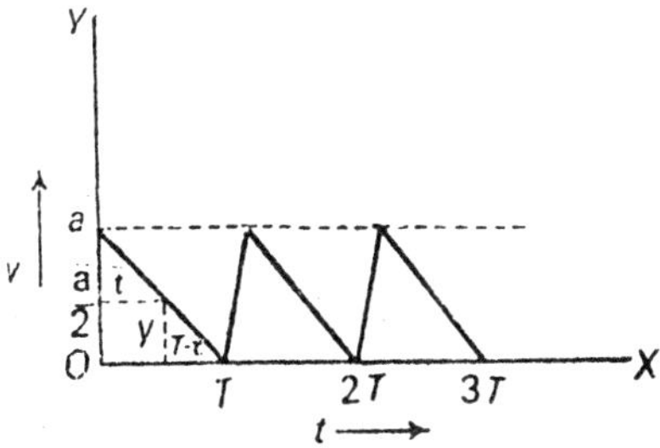


**Fig. 3.16**

Now, in the *Fourier series*

$$y = A_0 + A_1 \sin \omega t + \ldots + A_r \sin r\omega t + B_1 \cos \omega t + \ldots + B_r \cos r\omega t.$$

Since $A_0 = \frac{1}{T}\int_0^T y\,dt$, we have

$$A_0 = \frac{1}{t}\int_0^T a\left(1 - \frac{t}{T}\right)dt = \frac{a}{T}\left[t - \frac{t^2}{2T}\right]_0^T = \frac{a}{T}\left[T - \frac{T^2}{2T}\right] = \frac{a}{T}\left[\frac{T}{2}\right] = \frac{a}{2},$$

indicating that *the mean value of the displacement over a time-period is a/2 or that the axis of the displacement curve lies at a distance a/2 from the time-axis.*

And, clearly, $A_r = \frac{2}{T}\int_0^T y \sin r\omega t\,dt = \frac{2a}{T}\int_0^T \left(1 - \frac{t}{T}\right)\sin r\omega t\,dt$

$$= \frac{2a}{T}\int_0^T \sin r\omega t\,dt - \frac{2a}{T^2}\int_0^T t \sin r\omega t\,dt = \frac{2a}{T}\left[\frac{-\cos r\omega t}{r\omega}\right]_0^T$$

$$-\frac{2a}{T^2}\left[\frac{-t\cos r\omega t}{r\omega}\right]_0^T - \frac{2a}{T^2}\int_0^T \frac{-\cos r\omega t}{r\omega}dt$$

$$= \frac{2a}{Tr\omega}[-\cos r\omega T + \cos 0] - \frac{2a}{Tr\omega}[-\cos r\omega T + 0] - \frac{2a}{T^2}\left[\frac{-\sin r\omega T}{r^2\omega^2} + 0\right]$$

$$= \frac{2a}{Tr\omega}(-\cos r\omega T + 1) - \frac{2a}{Tr\omega}(-\cos r\omega T + 0) - 0$$

$$= \frac{2a}{Tr\omega}[(1 - \cos r\omega T) + 1] = \frac{2a}{Tr\omega}(2 - \cos r\omega T).$$

Since $\omega T = 2\pi$ and, therefore, $\cos r\omega T = 1$, we have

$$Ar = \frac{2a}{Tr\omega}(2-1) = \frac{2a}{Tr\omega} = \frac{2a}{T.r.2\pi/T} = \frac{a}{r\pi}.$$

And $$B_r = \frac{2a}{T}\int_0^T y \cos r\omega t\, dt = \frac{2a}{T}\int_0^T\left(1 - \frac{t}{T}\right)\cos r\omega t\, dt$$

$$= \frac{2a}{T}\int_0^T \cos r\omega t\, dt - \frac{2a}{T^2}\int_0^T t \cos r\omega t\, dt$$

$$= \frac{2a}{T}\left[\frac{\sin r\omega t}{r\omega}\right]_0^T = \frac{2a}{T^2}\left[\frac{t \sin r\omega t}{r\omega}\right]_0^T - \frac{2a}{T^2}\int_0^T \frac{\sin r\omega t}{r\omega} dt$$

$$= \frac{2a}{Tr\omega}[\sin r\pi - \sin 0] - \frac{2a}{Tr\omega}(\sin r\pi - \sin 0)$$

$$+ \frac{2a}{T^2}\left[\frac{\cos r\omega T - \cos 0}{r^2\omega^2}\right] = 0 - 0 + 0 = 0. \quad [\because \omega T = 2\pi.]$$

Thus $B_r$ being equal to *zero, all cosine terms of the Fourier series are absent.*

The series thus becomes

$$y = \frac{a}{2} + \frac{a}{\pi}\left(\sin \omega t + \frac{1}{2}\sin 2\omega t + \frac{1}{3}\sin 3\omega t + ...\right),$$

indicating that *both odd and even harmonics are present, with their amplitudes and frequencies proportional to \jr and r respectively, i.e., the frequencies are as 1 : 2 : 3 ...and amplitudes as* $1 : \frac{1}{2} : \frac{1}{3} : ...$

Here too, therefore, the superposition of ail these constituent harmonic waves should give the given saw-tooth wave. We find, however, that with the superposition of six constituent waves, we obtain a curve of the form shown in Fig. 1.17 (a), with only a partial resemblance with the given saw-tooth wave. But superposition of 10 or more constituent waves results in its acquiring a closer identity with the given form of the wave, as can be seen from Fig. 3.17 (b). If therefore, a still larger number of the constituent waves be superposed, we shall obtain the exact form of the saw-tooth wave analysed.

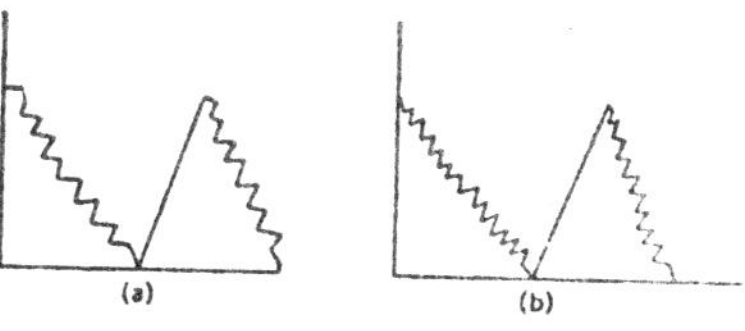


Fig. 3.17

## GROUP VELOCITY-ITS RELATIONSHIP WITH WAVE (OR PHASE) VELOCITY

We know that the equation of a plane progressive Harmonic wave, travelling in a medium along the positive direction of the x-axis is $y = a \sin \frac{2\pi}{\lambda}(vt - x)$, where v is the wave *velocity*, equal to $\frac{\lambda}{T}$.

Let us now put this equation in a slightly different form. We have from the above,

$$y = a \sin\left(\frac{2\pi vt}{\lambda} - \frac{2\pi}{\lambda}x\right)$$

Since $\frac{v}{\lambda} = \frac{1}{T} = n$, the *frequency* of the wave, and $\frac{2\pi}{\lambda} = k$, the *propagation constant*, the wave equation takes the form

$$y = a \sin (2\pi nt - kx).$$

Or, since 2pn = ω, the *angular frequency* of the wave, we have

$$y = a \sin (\omega t - kx),$$

showing that the constant phase (ωt – kx) of the wave travels along the positive direction of the x- axis with velocity dx/dt, *i.e.*, the *phase velocity of the wave* = dx/dt.

Now, since ωt – kx = *constant*, we have (on differentiating with respect to t), ω – k(dx/dt) = 0. Or, ω = k dx/dt, whence, *phase velocity of the wave* = dx/dt = ω/k.

But $\frac{\omega}{k} = \frac{2\pi n}{2\pi/\lambda} = n\lambda = \frac{\lambda}{T}$ = *wave velocity v.*

Thus, **for a single wave,** *in any given medium,*

**wave velocity = phase velocity** $= v = \frac{\lambda}{T} = \frac{\omega}{k}$.

Let us now consider the case of two (or more) wave trains; of slightly different wavelengths, $\lambda$ and $(\lambda + d\lambda)$, say, [Fig. 3.18 (a)], travelling simultaneously along the same path in a given medium, and, for the sake of simplicity, let their amplitudes be equal.

Then, if the medium be *non -dispersive, i.e.*, if the velocity of the wave in it does not depend upon the wavelength, both the waves travel through it with the same wave (or phase) velocity v.

Since they are both travelling *along the same path and in the same direction,* they get superposed, one over the other (although they have been shown separately in the figure).

This is a case, it will be recalled, exactly similar to that of the formation of beats in the case of sound waves. For, at certain points of the medium, where the positive or negative maximum displacements of the two waves come across each other, they get reinforced and we obtain the maxima of displacement of the resulting wave. And, at points where the positive maximum displacements of one wave come across the negative maximum displacements of the other, we obtain the minimum or zero displacement of the resulting wave.

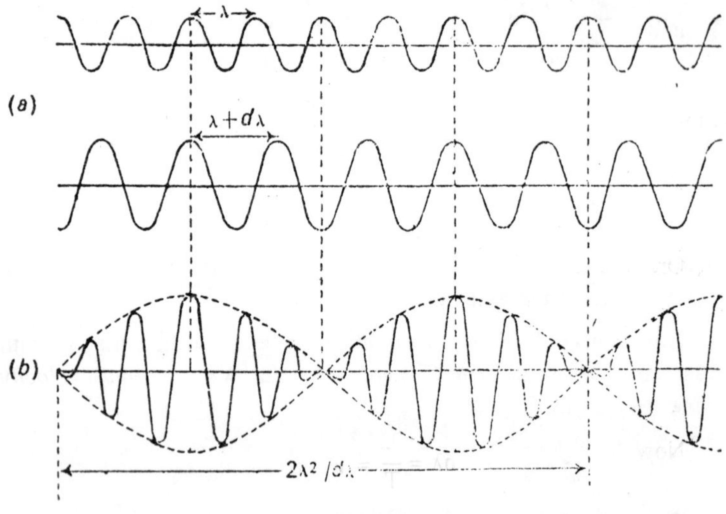


**Fig. 3.18**

In between these two extremes, the displacements in the resulting wave have intermediate values. The resulting wave or the '*beat wave*' thus consists of *groups of waves*, shown by the dotted curves in Fig. 9.18 (A), each group consisting of a number of waves, with the displacements in the wave at the centre of the group the maximum, with those of the others trailing off gradually to zero on either side, as shown. *The velocity with which the maxima of these groups of waves travel is called the* **group velocity** (u) *of the resultant wave.*

In the case we are considering, *i.e.*, in which the two waves are travelling in a *non-dispersive* medium so that their velocities are the same, the resultant wave, consisting of groups of waves, also travels with the same velocity as that of each constituent wave and the shape or profile of the groups, therefore, continues to be maintained throughout. In other words, *the group velocity u is, in this case, equal to the phase velocity v of each wave.* (We shall presently see this mathematically also).

In case, however, the medium be a dispersive one, so that the velocities of the two waves of wavelengths λ and (λ + dλ) be v and (v + dv) respectively, their wave equations are

$$y_1 = a \sin\frac{2\pi}{\lambda}(vt - x)$$

and
$$y_2 = a\sin\frac{2\pi}{(\lambda + d\lambda)}\,[(v + dv)t - x]$$

The equation of the resultant wave (consisting of groups of waves) is thus

$$y = y_1 + y_2 = a\sin\frac{2\pi}{\lambda}(vt - x) + a\sin\frac{2\pi}{(\lambda + d\lambda)}[(v + dv)t - x].$$

Or,
$$y = 2a \sin \pi\left\{\left(\frac{v}{\lambda} + \frac{v + dv}{\lambda + d\lambda}\right)t - x\left(\frac{1}{\lambda} + \frac{1}{\lambda + d\lambda}\right)\right\}$$

$$\times\cos \pi\left\{\left(\frac{v}{\lambda} - \frac{v + dv}{\lambda + d\lambda}\right)t - x\left(\frac{1}{\lambda} - \frac{1}{\lambda + d\lambda}\right)\right\} \qquad ...(A)$$

Now,
$$\frac{v}{\lambda} + \frac{v + dv}{\lambda + d\lambda} = \frac{v\lambda + vd\lambda + v\lambda - dv\lambda}{\lambda(\lambda + d\lambda)} = \frac{vd\lambda - \lambda dv}{\lambda^2} \text{ very nearly,}$$

$$\frac{1}{\lambda} + \frac{1}{\lambda + d\lambda} = \frac{\lambda + d\lambda + \lambda}{\lambda(\lambda + d\lambda)} = \frac{2\lambda}{\lambda^2} = \frac{2}{\lambda} \text{ very nearly,}$$

and $$\frac{1}{\lambda}-\frac{1}{\lambda+d\lambda}=\frac{\lambda+d\lambda-\lambda}{\lambda(\lambda+d\lambda)}=\frac{d\lambda}{\lambda^2} \text{ very nearly.}$$

Substituting these values in expression (A) above, we therefore have

$$y=2a\sin\pi\left(\frac{2vt}{\lambda}-\frac{2x}{\lambda}\right)\cos\pi\left\{\left(\frac{vd\lambda-\lambda dv}{\lambda^2}\right)t-\frac{xd\lambda}{\lambda^2}\right\}$$

Or, $$y=2a\sin\frac{2\pi}{\lambda}(vt-x)\cos\frac{2\pi}{2\lambda^2/d\lambda}\left\{\left(v-\lambda\frac{dv}{d\lambda}\right)t-x\right\}$$

If we put $2a\sin\frac{2\pi}{\lambda}(vt-x)=a'$, $\frac{2\lambda^2}{d\lambda}=\lambda'$ and $\left(v-\lambda\frac{dv}{d\lambda}\right)=u$, we have

$$y=a'\cos\frac{2\pi}{\lambda'}(ut-x).$$

This is clearly the equation of a wave having an *amplitude* $a=2a\sin\frac{2\pi}{\lambda}(vt-x)$, *wave length* $\lambda'=\frac{2\lambda^2}{d\lambda}$ and *velocity* $u=v-\lambda\frac{dv}{d\lambda}$.

This means, in other words, that (i) *the resultant wave or the beat wave (shown dotted) is divided up into groups, with each group consisting of waves represented by the equation* $y=2a\sin\frac{2\pi}{\lambda}(vt-x)$ *and its amplitude a' varying accordingly,* and (ii) *the wavelength of the resultant wave is* $2\lambda/d\lambda$ and *its velocity, i.e., the group velocity* $u = v - \lambda dv/d\lambda$.

Thus, group velocity (u) = wave velocity (v) $-\frac{\lambda dv}{d\lambda}$, which is the required relation between group velocity and wave (or phase) velocity *in a dispersive medium.*

Since in *a non-dispersive medium,* there is no change in velocity with wavelength, $dv/d\lambda = 0$. Hence, in such a medium, *group velocity = wave (or phase)* velocity and the wave groups continue to travel through the medium with their form or profile intact, as for example, in the case of sound waves in any homogeneous medium or light waves in vacuum.

In a *dispersive medium,* on the other hand, since velocity, in general, increases with wavelength, $dv/d\lambda$ is positive and, therefore, *the group velocity is, in general, less than the wave (or phase) velocity.* The wave groups in this case, therefore, do not retain their form or profile which keeps on changing as the wave advances.

*Alternatively,* we may also express group velocity u in another form. For, as we have just seen, $u = v - \lambda dv/d\lambda$. If, therefore, we denote the wavelengths of the two constituent waves by $\lambda_1$ and $\lambda_2$ and their velocities by $v_1$ and $v_2$ respectively.-we have

$$u = v_1 - \lambda_1 \frac{(v_2 - v_1)}{(\lambda_2 - \lambda_1)} = \frac{v_1\lambda_2 - v_1\lambda_1 - v_2\lambda_1 + v_1\lambda_1}{\lambda_2 - \lambda_1} = \frac{v_1\lambda_2 - v_2\lambda_1}{\lambda_2 - \lambda_1}$$

$$= \frac{\frac{v_1}{\lambda_1} - \frac{v_2}{\lambda_2}}{\frac{1}{\lambda_1} - \frac{1}{\lambda_2}}.$$

Or,

$$u = \frac{\left(\frac{2\pi}{\lambda_1}\right)v_1 - \left(\frac{2\pi}{\lambda_2}\right)v_2}{\frac{2\pi}{\lambda_1} - \frac{2\pi}{\lambda_2}}$$

Since $\frac{2\pi}{\lambda} = k$, we have $u = \frac{v_1 k_1 - v_2 k_2}{k_1 - k_2}$.

Again, since $\left(\frac{2\pi}{\lambda}\right)v = \left(\frac{2\pi}{\lambda}\right)n\lambda = 2\pi n = \omega$,

we have $u = \frac{\omega_1 - \omega_2}{k_1 - k_2}$,

which, in the event of the frequencies and propagation constants being really small, reduces, in the limit, to

$$u = \frac{d\omega}{dk}.$$

Thus, *whereas the wave (or phase) velocity v*, as we have seen before, *is given by ω/k, the group velocity u is given by its first derivative with respect to k.*

Again, *in a non-dispersive* medium, since wave (or phase) velocity v is the same for all wavelengths, $v = \omega/k$ = *constant* or $\omega = vk$ and, therefore, *group velocity* $u = \frac{d\omega}{dk} = v$, *i.e., group velocity = wave (or phase) velocity.*

*In a dispersive medium,* on the other hand, u may be greater or smaller than v.

Further, we may also put $v = \frac{d\omega}{dk} = \frac{d(vk)}{dk} = v + \frac{k\,dv}{dk}$, which again gives $u = v - \frac{\lambda\,dv}{d\lambda}$.

It may as well be pointed out here that whereas it is possible, in some cases, for the phase velocity to be higher than the velocity of light (c), the group velocity can never exceed c. Since it is the group velocity we actually measure by our experiments, the value obtained can never be higher than c.

Thus, taking the case of x-rays (or electromagnetic waves, including light waves, in general, their phase velocity through a carbon block works out to be greater than c but not their group velocity. This only shows that although the phase of the waves may travel faster in a medium than the velocity of light c, their velocity of propagation, *i.e.*, the velocity of the wave groups, in the form of which they actually travel through the medium, can never exceed c.

## SOLVED EXAMPLES

***Example 1:***

*Calculate the number of beats heard per second if three sources of sound of quencies 400, 401 and 402, of equal intensity, are sounded together.*

***Solution:***

This is a particular example of a general case cf three sources of sound simultaneously producing tones of frequencies (n – 1), n and (n + 1) of equal amplitudes their intensities being equal and intensity α (*amplitude*)$^2$}.

Let the displacements due to the three sound waves at a particular point in the medium through which they are passing simultaneously along the same direction be respectively.

$$y_1 = a \sin 2\pi (n - 1)t,$$

$$y_2 = a \sin 2\pi nt$$

and $$y_3 = a \sin 2\pi (n + 1)t.$$

Then, in accordance with the principle of superposition, the *resultant displacement at the point is* $y = y_1 + y_2 + y_2$ = a sin 2π (n – 1)t + a sin 2π nt + a sin 2π (n + 1)t.

Or, $y = [a \sin 2\pi (n_1 - 1)\, t + a \sin 2\pi (n + 1)t] + a \sin 2\pi\, nt = 2a \sin 2\pi\, nt \cos 2\pi t$

$+ a \sin 2\pi\, nt = \sin 2\pi\, nt\, (a + 2 \cos 2\pi t) = a\,(1 + 2 \cos 2\pi t) \sin 2\pi\, nt$,

indicating that the resultant displacement at the point varies cyclically with time.

Clearly, the *maximum resultant displacement or the amplitude of the displacement at the point,* is a (1 + 2cos 2πt).

This *amplitude will obviously have the maximum value* when cos 2πt = + 1, *i.e.*, when 2πt = 2rp, where r = 0, 1, 2, 3 etc. or, t = r = 0, 1, 2, 3 etc.

Thus, the maxima of amplitude, and hence also of intensity, occur at time-intervals of 1 sec each, or the frequency of maxima = 1 per sec,

And, *the amplitude, and hence also intensity, will be the minimum or zero when* 2 cos 2πt = – 1, or, cos 2πt = – 1/2, *i.e.*, when 2πt = (2rπ + 2π/3) or $t = \left(r + \frac{1}{2}\right)$, where r = 0, 1, 2, 3 etc. *i.e.*, when $t = \frac{1}{3}, \frac{4}{3}, \frac{7}{3}$ etc.

Thus, the minima of amplitude, and hence also of intensity, also occur at time-intervals of 1 sec each, or *their frequency too is* 1 per sec. though they alternate with the maxima.

Thus in all such cases, and hence also in the case given, one maximum and one minimum, *i.e., only one beat, is heard per second.*

### *Example 2:*

*A uniform circular hoop of string is rotating clockwise in the absence of gravity. The tangential speed is $v_0$. Find the speed of the wave travelling on this string.*

### *Solution:*

Since the circular hoop of the string preserves its shape and continues to rotate with a uniform speed $v_0$, there is *tension* along the string. Indeed, as we know, without any tension in the string, it would be simply impossible for a transverse wave to travel along it. The amplitude of the wave must, however, be very small or else, its presence would alter the tension in the string which would no longer remain constant.

Let us then consider a small circular hump AB of the wave, of length $\delta l$ of a very small amplitude so as to be almost coincident with the circumference of the hoop itself and hence of the same radius, *i.e.*, r, as shown in Fig. 3.19.

We know that the velocity of this hump (or the wave) along the string or the hoop is given by $\sqrt{T/\sigma}$, where T is the tension in the string acting tangentially to the hump or the hoop at A and B, as shown, and O, the *line density* or *mass per unit length* of the string. Let us obtain the value of $\sqrt{T/\sigma}$.

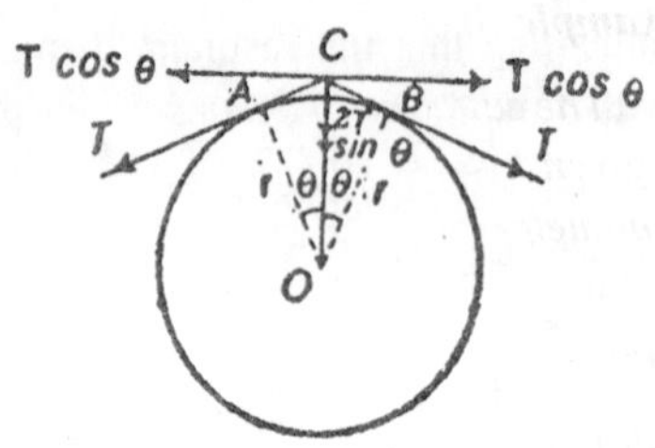


Fig. 3.19

Resolving the tension Tat A and B into its two rectangular components, along the horizontal and the vertical, we find that the horizontal components Teas θ are equal and opposite and thus cancel out; so that, the only effective components are the vertical ones along CO, each equal to T sin θ, where C is the *mid-point* of ABO, *the centre of the circular hoop* and θ, the angle AOC = angle COB. The *total force* acting on the portion AB of the string is thus 2T sin θ towards the centre 0 of the hoop and provides the necessary *centripetal force*, maintaining its rotation with speed $v_0$ in a circle of radius r.

But *centripetal force* acting on portion AB of the string is, as we know, given by *mass of portion AB of the stringy* $\times \dfrac{v_0^2}{r} = \sigma\delta l\dfrac{v_0^2}{r}$, because mass of portion AB of the string, of length $\delta l = \sigma\delta l$.

We, therefore, have $2T\sin\theta = \sigma\delta l\dfrac{v_0^2}{r}$.

Since θ is small, we may take sin θ = tan θ = θ (*in radian measure*).

So that, $2T\theta = \sigma\delta l\dfrac{v_0^2}{r}$.

Or, $$\frac{T\delta l}{r} = \sigma\delta l\frac{v_0^2}{r}. \qquad \left[\because 2\theta = \frac{\delta l}{r}\right]$$

Or, $\frac{T}{\sigma} = v_0^2.$

Or, $\sqrt{\frac{T}{\sigma}} = v_0.$

*i.e., the velocity of the transverse wave along the hoop of tiring is the same as the velocity of rotation of the hoop, viz.,* $v_0$.

***Example 3:***

*The equation of a transverse ware travelling along a stretched string is given by y = 10 sin π (2t – 0.01 x), where y and x are expressed in centimeters and t in seconds, (a) Find (be amplitude, frequency, velocity and wavelength of the ware. (b) Find the maximum transverse speed of a particle in the string.*

***Solution:***

(a) For any type of wave, transverse or longitudinal, we have the equation

$y = a \sin\frac{2\pi}{\lambda}(vt - x)$. This can be put in the form y – a sin p $\left(\frac{2vt}{\lambda} - \frac{2x}{\lambda}\right)$.

Or, since $\frac{v}{\lambda} = n$ and $\frac{1}{\lambda} = v$, it takes the form $y = a \sin \pi(2nt - 2vx)$.

Competing it with the given equation, we find it to be of the form, where a = 10 cm, 2nt = 2t or n = 1, 2vx = 0.01 x.

So that, $v = \frac{1}{\lambda} = \frac{0.01}{2} = 0.005/\text{cm},$

whence $\lambda = \frac{1}{0.005} = 200$ cm

and, therefore $v = n\lambda = 1 \times 200 = 200$ cm/sec.

Thus, the ***amplitude*** *of the wave* = 10 cm, *frequency* = – 1 cps, *velocity* = **200 cm/sec and** *wave length* = 200 cm.

**(b) Clearly,** ***particle*** *velocity* U = dy/dt = 10 × 2π cos (2t – 0.01x). **And, therefore if** ***maximum*** *value*, (*i.e.*, when cos (2t – 0.01 x) = 1), is 10 × 2π = **20π = 62.81 cm/sec.**

Since the particles of the string vibrate at right angles to (or transverse to) the length of the string, we have *maximum transverse speed of a particle of the strings 62.81 cm/sec.*

***Example 4:***

*One end of a string is continuously moved up and down through a distance of 1/2 cm, five times in one second, so as to send a transverse wave (rain along the string. If the due density of the string be 0.64 gm per cm and the tension in the string 1600 dynes, (a) obtain the value of the amplitude, frequency, velocity and wavelength of the wave; (b) taking y = 0 at x = 0 and f = 0, show that the equation of the wave is given by y = 0.25 sin (10πt – 0.2πx) and hence obtain the values of displacement, particle velocity and Wave Motion of the particle at a distance of 19 cm from the origin, after an interval of 0.4 second.*

***Solution:***

(a) Since rite end of the string is moved a total up and down distance of 1/2 cm, it is clear that it is moved 1/4 or 0.25 cm first up and then down from its mean or **equilibrium** position. *The amplitude of the wave generated is thus 6.25 cm.* **And** since the end is moved up and down five times in 1 second, the ***frequency*** *of the wave* = 5/sec.

Now, *velocity of the transverse wave along the string, i.e.,*

$$v = \sqrt{\frac{T}{\sigma}}.$$

Since, here, T = 1600 dynes and σ = 0.64 gm/cm, we have

$$v = \sqrt{\frac{1600}{0.64}} = \frac{40}{0.8} = 50 \text{ cm/sec.}$$

And, therefore, *wavelength of the wave,*

$$\lambda = \frac{v}{n} = \frac{50}{5} = 10 \text{ cm.}$$

(b) We have the *wave equation*

$$y = a \sin\frac{2\pi}{\lambda}(vt - x), \text{ with } y = 0 \textbf{ at } \mathbf{x} = 0 \text{ and } t = 0.$$

**Or,**

$$y = a \sin\left(\frac{2\pi vt}{\lambda} - \frac{2\pi x}{\lambda}\right).$$

Since $v = \frac{v}{\lambda} = n$ and $2\pi n = \omega$, we have $\frac{2\pi vt}{\lambda} = \omega t$ and since $\frac{2\pi}{\lambda} = k$, we have $\frac{2\pi x}{\lambda} = kx$. The equation of the wave thus takes the form $y = a \sin(\omega t - kx)$.

Here, $a = 0.025$ cm, $\omega = 2\pi n = 2\pi \times 5 = 10\pi$ and $k = \frac{2\pi}{\lambda} = \frac{2\pi}{10} = 0.2\pi$.

The *equation of the wave* is thus $y = 0.025 \sin(10\pi t - 0.2\pi x)$

For a distance 10 cm from the origin, we have $x = 10$ cm and $t = 0.4$ sec, and, therefore, displacement there is $y = 0.25 \sin(10\pi \times 0.4 - 0.2\pi \times 10) = 0.25 \sin 2\pi = 0$.

Thus, *the displacement of the particle at a distance 10 cm from the origin = 0,* indicating that *the particle is passing through its mean or equilibrium position.*

Since *particle velocity* $U = \frac{dy}{dt}$, we have $U = \omega a \cos(\omega t - kx)$,

*i.e.,* $U = 10\pi \times 0.25 \cos(10\pi t - 0.2\pi x)$

$= 2.5\pi \cos(10\pi \times 0.4 - 0.2\pi \times 10)$

$= 2.5\pi \cos 2\pi = 2.5\pi \times 1$

$= 2.5\pi = 7.852$ cm/sec.

This is obviously the *maximum particle velocity (the cosine term being equal to* 1), as is only to be expected, since the particle is passing through its mean or equilibrium position.

And, *particle Wave Motion,* $\frac{dU}{dt} = \frac{d^2y}{dt^2} = -\omega^2 a \sin(\omega t - kx) = -\omega^2 y$, the negative sign indicating that it is *oppositely directed to displacement and velocity.*

$\therefore$ *particle Wave Motion*, here, $= -(10\pi)^2 \times 0 = 0$

(Obviously, because the particle is passing through its mean or equilibrium position and has, therefore, its maximum velocity and hence *zero Wave Motion*).

***Example 5:***

*Show that the energy flux (or the energy passing through any point in unit time) in the case of a transverse wave along a string is given by the same general expression as deduced in 3.12 for a wave in any medium, with the difference that the volume density ($\rho$) of the medium is here replaced by the line density ($\sigma$) of the string.*

***Solution:***

Consider a small hump or a crest AB of a transverse wave travelling along a stretched string, as shown in Fig. 3.20. Then, the string is naturally inclined at a small angle $\theta$ to the x-axis or the equilibrium position of the string.

Since the tension T of the string acts tangentially to it at A, as shown, its downward component is clearly T sin $\theta$. So that, the force acting at point A of the string is $-$ T sin $\theta$, the negative sign indicating that the pull is *downwards (opposite io the direction of displacement or velocity).*

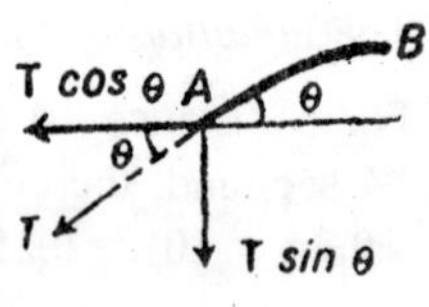


**Fig. 3.20**

If, therefore, U be the particle velocity at the point A, we have *work done or energy supplied by the string to point A, per unit time*

$$= -T \sin \theta . U.$$

$\therefore$ *energy supplied by the string to point A in time*

$$dt = -T \sin \theta .\ U.dt.$$

Now, $\theta$ being small, $\sin\theta = \tan\theta$ = slope of the curve or the hump at $A = \dfrac{dy}{dx}$.

So that, *energy passing through point A in time* $dt = -T.\left(\dfrac{dy}{dx}\right)U.dt$

Again, since $\quad y = a\sin\dfrac{2\pi}{\lambda}(vt - x),$

we have $\quad U = \dfrac{dy}{dt} = \dfrac{2\pi v}{\lambda} a\cos\dfrac{2\pi}{\lambda}(vt - x)$

and $\quad \dfrac{dy}{dx} = -\dfrac{2\pi}{\lambda} a\cos\dfrac{2\pi}{\lambda}(vt - x).$

Substituting these values of U and dy/dx is the **expression above,** we have

***energy** passing through point A in time* $dt = T\left[\dfrac{2\pi}{\lambda} a\cos\dfrac{2\pi}{\lambda}(vt - x)\right]$

$$\left[\frac{2\pi v}{\lambda} a\cos\frac{2\pi}{\lambda}(vt - x)\right]dt = \left(\frac{2\pi a}{\lambda}\right)^2 Tv\cos^2\frac{2\pi}{\lambda}(vt - x)dt.$$

∴ *energy passing through the point during the whole time-period, from* 0 to τ, (using the symbol τ for *time-period,* since T has been used for tension in the string)

$$= \int_0^{\tau}\left(\frac{2\pi a}{\lambda}\right)^2 Tv\cos^2\frac{2\pi}{\lambda}(vt - x)dt$$

$$= \left(\frac{2\pi a}{\lambda}\right)^2 Tv\int_0^{\tau}\frac{1}{2}\left[1 + \frac{\cos 4\pi}{\lambda}(vt - x)\right]dt$$

$$= \frac{4\pi^2 a^2}{\lambda^2}Tv\frac{\tau}{2} = 2\pi^2 a^2 nT\frac{\tau}{\lambda} \qquad \left[\because \frac{v}{\lambda} = n.\right]$$

∴ *overall energy passing through the point per second*

$$= \frac{2\pi^2 a^2 nT\tau}{\tau\lambda} = \frac{2\pi^2 a^2 nT}{\lambda}.$$

Now, $v = \sqrt{\frac{T}{\sigma}}$, whence, $T = v^2\sigma$.

∴ *average rate of transfer of energy through the point* or energy flax.

$$= \frac{2\pi^2 a^2 n}{\lambda}v^2\sigma = \frac{2\pi^2 a^2 nv}{\lambda}v\sigma = 2\pi^2 a^2 n^2 v\sigma, \qquad \left[\because \frac{v}{\lambda} = n.\right]$$

which is the same expression as obtained for the energy flux of a wave, in general, in a given medium, with only the volume density p of the medium replaced by the line density a of the string.

### *Example 6:*

*A sound wave of frequency 1000 and amplitude 2.5 × $10^{-5}$ cm is travelling through air at N.T.P. (a) Obtain the values of its velocity and wavelength. (b) Write down the equation of the wave. (c) Obtain the values of particle velocity amplitude and amplitude of pressure variation. (Density of air at N.T.P. – 1.293 gm/litre and γ for air = 1.4.*

### *Solution:*

(a) Since the wave is travelling through air at N T.P., we have

P = 76 cm of mercury column = 76 × 13.6 × 981 dynes/cm$^2$,

ρ = 1.293 gm/litre = 0.001293 gm/c.c. and γ = 1.4.

So that, *velocity of the wave,* $v = \sqrt{\gamma \frac{P}{\rho}}$

$\sqrt{1.4 \times 76 \times 13.6 \times 9.81 / 0.001293}$ = 33130 cm/sec.

And, therefore, *wavelength* $\lambda = \frac{v}{n} = \frac{33130}{1000}$ = 33.13 cm.

(b) The equation of the wave is, as we know, given by

$$y = a \sin \frac{2\pi}{\lambda}(vt - x).$$

Substituting the values of a, λ and v, therefore, we have *equation of the wave,*

$$y = 2.5 \times 10^5 \sin \frac{2\pi}{31.13}(33130\,t - x).$$

(c) Now, *particle velocity* $U = \frac{dy}{dt} = \frac{2\pi v}{\lambda} a \cos \frac{2\pi}{\lambda}(vt - x).$

Clearly, therefore, *particle velocity amplitude* $= \frac{2\pi v}{\lambda} a = 2\pi\, na$

$$= 2\pi \times 1000 \times 2.5 \times 10^{-5} = 0.157 \text{ cm/sec.}$$

And, since *pressure variation* $p = -K \frac{dy}{dx} = -K \frac{2\pi}{\lambda} a \cos \frac{2\pi}{\lambda}(vt - x).$

we have *amplitude of pressure variation* $= \frac{K.2\pi}{\lambda}$ (ignoring the sign)

$= v^2 \rho \frac{2\pi}{\lambda}$ a = 2π vr na

= 2π × 33130 × 0.001293 × 1000 × 2.5 × $10^{-5}$

= 6.726 dynes/cm². $\left[\because v = \sqrt{\frac{K}{\rho}}\right]$

**Example 7:**

***(a) Prove that at any point in the course of a plane progressive wave the ratio of the excess pressure to the velocity of the air particles at that place is constant, provided the temperature and pressure do not alter.***

***(b) In the faintest sound that can be beard at 1000 cps the pressure amplitude is 2.0 × $10^{-5}$ N/m². Find the corresponding displacement amplitude. (Atmospheric pressure = $10^5$ N/m², velocity of sound in air = 331 in/sec and ρ for air = 1.22 kg/m²).***

***Solution:***

(a) We know that particle velocity

$$U = \frac{dy}{dt} = v\left(\frac{dy}{dx}\right)$$ and that *excess pressure*

$$p = -K\left(\frac{dy}{dx}\right) - \gamma P\left(\frac{dy}{dx}\right)$$ because

K = *elasticity of air (adiabatic)* = $\gamma P$,

where P is the *atmospheric pressure.*

$$\therefore \quad \frac{p}{U} = \frac{\gamma P\left(\frac{dy}{dx}\right)}{v\left(\frac{dy}{dx}\right)} = \frac{\gamma P}{v}$$ [ignoring the sign.]

Since $\gamma$ is a constant for air and conditions of pressure and temperature of air remaining unaltered, P and r too are *constants*, we have

$$\frac{p}{U} = \frac{\text{excess pressure}}{\text{particle velocity}} = \text{constant}.$$

(b) We have for *excess pressure or pressure variation*, the relation

$$p = -\frac{K\,dy}{dx}.$$

And, since $$y = a \sin\frac{2\pi}{\lambda}(vt - x),$$

clearly, $$p = -K\frac{2\pi}{\lambda}a \cos\frac{2\pi}{\lambda}(vt - x).$$

Or, *amplitude of pressure variation*,

$$p_m = K\frac{2\pi}{\lambda}a = v^2\rho\frac{2\pi}{\lambda}a, \quad \left[\because v = \sqrt{\frac{K}{\rho}}\right]$$ ignoring the sign.

$\therefore$ *displacement amplitude*,

$$a = \frac{\lambda p_m}{2\pi v^2 \rho} = \frac{v}{n}\cdot\frac{p_m}{2\pi v^2 \rho} = \frac{p_m}{2\pi v n \rho}$$

$$= \frac{2.0\times10^{-5}}{2\pi\times331\times1000\times1.22} = 7.883 \times 10^{-12} \text{ metre.}$$

***Example 8:***

*(a) If the velocity of sound in hydrogen at 0°C is 4200 ft/sec, what will be the velocity of sound (at the same temperature) In a mixture of two parts by volume of hydrogen to one of oxygen? Density of oxygen is 16 times that of hydrogen.*

*(b) The planet Jupiter has an atmosphere of a mixture of ammonia and methane at a temperature – 130°C. Calculate the velocity of sound on this planet, assuming γ for the mixture to be 1.3 and the molecular weight of the mixture to be 16.5. (Gas constant 11 = 8.3 joules per deg. per gm. molecule).*

***Solution:***

(a) Let $\rho_H$ be the density of hydrogen and $\rho_M$ that of the mixture of hydrogen and oxygen. Then, if va and $v_M$ be the velocities of sound in hydrogen and the mixture respectively, we have

$$v_M = \sqrt{\frac{\gamma P}{\rho_M}} \text{ and } v_H = \sqrt{\frac{\gamma P}{\rho_H}}.$$

And therefore, $\dfrac{v_M}{v_H} = \sqrt{\dfrac{\rho_H}{\rho_M}}$.

Now, the mixture consists of two part by volume of hydrogen and oat part by volume of oxygen. So that, if each part by volume be V, we have

*mass of two parts by volume of hydrogen* 2V × 1 and *mass of one part by volume of oxygen* = V × 16.

∴ *mass of a volume 3V of the mixture*

$$= (2V + 16V) = 18V \text{ mass units.}$$

So that, *density of the mixture,*

$$\rho_M = \frac{\text{mass}}{\text{volume}} = \frac{18V}{3V} = 6 \text{ mass units/c.c.}$$

Hence $\dfrac{v_M}{v_H} = \sqrt{\dfrac{1}{6}}$.

Or, $v_M = v_H\sqrt{\dfrac{1}{6}} = 4200\sqrt{\dfrac{1}{6}} = 1714$ ft/sec.

Or, the *velocity of sound in the given mixture* = 1714 ft/sec.

(b) Here, molecular weight of the mixture of ammonia and methane or, the *mass of 1 gm molecule of the mixture* = 16.5. If, therefore, $\rho$ be the *density of the mixture*, the *volume of 1 gm molecule of it, i.e.*,

$$V = \frac{16.5}{\rho}.$$

Now, we have our *standard gas equation* PV = RT,

where, $V = \frac{16.5}{\rho}$, R = 8.3 joules or $8.3 \times 10^7$ *ergs per deg. C per gm molecule of the mixture* and T = (– 130 + 273) = 143° Abs.

So that, $$P\left(\frac{16.5}{\rho}\right) = 8.3 \times 10^7 \times 143,$$

whence, $$\frac{P}{\rho} = 8.3 \times 10^7 \times \frac{143}{16.5}.$$

∴ *velocity of sound in the mixture, i.e.. on the planet jupiter, is*

$$v = \sqrt{\frac{\gamma P}{\rho}}$$

$$= \sqrt{1.3 \times 8.3 \times 10^7 \times \frac{143}{16.5}}$$

$$= 30580 \text{ cm/sec} = 305.8 \text{ m/sec.}$$

***Example 9:***

*(a) Prove that v =* $v = C\sqrt{\frac{\gamma}{3}}$, *where v is the velocity of sound in a gas, C, the root mean square velocity of the molecules and $\gamma$ the ratio of the two specific heats.*

*(b) Show that the velocity of found to a gas, for which $\gamma$ = 1.41 is 0.68 C, where C is the root mean square velocity of the molecules.*

***Solution:***

(a) If P be the *pressure* and $\rho$, the *density* of a gas, we have *velocity of sound in the gas*,

$$v = \sqrt{\frac{\gamma P}{\rho}}.$$

Now, in accordance with the *kinetic theory* of gases,

$$P = \frac{1}{3}\rho C^2$$

or
$$\frac{P}{\rho} = \frac{C^2}{3},$$

where C is the *root mean square velocity* of the molecules of the gas.

$$\therefore \qquad v = \sqrt{\frac{\gamma P}{\rho}} = \sqrt{\frac{\gamma C^2}{3}} = C\sqrt{\frac{\gamma}{3}}.$$

(b) If y = 1.41 for a gas, we have

*velocity of sound in the gas,* $v = C\sqrt{\frac{1.41}{3}} = 0.6855\ C \approx 0.68\ C.$

***Example 10:***

(a) The speed of sound in a certain metal is V cm/sec. One end of a pipe of that metal of length 1 cm is truck a blow. A listener at the other cod bean two sounds, one from the wave that has travelled along the pipe and the other from the wave that has travelled through air. If v is the speed of the sound in air, what lime interval t elapses between the two sounds? (b) Suppose t = 1.4 sec. Young's modulus for the metal = $7.2 \times 10^{11}$ dynes/cm$^2$, density of the metal = 8 gm/c.c., atmospheric pressure = $10^6$ dynes/cm$^2$, density of air = 0.0012 gm/c.c. and $\gamma$ for air = 1.4, find the length 1.

***Solution:***

(a) We have *velocity of sound along the pipe,* $V = \sqrt{\frac{Y}{\rho}}$, where Y is the *Young's modulus* for the metal and $\rho$, its *density.*

And, *velocity of sound through air,* $v = \sqrt{\frac{\gamma P}{\rho'}}$, where P is the *atmospheric pressure* and $\rho'$, the *density of air*, with $\gamma = \frac{Cp}{Cv}$.

$\therefore$ *time taken by sound travelling along the pipe to cover the length I of the pipe* = $l$/B

and *time taken by sound travelling through air to cover the same distance l in air* $= \frac{l}{v}$.

Since V is very much greater than v ($\because$ the ratio $Y/\rho >> \gamma P/\rho'$), **we have** *time interval that elapses between the two sounds heard by the listener, i.e.,* t = $l/v - l/v$ ($l$ 1/v − 1/V) = $l(V - v)/vV$ sec.

(b) It is clear from the above that if t = 1.4 sec, we have

$$t = 1.4 = l\left(\frac{V - v}{vV}\right), \text{ whence, } l = \frac{1.4 \times v \times V}{V - v}.$$

Thus, $$V = \sqrt{\frac{Y}{\rho}} = \sqrt{\frac{7.2 \times 10^{11}}{8}} = 3 \times 10^5 \text{ cm/sec and}$$

$$v = \sqrt{\frac{\gamma P}{\rho'}} - \sqrt{\frac{1.4 \times 10^6}{0.0012}} = 3.415 \times 10^4 \text{ cm/sec.}$$

$$\therefore \quad l = \frac{1.4 \times 3.415 \times 10^4 \times 3 \times 10^5}{3 \times 10^5 - 3.415 \times 10^4} = 53940 \text{ cm} = 539.4 \text{ metres.}$$

***Example 11:***

*The lowest tone emitted by a cylindrical metal rod has a frequency 1000 cps and wavelength 300 cm. If the density of the metal be 7.5 gm/c.c., obtain the value of Young's modulus for it.*

***Solution:***

We have *velocity of the wave emitted by the rod,* v = nλ = 100 × 300 = 3 × $10^5$ cm/sec.

Now, $v = \sqrt{\frac{Y}{\rho}}$, where Y is *Young's modulus* for the material of the rod and ρ, its density.

$$\therefore \quad v^2 = \frac{Y}{\rho} \text{ or } Y = rv^2.$$

Substituting the values of ρ and v, therefore, we have *Young's modulus for the material of the rod*

$$= 7.5 \times (3 \times 10^5)^2 = 7.5 \times 9 \times 10^{10}$$

$$= 6.75 \times 10^{10} = 6.75 \times 10^{11} \text{ dynes/cm}^2.$$

***Example 12:***

*(a) Find the resultant of two plane simple harmonic waves of the mime period travelling in the same* **direction but** *differing in phase and*

*amplitude. What is the amplitude of the resultant wave if those of the component waves be 3.0 and 4.0 cm respectively and their phase difference π/2 radian?*

*(b) Explain why the interference fringes are hyperbolic when the prongs of a tuning fork are made to touch the surface of a liquid.*

**Solution:**

(a) Let the two simple harmonic waves, travelling along the same direction have amplitudes $a_1$ and $a_2$ and let the second wave lag a phase angle $\phi$ behind the first. Then, their equations may respectively be represented by

$$y_1 = a_1 \sin\frac{2\pi}{\lambda}(vt - x)$$

and $$y_2 = a_2 \sin\frac{2\pi}{\lambda}[(vt - x) - \phi].$$

The *resultant wave* is thus represented by

$$y = y_1 + y_2 = a_1 \sin\frac{2\pi}{\lambda}(vt - x) + a_2 \sin\frac{2\pi}{\lambda}[(vt - x) - \phi].$$

Or, $$y = a_1 \sin\frac{2\pi}{\lambda}(vt - x) + a_2 \sin\frac{2\pi}{\lambda}(vt - x)\cos\phi - a_2 \cos\frac{2\pi}{\lambda}(vt - x)\sin$$

$$= \sin\frac{2\pi}{\lambda}(vt - x)\,(a_1 + a_2\cos\phi) - \cos\frac{2\pi}{\lambda}(vt - x)\,(a_2 \sin\phi).$$

Putting $(a_1 + a_2 \cos\phi) = a\cos\theta$ and $a_2 \sin\phi = a\sin\theta$, we have

$$y = a\sin\frac{2\pi}{\lambda}(vt - x)\cos\theta - a\cos\frac{2\pi}{\lambda}(vt - x)\sin\theta$$

Or, $$y = a\sin\left[\frac{2\pi}{\lambda}(vt - x) - \theta\right],$$

indicating that *the resultant wave too is a simple harmonic wave, having the same frequency and wavelength as the two component waves but has an amplitude a and is a phase angle θ behind the first wave, where*

$$a = \sqrt{(a\cos\theta)^2 + (a\sin\theta)^2}$$
$$= \sqrt{a_1^2 + a_2^2\cos^2\phi + 2a_1a_2\cos\phi + a_2^2\sin^2\phi}$$

$$= \sqrt{a_1^2 + a_2^2 + 2a_1a_2\cos\phi}$$

and $$\tan\theta = \frac{a\sin\theta}{a\cos\theta} = \frac{a_2\sin\phi}{a_1 + a_2\cos\phi}$$

Obviously, for $\phi = 0$, *i.e.*, when the two component waves arrive at a point *in, phase*, we have $\cos\phi = +1$ and, therefore, $a = a_1 + a_2$ and for $\phi = \pi$, *i.e.*, when the two waves arrive at a point *out of phase*, we have $\cos\phi = -1$ and, therefore, $a = a_1 - a_2$. At all other points, the amplitude of the resultant wave lies between these two extremes.

If, therefore, $a_1 = 3.0$ cm,

$a_2 = 4.0$ cm

and $\phi = \frac{\pi}{2}$, we have

*amplitude of the resultant wave,*

$$a = \sqrt{3^2 + 4^2 + 2\times3\times4\cos\frac{\pi}{2}}$$

$$= \sqrt{9+16+0} = \sqrt{25} = 5.0 \text{ cm.}$$

(b) In Fig. 3.21, let the two dots, 1 and 2, represent the two points where the prongs of a vibrating tuning fork *just* touch a liquid surface. These points then obviously become the *sources of two identical, transversc, spherical waves, travelling onwards in the form of crests* and *troughs*, represented respectively by full and broken-line curves in the figure.

These two sets of waves thus get partially superposed and *interfere* with each other, *constructively at points where the crests or the troughs of the two waves arrive simultaneously and destructively at points where the crest of one wave arrives simultaneously with the trough of the other*. The former are thus points of maximum and the latter, of minimum, displacements and are marked in the figure by circles add crosses respectively.

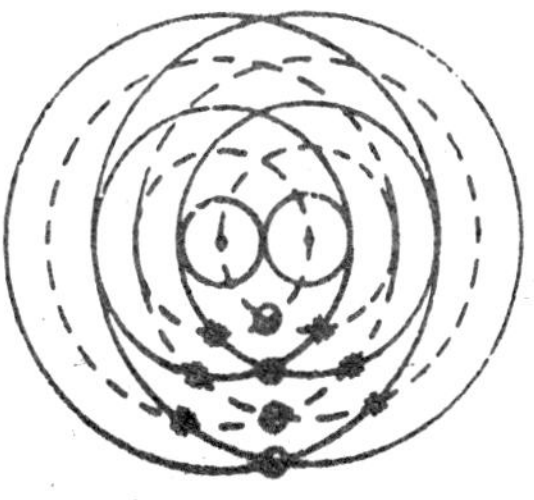

Fig. 3.21

Joining by a free band curve the points marked by circles and so also those marked by crosses, we obtain positions of maxima and minima of displacement in the resultant wave. These curves thus constitute the *interference* fringes formed by the two wave trains proceeding from the two dots.

Consider any point on one of these curves or interference fringes, such that its distances from dots 1 and 2 respectively are $x_1$ and $x_2$. Then, if the point lies on a curve representing a maximum of displacement, we have $(x_1 - x_2) = n\lambda$, *a constant*, and if it happens to be on a curve representing a minimum of displacement, we have $(x_1 - x_2) = (2n + 1)\lambda/2$, *again a constant* (where n = 0, 1, 2, 3 etc). In either case, therefore, the path difference for the point considered from the two fixed points or dots remains constant.

Now, *the locus of all points for which the path difference from two fixed points remains constant is a hyperbola, with the two fixed points as its two foci.*

This explains why the interference fringes in this case are so many hyperbolic curves, corresponding to n = 0, 1, 2, 3 etc, with their two foci at the two dots or points of disturbance.

***Example 13:***

*(a) A sound wave in air, having an amplitude of 0.005 cm and frequency 700 cps, and travelling along the positive direction of the x-axis with velocity 350 m/sec, suffers reflection at a free boundary. Obtain the values of (i) the displacement amplitude and (ii) the pressure amplitude in the resulting stationary wave at a point x = 50 cm. (Density of air =1.293 gm/litre).*

*(b) Also obtain the distance between (i) two successive displacement nodes and antinodes, (ii) two successive pressure nodes and antinodes, (iii) a displacement node and an adjacent pressure node.*

***Solution:***

(a) (i) We have the equation of a stationary wave formed by reflection at a free boundary given by

$$y = 2a\cos\frac{2\pi x}{\lambda}\sin\frac{2\pi vt}{\lambda},$$

whence, clearly, *displacement amplitude* $= 2a\cos 2\pi\dfrac{x}{\lambda}$.

Here, $a = 0.005$ cm, $x = 50$ cm

and $\lambda = \frac{v}{n} = 350 \times \frac{100}{700} = 50$ cm.

$\therefore$ *displacement amplitude* $= 2 \times 0.005 \cos 2p \times \frac{50}{50} = 0.01$ cm,

*i.e.*, the *maximum* (the cosine term being equal to 1).

*The point is thus a displacement antinode.*

(ii) We have *excess pressure* $p = \rho v^2 \frac{4\pi a}{\lambda} \sin\frac{2\pi x}{\lambda} \sin\frac{2\pi vt}{\lambda}$

So that, *pressure amplitude*

$$= \rho v^2 \frac{4\pi a}{\lambda} \sin\frac{2\pi x}{\lambda} = \frac{0.001293 \times (3500)^2 \times 4\pi 0.005}{50} \sin\frac{2\pi \times 50}{50} = 0,$$

*i.e., the point is a pressure node* (naturally because it is a displacement antinode).

(b) We know that the distance between two successive displacement nodes or antinodes is $\lambda/2$ and, therefore, the distance between a displacement node and an adjacent displacement antinode is $\lambda/4$, and also that a displacement node coincides with a pressure antinode and vice versa.

It will thus be easily seen that, in the case above, *distance between two successive displacement nodes = distance between two successive displacement antinodes*

$$= = \frac{\lambda}{2} = \frac{50}{2} = 25.0 \text{ cm.}$$

And, therefore, *distance between two successive nodes or antinodes is also equal to* $\frac{\lambda}{2} = \frac{50}{2}$ *= 25.0 cm*, and *distance between a displacement node (or pressure antinode) and an adjacent pressure node (or displacement antinode)* $= \frac{\lambda}{4} = \frac{50}{4} = 12.5$ cm.

***Example 14:***

***The equation** of a stationary wave in a metal rod is given by* $y = 0.002 \sin\left(\frac{\pi}{3}x\right) \sin(1000\,t)$. *Obtain the values of (i) amplitude of* *particle velocity, (ii) maximum tensile stress at a point x = 2.0 cm. (Y for the material of the rod = 8 × $10^{11}$ dynes/cm$^2$).*

**Solution:**

(i) We have, *particle velocity*

$$U = \frac{dy}{dt} = 0.002 \times 1000 \sin\left(\frac{\pi}{3}x\right) \cos(1000\,t).$$

So that, *amplitude of particle velocity*

$$= 0.002 \times 1000 \sin\left(\frac{\pi}{3}x\right) = 2\sin\left(\frac{\pi}{3}\times 2\right)$$

$$= 2\sin\frac{2\pi}{3} = 2\sin 120° = 2\sqrt{3/2} = 1.732 \text{ cm/sec}$$

(ii) *Since Young's modulus*

$$Y = \frac{\text{tensile stress}}{\text{tensile strain}} = \frac{\text{tensile stress}}{dy/dx},$$

we have *tensile stress* = $Y\left(\frac{dy}{dx}\right)$

Here, Y = 8 × $10^{11}$ dynes/cm$^2$ and

$$\frac{dy}{dx} = 0.002\left(\frac{\pi}{2}\right)\cos\left(\frac{\pi}{3}x\right) \sin(1000\,t).$$

Clearly, *maximum value of strain*

$$\left(\frac{dy}{dx}\right) = 0.002\left(\frac{\pi}{3}\right)\cos\left(\frac{\pi}{3}x\right).$$ [*i.e.* when sin (1000 t) = 1.]

∴ *maximum value of stress*

$$= Y \times 0.002 \times \left(\frac{\pi}{3}\right)\cos\left(\frac{\pi}{3}x\right)$$

$$= Y \times 0.002 \times \left(\frac{\pi}{3}\right) \cos 60°$$

$$= 8 \times 10^{11} \times 0.002 \times \left(\frac{\pi}{3}\right) \times \frac{1}{2}$$

$$= 8 \times 10^{8} \times \frac{\pi}{3} = \mathbf{8.377 \times 10^{8}\ dynes/cm^{2}}.$$

***Example 15:***

*(a) A loaded string, one metre in length and weighing 0.5 gm is banging from a tuning fork of frequency 200 and is vibrating in 4 loops. Calculate the tension in the string.*

*(b) In Melde's experiment when the tension is 100 gm wt and the fork vibrates at right angles to the direction of the string, the latter is thrown into four segments. If now the fork is let to vibrate along the string, find what additional load will make the string vibrate in one segment.*

***Solution:***

(a) Here, obviously, two cases arise: (i) when the prongs of the fork vibrate *transversely* (*i.e.*, at right angles) to the length of the string, as shown in Fig. 3.22 (a), called the *transverse mode* and (ii) when the prongs vibrate along the length of the string, as shown in Fig. 3.22 (b), called the *longitudinal mode.*

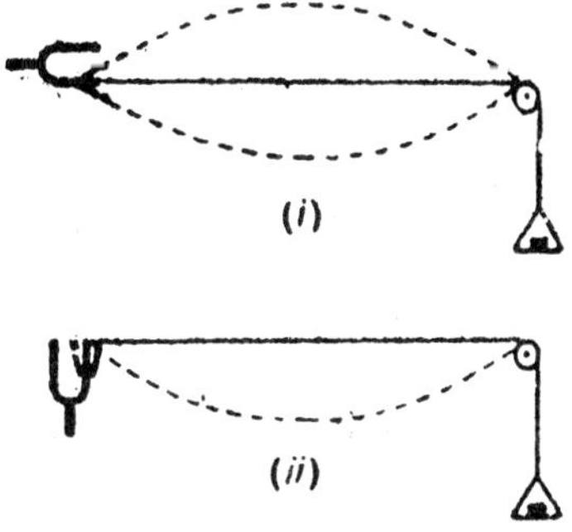


Fig. 3.22

In case (i) or the *transverse mode*, with the prong in its normal position, the string is horizontal and when the prong moves us or down the string also does so; so that, the frequency of vibration of the string in its *fundamental mode of vibration*, *i.e.*, when it vibrates as a whole, in one single segment, is the same as that of the prong or the tuning fork, say, n, and we, therefore, have

$$n = \frac{1}{2l}\sqrt{\frac{T}{\sigma}}.$$

And, if the string vibrates in *p segments,* the frequency of the note emitted by it is given by $n = \frac{p}{2l}\sqrt{\frac{T_p}{\sigma}}$, where $T_p$ is now the tension in the string.

$$\therefore\ T_p = \frac{4l^2n^2\sigma}{p^2} = 4 \times (100)^2 \times (200)^2 \times \frac{0.5}{(4)^2} \times 100$$

$$= 5 \times 10^5 \text{ dynes.} \qquad \left[\because \sigma = \frac{0.5}{100} \text{ gm/cm}\right]$$

Thus, *the tension required in the string* = $5 \times 10^5$ dynes.

Incase (ii), or the *longitudinal mode,* when the prong of the fork is in its extreme right position, the string takes the lower dotted position shown and when the prong has moved to its extreme left position, the string comes to its normal horizontal position, so that, by the time the prong has completed half its vibration, the string has completed only one-fourth of its vibration. The frequency of the string, is thus *half* that of the fork and we, therefore, have

$$\frac{n}{2} = \frac{1}{2l}\sqrt{\frac{T}{\sigma}}.$$

And, if the string vibrates in p segments, the frequency of the note emitted by it is given by $\frac{n}{2} = \frac{p}{2l}\sqrt{\frac{T_p'}{\sigma}}$, where $T_p'$ is now the tension in the string.

$$\therefore\ T_p' = 4l^2n^2\,\frac{\sigma}{4}\,p^2 = l^2n^2\,\frac{\sigma}{p^2} = \frac{1}{4}\text{th of the tension } T_p \text{ in case (i)}$$

$$= \frac{1}{4} \times 5 \times 10^5 = 1.25 \times 10^5 \text{ dynes.}$$

Thus, *the tension required in the string, in this rase,*

$$= 1.25 \times 10 \text{ dynes.}$$

(b) It will be seen from the discussion under (a) above that in the transverse mode,

$$p^2 \times T_p = \text{a } constant\ k. \text{ Or, } p \propto \sqrt{\frac{1}{T_p}}$$

and in the *longitudinal mode,*

$$p^2 \times T_p' = a\ constant\ k'. \text{ Or, } p \propto \sqrt{\frac{1}{T_p'}},$$

*i.e.*, in either case, *the number of segments or loops into which the string is thrown into vibration is inversely proportional to the square root of the stretching force.*

Since, however, $T_p$ is four times $T_p'$, it is clear that the stretching force required to set the string vibrating into a given number of segments in the transverse mode is four times that required to set it vibrating into the same number of segments in the longitudinal mode.

If, therefore, we apply the same tension $T_p = 4T_p'$, in either case, we shall find that whereas in the transverse mode the string vibrates in p segments, in the longitudinal mode, the number of segments is reduced to p/2 because the number of segments is inversely proportional to the square root of the stretching force which is now 4 times its previous value. In other words, *for the same tension in the string, the number of segments in the longitudinal mode is half that in the transverse mode.*

Now, in the given problem, a stretching force of 100 gm wt produces 4 loops in the string *in the transverse mode*. In the longitudinal mode, therefore, this stretching force will produce only half the number of loops, *i.e.* , only 2 loops. The problem thus reduces to this that if with a stretching force of 100 gin wt, the string vibrates in 2 loops in the longitudinal mode, what additional force will be needed to make it vibrate in only one loop?

In other words, p = 2 when $T_p'$ = 100 gm wt.. What should be the value of $T_p'$ when p = 1?

As we know, $p^2 \times T_p'$ = constant. And, therefore,

$$2^2 \times 100\ (1)^2 \times T_p',$$

whence $T_p'$ = 400 gm wt.

Therefore, *additional load required to make the string vibrate in*

*1 loop* = 400 – 100 = 300 gm wt.

***Example 16:***

*One end of the string in Melde's experiment is attached to a vibrating tuning fork while its other end carries a piece of stone. The string shows eight vibrating loops. When me stone is immersed in water, 10 loops are formed. Calculate the specific gravity of the stone.*

***Solution:***

Although it is not mentioned here whether it is a case of transverse or of longitudinal mode, in either case, as we know,

*(number of segments)*$^2$ × *stretching force* = *constant*.

∴ if $T_1$ and $T_2$ be the tensions in the string when the stone is in air and in water respectively, we have

$$8^2 \times T_1 = 10^2 \times T_1,$$

whence $T_2 = 8^2 \times T_1/10^2 = (16/25)T_1$.

∴ *loss of weight of in water* $= T_1 - T_2 = T_1 - \left(\frac{16}{25}\right)T_1 = \left(\frac{9}{25}\right)T_1$.

And, therefore, *specific gravity of the stone* $= \frac{T_1}{(9/25)T_1} = \frac{25}{9} = 2.78$.

***Example 17:***

*(a) A tuning fork of frequency 200 is in unison with a sonometer wire. How many beats per second will be heard if the tension of the wire be increased by one per cent?*

*(b) A stretched sonometer wire gives 2 beats per second with a tuning fork when the length of the wire is 143 cm and also when its lengths is 145 cm. What Is the frequency of the tuning fork?*

***Solution:***

(a) We know that other factors remaining the same, the frequency of the note emitted by a sonometer wire is directly proportional to the square root of the stretching force applied to it or the tension in it.

If, therefore, $n_1$ and $n_2$ be the frequencies of the notes emitted by the wire with tension T and $\left(T + \frac{T}{100}\right)$ respectively, we have

$$\frac{n_2}{n_1}\sqrt{\frac{T + \frac{1}{100}}{T}} = \left(1 + \frac{1}{100}\right)^{1/2} = \left(1 + \frac{1}{200} + \ldots\right).$$

Or, $$\frac{n_2}{n_1} = \frac{201}{200}$$

Since the first note ($n_1$) of the sonometer wire is in unison with the tuning fork, we have $n_1 = 200$ and, therefore, $n_2 = 201$. So that, $(n_2 - n_1) = 201 - 200 = 1$.

Thus, the difference between the frequencies of the second note of the wire and the tuning fork being 1, *only 1 be at per second will be heard.*

(b) Here, the tension and mass per unit length of the sonometer wire remaining the same, the frequency of the note emitted by it is inversely proportional to the length of the wire. So that, if $n_1$ and $n_2$ be the frequencies of the notes when the lengths of the wire are $l_1$ and $l_2$ respectively, we have

$$\frac{n_1}{n_2} = \frac{l_1}{l_2}.$$

Or, $n_1 l_1 = n_2 l_2$. And, therefore, 143 $n_1$ = 145 $n_2$, whence,

$$n_2 = 143\frac{n_1}{145}.$$

If the frequency of the tuning fork be n, it must be 2 lower than $n_1$ and 2 higher than $n_2$ in order to produce 2 beats per second in either case.

We, therefore, have $n_1 - n = 2$ ...(i)

and $$n - n_2 = n - \frac{143}{145}n_1 = 2$$ ...(ii)

Adding relations (i) and (ii), we have $n_1 - \frac{149}{145}n_1 = 4$

Or, $2n_1 = 4 \times 145$, whence, $n_1 = 4 \times \frac{145}{2} = 290.$

Substituting this value of $n_1$ in relation (i) above, we have

$$290 - n = 2.$$

Or, $n = 288.$

Thus, the *frequency of the tuning fork* = 288 cps.

***Example 18:***

*(a) A sonometer wire of non-magnetic material is stretched across two permanent bridges one metre apart and a permanent horse-shoe magnet M placed athwart (i.e., across) it at its mid-point, so that the two poles of the magnet lie on either side of the wire, as shown in Fig. 3.23 (a). The tension in the wire is so adjusted that on passing an alternating current under a low voltage (to avoid undue heating of the wire), the wire vibrates up and down with its maximum amplitude. If the*

*line density of the wire be 0.049 gm/cm and the stretching force necessary be f kg wt, calculate the frequency of the alternating current. (g = 980 cm/sec²).*

*(b) If the sonometer wire be of a magnetic material, like iron, it can be made to vibrate up and down with maximum amplitude by bringing close to it an electromagnet E.M., energised by the alternating current, [Fig. 3.23 (b)], and suitably adjusting the length of the wire and the tension in it. Calculate the frequency of the alternating current in this case if the line density of the wire and the tension in it be the same as in case (a) but the length of the wire is now 50 cm.*

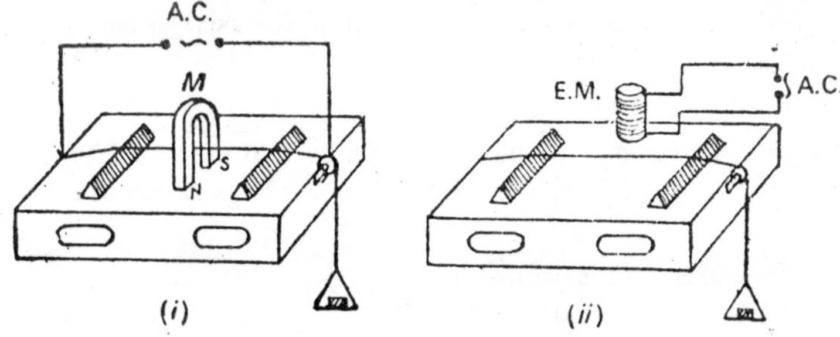


Fig. 3.23

***Solution:***

(a) Here, the magnetic field due to the magnet (M) being perpendicular to the length of the wire, in the direction NS [Fig. 3.23 (a)], we have the case of a current-carrying conductor lying in a magnetic field. The conductor or the wire, therefore, gets deflected in accordance with *Fleming's left hand rule, i.e.*, upwards during one half cycle of the current and downwards during the other half. It thus executes a forced vibration in one single segment with the frequency of the current.

With its length and tension properly adjusted, the wire vibrates with *maximum amplitude*, showing that *now its natural frequency n is the same as the frequency f of the alternating current, i.e.*, it now vibrates in resonance with the alternating current.

Clearly, the natural frequency of the wire is $n = \frac{1}{2l}\sqrt{\frac{T}{\sigma}}$.

Or, since T = 5 kg wt = 5 × 1000 × 980 dynes

and $\sigma$ = 0.049 gm/cm, we have

$$n = \frac{1}{2\times100}\sqrt{\frac{5\times1000\times980}{0.049}} = \frac{10^4}{2\times100} = 50 \text{ cps.}$$

Since the wire is in resonance with the alternating current, *the frequency of the alternating current too is 50 cps.*

(b) In this case, as the alternating current is passed around the electromagnet, the pole close to the wire acquires alternately north and south polarities (due to the current flowing in one direction during one half cycle and in the opposite direction during the other half). In either case, however, the wire (being of a magnetic material) is attracted towards it. So that, although the current passing through the electromagnet changes direction only once in one cycle, the wire is attracted upwards twice in that one cycle. In other words, the wire is forced to vibrate in one segment with twice the frequency of the alternating current. The length of the wire and its tension have been so adjusted that the wire vibrates with its maximum amplitude.

Again, the *frequency of the wire,*

$$n = \frac{1}{2l}\sqrt{\frac{T}{\sigma}} = \frac{1}{2\times50}\sqrt{\frac{5\times100\times980}{0.049}}$$

$$= \frac{104}{2\times50} = 100 \text{ cps.}$$

Since the frequency of the wire is twice that of the alternating current, we have

$$\textit{frequency of the alternating current} = \frac{100}{2} = 50 \text{ cps.}$$

***Example 19:***

*A one-metre long sonometer wire is stretched with a force of 4 kg wt. Another wire of the same material and diameter is arranged alongside the first and stretched with a force of 16 kg wt. What should be the length of the second wire so that its second harmonic is the same as the fifth harmonic of the first?*

***Solution:***

The two wires being of the same material and diameter, their *line density* $\sigma$ is the *same*. If, therefore, $n_5$ be the *fifth harmonic of the first wire,* $T_1$, *the stretching force* applied to it and $l_1$, its length, we have

$n_5 = \frac{5}{2l}\sqrt{\frac{T_1}{\sigma}}$. And if $n_2$ be the *second harmonic of the second wire*, $T_2$, the *stretching force* applied to it and $l_2$, its *length*, we have

$$n_2' = \frac{2}{2l_2}\sqrt{\frac{T_2}{\sigma}} = \frac{1}{l_2}\sqrt{\frac{T_2}{\sigma}}.$$

Since $n_2' = n_5$, we have $\frac{1}{l_2}\sqrt{\frac{T_2}{\sigma}} = \frac{5}{2l_1}\sqrt{\frac{T_1}{\sigma}}$

Or, $\frac{\sqrt{T_2}}{l_2} = 5\sqrt{\frac{T_1}{2l_1}}$.

Or, $l_2 \times 5\sqrt{T_1} = 2l_1\sqrt{T_2}$.

Or, $l_2 \times 5\sqrt{4} = 2l_1\sqrt{16}$.

Or, $10l_2 = 8l_1$.

whence, $l_2 = \frac{8l_1}{10} = 8 \times \frac{100}{10} = 80$ cm.

Thus, *the length of the second wire must be 80 cm.*

**Example 20:**

*Calculate the frequency of the lowest note, or the first harmonic, emitted by a metal rod set in longitudinal vibration when the rod is (i) clamped nowhere (free-free rod), (ii) clamped in the middle and (iii) clamped at both ends (fixed-fixed rod), if the length of the rod be 100 cm, its density 8 gm/c.c. and the value of Young's modulus for its material, $10^{11}$ dynes/cm$^2$.*

*Is it possible to obtain a note of a lower frequency from this rod? If so, bow? What Is the value of this frequency?*

**Solution:**

In all the cases, (i), (ii) and (iii), the frequency of the lowest or the fundamental note, or the first harmonic, is given by

$$n = \frac{1}{2l}\sqrt{\frac{Y}{\rho}}.$$

So that, substituting the given value of $l$, Y and $\rho$, we have *frequency of the lowest note,*

$$n = \frac{1}{2\times 100}\sqrt{\frac{10^{11}}{8}} = 5591 \text{ cps.}$$

It is possible to obtain a note of a lower frequency than this (in fact, half of this) if the rod be fixed at one end and be free at the other (fixed-free rod), when the frequency of the lowest note or the first harmonic is given by

$$n' = \frac{1}{4l}\sqrt{\frac{Y}{\rho}} = \frac{1}{4\times 100}\sqrt{\frac{10^{11}}{8}} = 2796 \text{ cps.}$$

***Example 21:***

*The frequency of the note next higher to the fundamental, as given by a rod of an alloy, 100 cm long and clamped at its mid-point is 1000. If the density of the alloy be 7.5 gm/c.c.. calculate the value of Young's modulus for it.*

***Solution:***

We know that in the case of a rod, clamped at its mid point, only odd harmonics can be excited. So that, the note next to the fundamental has thrice the frequency of the fundamental. If, therefore, n be the frequency of the note we have

$$n = \frac{3}{2l}\sqrt{\frac{Y}{\rho}}.$$

Substituting the given values, we have

$$1000 = \frac{3}{2\times 100}\sqrt{\frac{Y}{7.5}}.$$

Or, $$\sqrt{\frac{Y}{7.5}} = \frac{1000\times 2\times 100}{3},$$

whence, $$Y = \frac{4\times 10^{10}\times 7.5}{9} = 3.33 \times 10^{10}$$

Thus, *Young's modulus for the alloy of the rod*

$= 3.33 \times 10^{10}$ dynes $cm^2$.

***Example 22:***

*A wire, of area of cross section 0.04 sq. cm, is fixed at one end and is stretched by a weight of 20 kg suspended from the other. Obtain (he* *ratio between the frequencies of the notes emitted by it when it is rubbed longitudinally at its mid-point by a resined cloth and when it is plucked at its mid-point. (Y for the material of the wire = 4.9 × $10^{12}$ dynes/$cm^2$).*

**Solution:**

In the *first case*, when the wire is rubbed at its mid-point, it functions as a rod fixed at both ends and is set in longitudinal vibration, giving out Bs fundamental note of frequency n, say, where

$$n = \frac{1}{2l}\sqrt{\frac{Y}{\rho}}, \quad \text{...(i)}$$

if being the (volume) density of the wire.

And, in the second case, it functions as a string fixed at both ends which, when plucked at its mid-point gives its fundamental note of frequency n', say, where

$$n' = \frac{1}{2l}\sqrt{\frac{T}{\sigma}}, \quad \text{...(ii)}$$

σ being is the *line density (or mass per unit length)* of the wire = (area of cross-section of the wire × 1) × ρ = 0.04 ρ gm/cm.

So that, $$n' = \frac{1}{2l}\sqrt{\frac{T}{0.04\rho}}. \quad \text{...(iii)}$$

∴ *ratio of the two frequencies*

$$= \frac{n}{n'} = \frac{\frac{1}{2l}\sqrt{\frac{Y}{\rho}}}{\frac{1}{2l}\sqrt{\frac{T}{0.04\rho}}} = \sqrt{\frac{Y \times 0.04}{T}}.$$

Since Y = 4.9 × $10^{12}$ dynes/$cm^2$, T = 20 kg wt = 20 × 1000 × 980 = 1.96 × $10^7$ dynes, we have

$$\frac{n}{n'} = \sqrt{\frac{4.9 \times 10^{12} \times 0.04}{1.97 \times 10^7}} = 100.$$

Thus, *the ratio between the two notes is 100 : .*

**Example 23:**

*A tube 100 cm long is closed at one end. A stretched wire is placed near the open end. The wire is 30 cm long and has a mass of 10 gm.*

*It is held fixed at both ends and vibrates In its fundamental mode. It sets the air column In the tube into vibration at its fundamental frequency by resonance. Find (a) the frequency of oscillation of the air column and (b) the tension in the wire. (Velocity of sound in air = 340 m/sec). Neglect end-correction of the tube.*

***Solution:***

(a) We know that the fundamental frequency of the air column in a closed pipe is given by $n = \frac{1}{4l}\sqrt{\frac{K}{\rho}} = \frac{v}{4l}$, neglecting the end-correction.

Here, v = 340 × 100 = 34000 cm/sec and $l$ = 100 cm. And, therefore,

$$n = \frac{34000}{4} \times 100 = 85.$$

Thus, the *frequency of oscillation of the air column* = 85 cps..

(b) Since the air column has been set into resonant oscillation by toe fundamental frequency of the wire the latter too is n = 85.

Now, with the wire fixed at both ends, its *fundamental frequency* $= \frac{1}{2l}\sqrt{\frac{T}{\sigma}}$.

We, therefore, have $n = 85 = \frac{1}{2l}\frac{T}{\sigma}$.

Or, $$\sqrt{\frac{T}{\sigma}} = 85 \times 2l = 85 \times 2 \times 30.$$

Or, $$\frac{T}{\sigma} = (85 \times 2 \times 30)^2.$$ And, therefore,

$$T = (85 \times 2 \times 30)^2 \times \sigma.$$

Clearly, $\sigma$ = *mass per unit length of the wire* $= \frac{10}{30} = \frac{1}{3}$ *gm/cm.*

∴ *tension in the wire,* $T = (85 \times 2 \times 30)^2 \times \frac{1}{3} = 8.672 \times 10^6$ dynes.

***Example 24:***

*(a) It two pipes, one closed at one end and the other open at both ends, have their first overtones of the same frequency, what is the ratio of their respective lengths?*

*(b) A tuning fork is held above a column of air in a glass tube. The positions of the first and second resonance are 33.0 cm and 100.5 cm respectively. Calculate the end correct.*

***Solution:***

(a) Let the lengths of the closed and open pipes be $l$ and $l'$ respectively.

Then, the *first overtone of the former has a frequency* $\frac{3}{4l}\sqrt{\frac{K}{\rho}} = \frac{3v}{4l}$ and the *first overtone of the latter has a frequency* $\frac{2}{2l'}\sqrt{\frac{K}{\rho}} = \frac{2}{2l'} = \frac{v}{l'}$ where v is the velocity of sound in air.

Since the frequencies of the two overtones are the same, we have

$$\frac{3v}{4l} = \frac{v}{l'}.$$

Or, $4l = 3l'$ or $l : l' : : 3 : 4$

Thus, *the ratio of the lengths of the closed and open pipes is 3 : 4.*

(b) Here, the glass tube is, obviously, serving as a *closed pipe.* Therefore, $l_1$ be the length of the air column at the *first resonance* and x, the *end-correction* of the tube, we have

$$(l_1 + x) = \frac{3\lambda}{4} \quad \text{...(ii)}$$

And, if $l_2$ be the length of the air column at the *second resonance,* we have

$$(l_2 + x) = \frac{3\lambda}{4} \quad \text{...(ii)}$$

$\therefore$ multiplying relation (i) by 3, we have $3l_1 + 3x = \frac{3\lambda}{4}$ ...(iii)

From relations (ii) and (iii), therefore, we have $3l_1 + 3x = l_1 + x$.

Or, $\quad 3x - x = l_2 - 3l_1$.

Or, $\quad 2x = l_2 - 3l_1$, whence, $x = \frac{(l_2 - 3l_1)}{2}$.

Here, $l_1 = 33.0$ cm and $l_2 = 100.5$ cm. So that, *end correction of the tube,* $x = \frac{(100.5 - 3 \times 33.0)}{2} = \frac{1.5}{2} = 0.75$ cm.

***Example 25:***

*It is found that 4 beats are heard per second when the fundamental note of a closed pipe, 25 cm long, is sounded simultaneously with this second harmonic of an open pipe. What can be the possible lengths of the open pipe? The end-correction may be neglected in either case. (Velocity of sound in air = 340 m/sec)*

***Solution:***

Here, clearly, the length of the closed pipe being given to be 25 cm, the frequency of its fundamental note is fixed and is equal to

$$n = \frac{v}{4l} = 340\times\frac{100}{4}\times 25 = 340.$$

In order, therefore, that 4 beats may be heard per second, the frequency of the second harmonic of the open pipe must either be 340 + 4 = 344 or 340 – 4 = 336.

If, therefore, $l'$ be the length of the open pipe which gives a second harmonic of frequency 344, we have

$$344 = \frac{2v}{2l'} = \frac{v}{l'} = \frac{34000}{l'},$$

whence, $$l' = \frac{34000}{344} = 98.83 \text{ cm}.$$

And, if $l''$ be the length of the open pipe which gives a second harmonic of frequency 336, we have

$$336 = \frac{2v}{2l''} = \frac{v}{l''} = \frac{34000}{l''},$$

whence, $$l'' = \frac{34000}{336} = 101.2 \text{ cm}$$

Thus, *the two possible lengths of the open pipe are* 98.83 cm and 101.2 cm.

***Example 26:***

*Fig. 3.24 gives a periodic function. Obtain a Fourier analysis of this up to the fifth harmonic.*

***Solution:***

Here, as we can see, $\xi = f(x) = 0$ from

$x = 0$ to $x = \frac{a}{2}$ and $\xi = f(x) = h$ from $x = \frac{a}{2}$ to $x = a$.

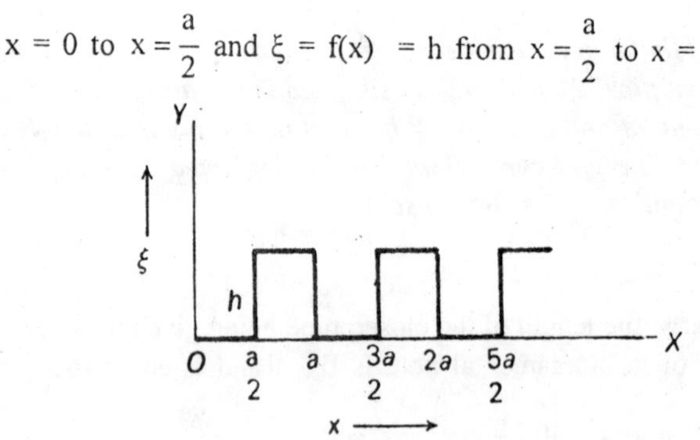


Fig. 3.24

Now, the *Fourier series* gives $x = A_0 + A_1 \sin \omega x + ... + A_r \sin r\omega x + B_1 \cos \omega x + ... + B_r \cos r\omega x$.

Since $A_0 = \frac{1}{a}\int_0^a \xi dx$, we have

$$A_0 = 0 + \frac{h}{a}\int_{a/2}^{a} h\,dx = \frac{1}{a}[x]_{a/2}^{a}$$

$$= \frac{1}{a}\left(ha - \frac{ha}{2}\right) = \frac{h}{2},$$

indicating that *the axis of the curve lies at distance h/2 from the axis of x.*

And, because $A_r = \frac{2}{a}\int_0^a \xi \sin r\omega x\, dx$, we have

$$A_r = 0 + \frac{2}{a}\int_{a/2}^{a} h \sin r\omega x\, dx = \frac{2h}{a}\left[-\frac{1}{r\omega}\cos r\omega x\right]_{a/2}^{a}$$

$$= \frac{2h}{a}\left[-\frac{a}{2\pi r} + \frac{a}{2\pi r}\cos r\pi\right] = \frac{h}{\pi r}(-1 + \cos r\omega) \qquad \left[\because \omega = \frac{2\pi}{a}\right]$$

*For even values of r,* therefore, $A_r = 0$ and *for odd values or r,* $A_r = -\frac{2h}{\pi r}$.

So that, $A_1 = -\frac{2h}{\pi}$, $A_2 = 0$,

$$A_3 = -\frac{2h}{3\pi},\ A_4 = 0$$

and $$A_5 = -\frac{2h}{5\pi}.$$

Again, since $B_r = \frac{2}{a}\int_0^a \xi \cos r\omega x\,dx$, we have

$$B_r = \frac{2}{a}\int_{a/2}^{a} h\cos r\omega x\,dx = \frac{2h}{a}\left[\frac{a}{2\pi r}\sin r\omega x\right]_{a/2}^{a}$$

$$= \frac{2h}{a}\left[\frac{a}{2\pi r}(\sin r\pi - \sin r\pi)\right] = 0.$$

And, therefore, $B_1$, $B_2$, $B_3$ *are all zero, i.e., the cosine terms in the Fourier series are all absent.*

The series, in this case, thus becomes

$$\xi = \frac{h}{2} - \frac{2h}{\pi}\left(\sin\omega x + \frac{1}{3}\sin 3\omega x + \frac{1}{5}\sin 5\omega x + \ldots\right)$$

Or, $$\xi = \frac{h}{2} - \frac{2h}{\pi}\left(\sin\frac{2\pi x}{a} + \frac{1}{3}\sin\frac{6\pi x}{a} + \frac{1}{5}\sin\frac{10\pi x}{a} + \ldots\right)$$

***Example 27:***

Obtain the Fourier series for the complex periodic function represented by Fig. 3.25, when y = f(t) = at t = 0 and t = T and y = f(t) = a at $\frac{T}{2}$.

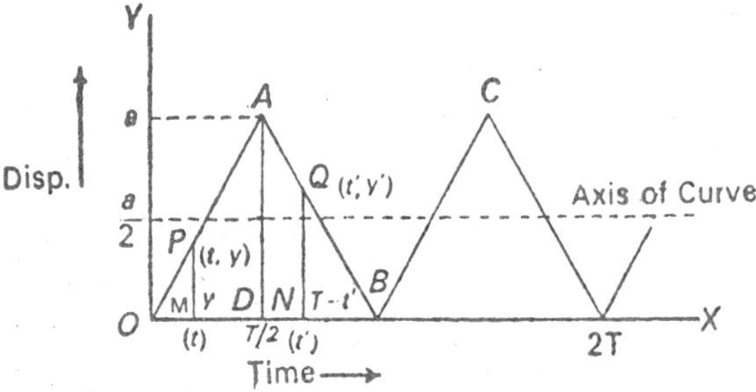


Fig. 3.25

***Solution:***

Let us take a point P on OA, with coordinates t and y and a point Q on AB, with coordinates t', y' and drop perpendiculars PM and QN from P and Q respectively on to the time-axis. Then, from similar right-angled triangles OMP and ODA, we have

$$\frac{PM}{AD} = \frac{OM}{OD}.$$

Or, $$\frac{y}{a} = \frac{1}{T/2},$$

whence, $$y = \frac{Zat}{T} \text{ from } 0 \text{ to } \frac{T}{2}$$

Similarly, from similar right-angled triangles BNQ and BDA, we have

$$\frac{QN}{AD} = \frac{NB}{DB}.$$

Or, $$\frac{y'}{a} = \frac{T-t'}{T/2},$$

whence $$y = 2a\left(1 - \frac{t'}{T}\right) \text{ from } \frac{T}{2} \text{ to } T.$$

Now, the coefficient $A_0$ of the Fourier series is given by $A_0 = \frac{1}{T}\int_0^T y\, dt$.

We, therefore, have $A_0 = \frac{1}{T}\left[\int_0^{T/2} \frac{2at}{T} dt + \int_{T/2}^{T} 2a\left(1 - \frac{t'}{T}\right) dt'\right]$

$$= \frac{a}{T}\left[\frac{2t^2}{2T}\right]_0^{T/2} + \frac{2a}{T}\left[t' - \frac{t'^2}{2T}\right]_{T/2}^{T} = \frac{a}{4} + a - \frac{3a}{4} = \frac{a}{2}.$$

*indicating that the axis of the curve lies at a distance a/2 from the time axis.*

The coefficient Ar of the series is given by

$$A_r = \frac{2}{T}\left[\int_0^{T/2} \frac{2at}{T} \sin r\omega t\, dt + \int_{T/2}^{T} 2a\left(1 - \frac{t'}{T}\right) \sin r\omega t'\, dt'\right]$$

Or, $$A_r = \frac{4a}{T^2}\int_0^{T/2} t \sin r\omega t\, dt + \frac{4a}{T}\int_{T/2}^{T} \sin r\omega t' - \frac{4a}{T^2}\int_{T/2}^{T} t' \sin r\omega t'\, dt'.$$

$$= \frac{4a}{T^2}\left[-t\frac{\cos r\omega t}{r\omega}\right]_0^{T/2} + \frac{4a}{T^2}\int_0^{T/2} -\frac{\cos r\,\omega t}{r\omega}dt$$

$$+\frac{4a}{T^2}\left[-\frac{\cos r\,\omega t'}{r\omega}\right]_{T/2}^{T} - \frac{4a}{T^2}\left[-t'\frac{\cos r\omega t'}{r\omega}\right]_{T/2}^{T}$$

$$= \frac{4a}{T^2}\left(-\frac{T}{2}\frac{\cos r\,\omega t'\,T/2}{r\omega}+0\right) + \frac{4a}{T^2}\left[-\frac{\sin r\omega\,T/2}{r^2\omega^2}\right]$$

$$+\frac{4a}{T}\left[-\frac{\cos r\,\omega t}{r\omega}+\frac{\cos r\omega\,T/2}{r\omega}\right] - \frac{4a}{T^2}\left(-\frac{T\cos r\,\omega T}{r\omega}\right)$$

$$+\frac{(T/2)\cos r\omega\,T/2}{r\omega} - \frac{4a}{T^2}\left[-\frac{\sin r\,\omega T}{r^2\omega^2}+\frac{\sin r\omega\,T/2}{r^2\omega^2}\right]$$

Since $\omega T = 2\pi$ and, therefore, $\frac{\omega T}{2} = \pi$, we have

$$A_r = \frac{4a}{Tr\omega}\left(-\frac{1}{2}\cos r\pi\right) + 0 + \frac{4a}{Tr\omega}(-1+\cos r\pi)$$

$$-\frac{4a}{Tr\omega}\left(-1+\frac{1}{2}\cos r\pi\right) + \frac{4a}{T^2}(0)$$

$$= \frac{4a}{Tr\omega}\left(-\frac{1}{2}\cos r\pi - 1 + \cos r\pi + 1 - \frac{1}{2}\cos r\pi\right) = 0.$$

Thus, $A_r$ *being zero, all the sine terms are absent from the series.*

And, coefficient

$$B_r = \frac{2}{T}\int_0^{T/2}\frac{2at}{T}\cos r\omega t\,dt + \frac{2}{T}\int_{T/2}^{T} 2a\left(1-\frac{t'}{T}\right)\cos r\omega t'\,dt'.$$

$$= \frac{4a}{T^2}\int_0^{T/2} t\cos r\omega t\,dt + \frac{4a}{T}\int_{T/2}^{T}\cos r\,\omega t'\,dt' - \frac{4a}{T^2}\int_{T/2}^{T} t'\cos r\,\omega t'\,dt'.$$

$$= \frac{4a}{T^2}\left[\frac{t\sin r\,\omega t}{r\omega}\right]_0^{T/2} - \frac{4a}{T^2}\int_0^{T/2}\frac{\sin r\,\omega t}{r\omega}dt + \frac{4a}{T}\left[\frac{\sin r\,\omega t'}{r\omega}\right]_{T/2}^{T}$$

$$-\frac{4a}{T^2}\left[\frac{t'\sin r\,\omega t'}{r\omega}\right]_{T/2}^{T} + \frac{4a}{T^2}\int_{T/2}^{T}\frac{\sin r\,\omega t'}{r\omega}dt'.$$

$$= \frac{4a}{T}\left(\frac{1}{2} - \frac{\sin r\pi}{r\omega}\right) - \frac{4a}{T^2}\left[\frac{-\cos r\omega\, T/2 + \cos 0}{r^2\omega^2}\right] + \frac{4a}{T}(0) - \frac{4a}{T^2}(0)$$

$$+ \frac{4a}{T^2}\left[\frac{-\cos r\omega T + \cos r\omega T/2}{r^2\omega^2}\right]$$

$$= 0 - \frac{4a}{T^2 r^2 \omega^2}(-\cos r\pi + 1) + 0 - 0 + \frac{4a}{T^2 r^2\omega^2}(-1 + \cos r\pi).$$

$$= \frac{4a}{T^2 r^2\omega^2}(\cos r\pi - 1) + \frac{4a}{T^2 r^2\omega^2}(\cos r\pi - 1)$$

$$= \frac{4a}{T^2 r^2\omega^2}\left[2(\cos r\pi - 1)\right] = \frac{8a}{T^2 r^2\omega^2}(\cos r\pi - 1).$$

Or, since $\omega = \frac{2\pi}{T}$, we have

$$B_r = \frac{2a}{r^2\pi^2}(\cos\pi - 1).$$

So that, *for even values of r*, $\cos r\pi = 1$ and, therefore, $B_r = 0$ and, *for odd values or r*, $\cos r\pi = -1$ and, therefore,

$$B_r = -\frac{4a}{r^2\pi^2}.$$

Thus, *the complete Fourier series for the given periodic function is*

$$y = \frac{a}{2} - \frac{4a}{\pi^2}\left(\cos\omega t + \frac{\cos 3\omega t}{3^2} + \frac{\cos 5\omega t}{5^2} + \ldots\right).$$

***Example 28:***

*The velocity of water waves, under the combined action of gravity and surface tension is given by the relation* $v = \sqrt{\lambda g/2\pi + 2\pi T/\lambda\rho}$, *where T stands for surface tension, and ρ for the density of water. Obtain an expression for the group velocity of the wave and show that the group velocity of surface waves or ripples (due in main to surface tension) is thrice the group velocity of gravity waves (due in main to gravity).*

***Solution:***

We know that the wave velocity of the water waves is given by $v = \omega/k$, where $k = 2\pi/\lambda$. We, therefore, have

$$v = \frac{\omega}{k} = \sqrt{\frac{1}{k}\left(g + \frac{k^2 T}{\rho}\right)},$$

whence, $$\omega = \sqrt{kg + k^3 \frac{T}{\rho}}.$$

∴ *group velocity of the wave*

$$u = \frac{d\omega}{dk} = \frac{1}{2} \cdot \frac{g + 3k^2 \frac{T}{\rho}}{\sqrt{kg + k^3 \frac{T}{\rho}}} = \frac{1}{2} \frac{g + 3k^2 \frac{T}{\rho}}{\omega}$$

Or, dividing the numerator and denominator by k, we have

$$u = \frac{g/2k + 3kT/2\rho}{\omega/k} = \frac{g/2k + 3kT/\rho}{v}$$

And ∴ $$\frac{u}{v} = \frac{g/2k + 3kT/2\rho}{v^2} = \frac{g/2k + 3kT/2\rho}{g/k + kT/\rho}$$

Or, *group velocity*, $$u = \frac{g/2k + 3kT/2\rho}{g/k + kT/\rho} v. \quad ...(1)$$

Now, if we consider the case of *ripples* in which the surface tension alone plays the dominant part, we obtain the group velocity of ripples by neglecting the terms involving g in expression I above. Thus,

*group velocity of ripples*, $$u_{(ripples)} = \frac{3kT/2\rho}{kT/\rho} v = \frac{3}{2} v,$$

i.e., *the group velocity of ripples is 3/2 times the wave velocity.*

If, on the other hand, we are interested in the waves in deep water or the *gravity waves*, in which gravity plays the major part, we obtain their group velocity by neglecting the terms involving surface tension in expression I above. Thus

*group velocity of gravity waves*, $$u_{(gravity\ wave)} = \frac{\frac{g}{2k}}{\frac{g}{k}} v = \frac{1}{2} v$$

*i.e., the group velocity of gravity wives (if the water be deep enough) is half the waves velocity.*

It will thus be easily seen that the *group velocity of ripples* $\left(\frac{3}{2}v\right)$ *is thrice the group velocity of gravity waves* $\left(\frac{1}{2}v\right)$.

***Example 29:***

*Show that the group velocity of de Broglie waves associated with a moving particle is the same as the velocity of the particle on both non-relativistic and relativistic considerations but their phase velocity is half the particle velocity in the former, and much higher than the velocity of light (c) in the latter, case.*

***Solution:***

*Ignoring first any relativistic considerations*, let m be the mass of a moving particle and v, its *velocity*, so that its *momentum* p = mv.

Then, the *de Broglie wavelength of the wave associated with the particle*, say, λ, is given by λ = h/p, where h is the *Planck's constant.*

The *frequency of the wave* $\nu = \omega/2\pi$ and the *energy associated with* it = hν.

We thus have energy $E = h\nu = \frac{h\omega}{2\pi}$. Also

$$E = \frac{1}{2}mv^2 = \frac{p^2}{2m} = \frac{h^2}{2m\lambda^2}, \left(\text{because } p = \frac{h}{\lambda}\right).$$

Or, putting $\frac{2\pi}{\lambda} = k$ and, therefore,

$$\lambda = \frac{2\pi}{k}, \text{ we have}$$

$$E = \frac{h^2k^2}{8\pi^2 m}.$$

Equating the two values of E, we have

$$\frac{h\omega}{2\pi} = \frac{h^2k^2}{8\pi^2 m}, \text{ whence}$$

$$\omega = \frac{hk^2}{4\pi m}.$$

Hence, *group velocity of de Broglie waves,*

$$u = \frac{d\omega}{dk} = \frac{2hk}{4\pi m} = \frac{hk}{2\pi m} = \frac{h}{\lambda m} \qquad \left[\because k = \frac{2\pi}{\lambda}\right]$$

Since $p = mv = \frac{h}{\lambda}$, we have *particle velocity* $v = \frac{h}{\lambda m}$.

Thus, *group velocity of de Broglie waves associated with the p particle = velocity of the particle itself.*

Now, *under relativistic considerations*, the mass m of the particle is its *relativistic mass* = $m_0/\sqrt{1-v^2/c^2}$, where $m_0$ is its *rest mass* and c, the *velocity of light in free space.*

We, therefore, have $E = h\nu = \frac{h\omega}{2\pi}$, whence,

$$\omega = \frac{2\pi E}{h}.$$

Since $E = mc^2$, we have $\omega = \frac{2\pi\, mc^2}{h} = \frac{2\pi\, m_0 c^2}{h}\sqrt{1-\frac{v^2}{c^2}}$. ...(i)

And, clearly, $k = \frac{2\pi}{\lambda} = \frac{2\pi}{h/p} = \frac{2\pi\, p}{h} = \frac{2\pi\, mv}{h} = \frac{2\pi\, m_0 v}{h}\sqrt{1-\frac{v^2}{c^2}}$ ...(ii)

From relation (i) we have $\frac{d\omega}{dv} = \frac{2\pi m_0 v}{h\left(1-\frac{v^2}{c^2}\right)^{3/9}}$ ...(iii)

and from relation (ii) we have $\frac{dk}{dv} = \frac{2\pi m_0}{h\left(1-\frac{v^2}{c^2}\right)^{3/2}}$. ...(iv)

From relations (iii) and (iv), therefore, we have *group velocity of de Broglie waves,*

$$u = \frac{d\omega}{dk} = \frac{\frac{d\omega}{dv}}{\frac{dk}{dv}} = \frac{\frac{2\pi m_0 v}{h\left(1-v^2/c^2\right)^{3/2}}}{\frac{2\pi m_0}{h\left(1-v^2/c^2\right)^{3/2}}} = v$$

Thus, *the group velocity of de Broglie waves here too is equal to the velocity of the particle itself.*

Let us now calculate the *phase velocity* in the two cases:

*Under non-relativistic considerations*, we have

$$\lambda = \frac{2\pi}{k} = \frac{h}{mv}, \text{ whence,}$$

$$k = \frac{2\pi}{h} mv.$$

And, $$E = \frac{1}{2} mv^2 = hv = \frac{h\omega}{2\pi},$$

whence, $$\omega = \frac{2\pi}{h} hv = \frac{2\pi}{h} \times \frac{1}{2} mv^2 = \frac{\pi mv^2}{h}.$$

∴ *phase velocity of de Brogile waves*

$$= \frac{\omega}{h} = \frac{\pi mv^2}{h} \times \frac{h}{2\pi mv} = \frac{1}{2} mv^2 / mv = \frac{1}{2} v,$$

*i.e., the phase velocity here is equal to half the particle velocity.*

*Under relativistic considerations*, we have ω given by expression (i) and k by expression (ii) above. And, therefore,

*phase velocity of de Broglie waves*

$$= \frac{\omega}{k} = \frac{2\pi m_0 c^2 / h\sqrt{1 - v^2/c^2}}{2\pi m_0 v / h\sqrt{1 - v^2/c^2}} = \frac{c^2}{v}.$$

Since $v << c^2$, *the phase velocity, in the case, in very much greater than c, the velocity of light in free space.*

**_Example 30:_**

*A train of simple harmonic waves is travelling in a gas along the positive direction of the x-axis, with an amplitude equal to 2 cm, velocity 300 metres/sec and frequency 400. Calculate-the displacement, particle velocity and particle Wave Motion at a distance of 4 cm from the origin after an interval off seconds.*

**_Solution:_**

(i) Displacement. As we know, in a simple harmonic wave, the displacement of a particle at a distance x from the origin, after time t, is given by $y = a\sin\frac{2\pi}{\lambda}$ (vt – x), where a is the *amplitude* of the wave, λ, its *wavelength* and v, its *velocity* in the given medium.

Here, a = 2 cm,

v = 300 m/sec = 30000 cm/sec,

*frequency* n = 400 and, therefore, wave length

λ = v/n = 30000/400 = 75 cm,

x = 4 cm

and t = 5 sec. So that,

$$y = 2\sin\frac{2\pi}{75}(3000\times5-4) = 2\sin\left(\frac{2\pi}{75}\times149996\right)$$

$= 2 \sin (2\pi \times 1999.9)$

$= 2 \sin (1999 \times 2\pi + 0.9 \times 2\pi)$

$= 2 \sin (1.8\,\pi)$

$= 2 \sin (\pi + 0.8\,\pi)$

$= -2 \sin (0.8\,\pi) = -2 \sin (0.8\pi \times 180/\pi)^{\circ}$

$= -2 \sin 144^{\circ}$

$= -2 \sin (180 - 144)^{\circ} = -2 \sin 36^{\circ}$

$= -2 \times 0.5878$

$= -1.1756$ cm.

Thus, the *displacement of the particle at a distance of 4 cm from the origin, after an internal of 5 seconds is* – 1.1756 cm.

(ii) **Particle velocity:** We know that *particle velocity* U

$$= \frac{dy}{dt} = \frac{2\pi v}{\lambda} a \cos\frac{2\pi}{\lambda}(vt - x).$$

Now, as we have seen under (i) above,

$$\sin\frac{2\pi}{\lambda}(vt - x) = \sin 36^{\circ},$$

so that, $$\frac{2\pi}{\lambda}(vt - x) = 36^{\circ}$$

and, therefore, $\cos^2\frac{\pi}{\lambda}(vt - x) = \cos 36^{\circ} = 0.8090.$

$\therefore$ *particle velocity* $U = \frac{2\pi v}{\lambda} a \cos 36° = \frac{2\pi \times 3000}{75} \times 2 \times 0.8090$

$= 4068$ cm/sec $= 40.68$ m/sec.

(iii) **Particle Wave Motion:** Clearly, *particle Wave Motion*

$$= \frac{dv}{dt} = \frac{d^2y}{dt^2}$$

$$= -\frac{4\pi^2 v^2}{\lambda^2} a \sin\frac{2\pi}{\lambda}(vt - x) = -\left(\frac{4\pi^2 v^2}{\lambda^2}\right)y.$$

Since $y = -1.1756$ cm, we have *particle Wave Motion*

$$= -\frac{4\pi^2 (30000)^2}{75^2}(-1.1756) = 7.429 \times 10^6 \text{ cm/sec}^2.$$

***Example 31:***

*A wave of frequency 500 cycles/sec has a phase velocity of 360 metres/sec. (a) How far apart are two points 60° out of phase? (b) What is the phase difference between two displacements at a certain point at times* $10^{-3}$ *sec apart?*

***Solution:***

(a) We have the equation $y = a \sin\frac{2\pi}{\lambda}(vt - x)$ representing a simple harmonic wave, where a is the *amplitude* and $\frac{2\pi}{\lambda}(vt - x)$, the *phase angle* of a point distant x from the origin at time t.

$\therefore$ phase angle of a point distant $x_1$ from the origin at time $t = \frac{2\pi}{\lambda}(vt - x_1)$ and phase angle of a point distant $x_2$ from the origin at time $t = \frac{2\pi}{\lambda}(vt - x_2)$.

Hence *phase difference between the two points*

$$= \frac{2\pi}{\lambda}(vt - x_1) - \frac{2\pi}{\lambda}(vt - x_2)$$

$$\frac{2\pi}{\lambda}(x_2 - x_1) = \frac{2\pi v}{\lambda}\frac{(x_2 - x_1)}{v} = 3\pi l\left(\frac{x_2 - x_1}{v}\right),$$

because, as we know, $v = n\lambda$ and, therefore $\frac{v}{\lambda} = \lambda$.

Here, the phase difference between the two points is given to be 60°

$$= \frac{60 \times \pi}{180} = \frac{\pi}{3} \text{ radian.}$$

We, therefore, have $2\pi n \frac{(x_2 - x_1)}{v} = \frac{\pi}{3}$.

Or, $2n \frac{(x_2 - x_1)}{v} = \frac{1}{3}$.

Since n = 500 and v = 360 m/sec = 36000 cm/sec, we have

$$\frac{2 \times 500(x_2 - x_1)}{36000} = \frac{1}{3}.$$

Or, $(x_2 - x_1) = \frac{36}{3} = 12.0$ cm.

Thus, *the two points 60° out of phase are 12.0 cm apart.*

(b) Again, phase angle at a point distant x from the origin at time $t_1 = \frac{2\pi}{\lambda}(vt_1 - x)$ and phase angle at the same point at time $t_2 = \frac{2\pi}{\lambda}(vt_2 - x)$

∴ phase differencc at thc point at times $(t_2 - t_1)$ sec apart

$$= \frac{2\pi}{\lambda}(vt_2 - x)\frac{2\pi}{\lambda}(vt_1 - x) = \frac{2\pi v}{\lambda}(t_2 - t_1) = 2\pi n(t_2 - t_1).$$

Here, $(t_2 - t_1) = 10^{-3}$ sec and, therefore,

*phase difference between displacement M the point at times* $10^{-3}$ sec *apart*

$$= 2\pi \times 500 \times 1/1000 = \pi \text{ radian} = 180°.$$

***Example 32:***

*(a) Obtain the value of (i) frequency, (ii) time-period, (iii) wave number (in the sense of waves per cm), (iv) propagation constant and (v) angular frequency for light waves of wavelength 5000 A.U., travelling in free space.*

*(b) The propagation constant of a wave is 120/cm and its velocity 360 in/sec. Obtain the values of*

*(i) wave number (i.e., number of waves per cm),*

*(ii) wavelength, and*

*(iii) frequency of the wave.*

***Solution:***

(a) The *velocity* of light in free space is $c = v\lambda = 3 \times 10^{10}$ cm/sec, where v is its *frequency* and $\lambda$, its *wavelength.*

(i) Here, since $\lambda = 5000$ A.U. $= 5000 \times 10^{-8}$ cm, we have *frequency of the light wave,* $v = \frac{c}{\lambda} = 3 \times 10^{10}/5000 \times 10^{-8} = 6 \times 10^{14}$ cycles/sec.

(ii) Hence, *time-period* $T = \frac{1}{v} = \frac{1}{6} \times 10^{14} = 1.7 \times 10^{-15}$ sec.

(iii) *Wave number* $v = \frac{1}{\lambda} = \frac{1}{5000} \times 10^{-8} = 2 \times 10^{4}$/cm.

(iv) *Propagation constant* of the wave (k)

$$= \frac{2\pi}{\lambda} = 2\pi \times 2 \times 10^{4}$$

$$= 4\pi \times 10^{4} = 1.257 \times 10^{5}\text{/cm. and}$$

(v) *angular frequency* $\omega = 2\pi v = 2\pi \times 6 \times 10^{14}$

$$= 12\pi \times 10^{14} \text{ rad/sec}$$

$$= 3.77 \times 10^{15} \text{ rad/sec.}$$

(b) We know that *propagation constant* $k = 2\pi/l$ and *wave number* $v = \frac{1}{\lambda}$.

(i) Here, $\frac{2\pi}{\lambda} = 120$ and, therefore, *wave number*

$$v = \frac{1}{\lambda} = \frac{120}{2\pi} = 19.1 \text{ per cm.}$$

(ii) $\therefore$ *wavelength* $\lambda = \frac{1}{v} = \frac{1}{19.1} = 0.05238$ cm.

(iii) Since $v = n\lambda$, we have *frequency* $n = \frac{v}{\lambda} = v \times \left(\frac{1}{\lambda}\right) = v$

Here, $v = 360$ m/sec $= 36000$ cm/sec and, therefore,

*frequency n of the wave* $= 36000 \times 19.1 = 6.876 \times 10^{5}$ c.p.s.

***Example 33:***

*Show that (i)* $y = x^2 + v^2t^2$, *(ii)* $y = (x + vt)^2$, *(iii)* $y = (x - vt)^2$ *and (iv)* $y = 2 \sin x \cos vt$ *are each a solution of the one-dimensional wave equation but not (v)* $y = x^2 - v^2t^2$ *and (vi)* $y = \sin 2x \cos wt$.

***Solution:***

(i) Differentiating expression (i) with respect to t, we have dy/dt = 2 $v^2t$ and, therefore, $d^2y/dt^2 = 2v^2$.

And, differentiating expression (i) with respect to x, we have

$$dy/dx = 2x$$

and, therefore, $d^2y/dt^2 = 2v^2$.

Clearly, $2v^2 = (v^2)2$, *i.e.*, $\frac{d^2y}{dt^2} = v^2\left(\frac{d^2y}{dx^2}\right)$

*Expression (i) is therefore a solution of the one-dimensional wave equation,*

$$\frac{d^2y}{dt^2} = v^2\left(\frac{d^2y}{dx^2}\right).$$

(ii) and (iii). Differentiating each of the expressions (ii) and (iii) *twice* with respect to t, we have

$$\frac{d^2y}{dt^2} = 2v^2$$

And differentiating the two expressions with respect to x, we have

$\frac{d^2y}{dt^2} = 2$. Again, therefore.

$2v^2 = (v^2)2$.

Or, $$\frac{d^2y}{dt^2} = v^2\left(\frac{d^2y}{dx^2}\right).$$

*Both expression (ii) and (iii) are therefore, a solution of the one-dimensional wave equation.*

(iv) Differentiating expression (iv) twice with respect to t, we have

$$\frac{d^2y}{dt^2} - v^2y$$

and differentiating it twice with respect to x, we have

$$\frac{d^2y}{dx^2} = -y.$$

Now, $-v^2y = v^2(-y).$

Or, $$\frac{d^2y}{dt^2} = v^2\left(\frac{d^2y}{dx^2}\right)$$

So that *expression (iv) too is a solution of the one-dimensional wave equation*

(v) Differentiating expression (v) twice with respect to t, we have

$$\frac{d^2y}{dt^2} = -v^2,$$

and differentiating it twice with respect to x, we have

$$\frac{d^2y}{dx^2} = 2.$$

Now $-2v^2 \neq v^2(2)$ and, therefore, here,

$$\frac{d^2y}{dt^2} \neq v^2\left(\frac{d^2y}{dx^2}\right)$$

*Expression (v) is not thus a solution of the one-dimensional wave equation,*

(vi) Differentiating expression (vi) with respect to t, we have

$$\frac{d^2y}{dt^2} = -v^2y$$

and differentiating it twice with respect to x, we have

$$\frac{d^2y}{dx^2} = -2y.$$

Again, clearly $-v^2y \neq v^2(-2y)$ and, therefore,

$$\frac{d^2y}{dt^2} \neq v^2\left(\frac{d^2y}{dx^2}\right).$$

*Expression (vi) too, therefore, is not a solution of the one-dimensional wave equation.*

***Example 34:***

*(a) A continuous sinusoidal longitudinal wave is sent along a coil spring from a vibrating source attached to it. The frequency of the source is 25 vibrations per second and the distance between successive*

*rarefactions in the spring is 24 cm. Find the wave speed. If the maximum longitudinal displacement of a particle in the spring is 3.0 cm and the wave moves in the—x direction, write down the equation of the wave. Let the Source be at x = 0 and the displacement at x = 0 and t = 0 be zero.*

*(b) Write down the equation of a harmonic wave travelling along the +x direction is terms of a = 5 cm, ω = 100π radian/sec and v = 250 cm/sec with y = 0 at x = 0 and t = 0, and obtain the values of λ, v and k.*

**Solution:**

(a) We know that in a sinusoidal (or harmonic) longitudinal wave, the wavelength is equal to the distance between two successive condensations or rarefactions. So that, here, λ = 24 cm and frequency n, the same as that of the source of vibrations, *i.e.*, 25 cps.

∴ *velocity (or speed) of the wave,* v = nλ = 25 × 24 = 600 cm/sec.

Now, the *maximum longitudinal displacement of a particle or amplitude of the* wave is 0 = 3.0 cm.

Then, since displacement y = 0 at x = 0 and t = 0 and the wave is travelling along the – ve direction of the x-axis, its equation is

$$y = a\sin\frac{2\pi}{\lambda}(vt + x)$$

*i.e.*, $$y = 3\sin\frac{2\pi}{24}(600t - x),$$

or, $$y = 3\sin\frac{\pi}{12}(600t + x).$$

(b) For a harmonic wave travelling along the +x direction, with y = 0 at x = 0 and t = 0, we have the equation

$$y = a\sin\frac{2\pi}{\lambda}(vt - x).$$

We may put it as $$y = a\sin\frac{2\pi v}{\lambda}\left(t - \frac{x}{v}\right)$$

Or, since $\frac{v}{\lambda} = n\ 2\pi n = \omega$, it takes the form $y = a\sin\omega\left(t - \frac{x}{v}\right)$

Here, a = 5 cm, ω = 100π rad/sec and v = 250 m/sec. So that , the equation becomes

$$y = 5 \sin 100\pi \left(t - \frac{x}{250}\right).$$

Now, $\omega = 2\pi n = 100\pi$, and, therefore, $n = \dfrac{100\pi}{2\pi} = 50$ cps.

And, since $v = n\lambda = 250$ cm/sec, we have $\lambda = \dfrac{v}{n} = \dfrac{250}{50} = 5.0$ cm.

Hence, $\quad v = \dfrac{1}{\lambda} = \dfrac{1}{5} = 0.2/\text{cm}$

and $\quad k = \dfrac{2\pi}{\lambda} = 2\pi \times 0.2 = 0.4\pi = 1.257/\text{cm}.$

***Example 35:***

*Plane harmonic waves of frequency 500 sec*$^{-1}$ *are produced in air with displacement amplitude* $1.00 \times 10^{-3}$ *cm. Deduce (i) the pressure amplitude, (ii) the energy density, (iii) energy flux in the wave. (Density of air = 1.29 gm/litre, speed of sound in air = 300 m/sec.)*

***Solution:***

We know that in the case of a plane harmonic wave,

$$y = a \sin \frac{2\pi}{\lambda}(vt - x)$$

and
$$p = -\frac{K dy}{dx} = K\frac{2\pi}{\lambda} a \cos \frac{2\pi}{\lambda}(vt - x).$$

Obviously, p will have its *maximum* value $K\dfrac{2\pi}{\lambda}a$ when $\cos\dfrac{2\pi}{\lambda}(vt - x) = 1.$

So that, the *maximum value of p or the pressure amplitude* $= K\dfrac{2\pi}{\lambda}a.$

Or, since $v = \sqrt{\dfrac{K}{\rho}}$ and, therefore, $K = \rho v^2$, we have *pressure amplitude* $= \rho v^2 \dfrac{2\pi}{\lambda} a.$

Again, because $v = n\lambda$, we have $\dfrac{1}{\lambda} = \dfrac{n}{v}$ and, therefore,

*pressure amplitude* $= 2\pi\ \rho v^2 a\ \dfrac{n}{v} = 2\pi\rho v a n.$

Here, $\rho$ = .00129 gm/c.c.,

v = 34000 cm/sec,

a = $10^{-3}$ cm and n = 500.

$\therefore$ *pressure amplitude* = $2\pi \times .00129 \times 34000 \times 10^{-3} \times 500$

= 135.8 dynes/cm$^2$

(ii) *Energy density* E = $2\pi^2 n^2 a^2 \rho$.

So that, $E = 2\pi^2 \times (500)^2 \times (10^{-3})^2 \times .000129$

= $6.365 \times 10^{-3}$ erg/c.c.

(iii) *Energy flux or energy current*

$I = Ev = 6.365 \times 10^{-3} \times 3400$

= $2.164 \times 10^2$ erg sec$^{-1}$ cm$^{-2}$

$$= \frac{2.164 \times 10^2}{4.2 \times 10^7}$$

= $5.153 \times 10^{-4}$ watt cm$^{-2}$.

[$\because$ $4.2 \times 10^7$ crg/scc = 1 watt.]

**Example 36:**

*(a) Spherical waves are emitted from a 1.0 watt source in an isotropic non-absorbing medium. What is the ware intensity 1.0 metre from the source.*

*(b) A line source emits a cylindrical expanding wave. Assuming the medium absorbs no energy, find how the intensity and amplitude of the ware depend on the distance from the source.*

**Solution:**

(a) We know that the *intensity (I) of a wave is the quantity of incident energy per area of the wave front per unit time.*

Here, the *energy incident on the wave front per second* = 1 joule.

And, the *wave front being spherical, its surface area at distance r from the* sources = $4\pi r^2 = 4\pi(1)^2 = 4\pi$ sq.m.

$\therefore$ *wave intensity at a distance 1 metre from the source, i.e.,* $I = 1/4\pi$ = *0.0796 joules per metre$^2$ per sec = 0.0796 watt/metre$^2$.*

(b) Here, *the wave front being cylindrical, its surface area at a distance r from the line source* = $2\pi rl$, where $l$ is *the length of the cylindrical wave front or the line source.*

If, therefore, E be the *energy emitted by the source per second*, we have *intensity of the wave at a distance r from the source, i.e.,*

$$I = \frac{E}{2\pi rl}.$$

Or, $$I \propto \frac{1}{r}$$

or, $$I \propto r^{-1},$$

*i.e., the intensity of the wave is inversely proportional to the distance from the source.*

Now, as we know, the intensity of a wave is directly proportional to the square of its amplitude, *i.e.*, $I \propto a^2$. We therefore, have

$$a^2 \propto I \propto r^{-1}. \text{ So that, } a \propto r^{-1/2}$$

*i.e., the amplitude of the wave is inversely proportional to the square root of the distance from the source.*

***Example 37:***

Explain analytically the forı ıation of beats when two sound wares of nearly the same frequencies travel along the same path and in the same direction through air and show that the number of beats produced per second is equal to the difference between the frequencies of the two wares.

***Solution:***

Let two sound waves, or plane harmonic waves, of *amplitudes* $a_1$ and $a_2$ (where $a_1 > a_2$) and of nearly equal frequencies $n_1$ and $n_2$ (where $n_1 > n_2$) be travelling along the *same path* and in the *same direction* through air.

Considering any one point in the medium (air) through which the waves are passing, the displacements there due to the two waves which, as we know, very *cyclically with time*, are given by $y_1 = a_1 \sin 2\pi\, n_1 t$ and $y_2 = a_2 \sin 2\pi n_2 t$.

In accordance with the principle of superposition, therefore, the resultant displacement at the point will be

$$y = y_1 + y_2 = a_1 \sin 2\pi n_1 t + a_2 \sin 2\pi n_2 t.$$

Or, $$y = a_1 \sin 2\pi n_1 t + a_2 \sin 2\pi [n_1 - (n_1 - n_2)]t$$

$$= a_1 \sin 2\pi n_1 t + a_2 \sin 2\pi n_1 t \cos 2\pi (n_1 - n_2)t - a_2 \cos 2\pi n_1 t \sin 2\pi (n_1 - n_2)t.$$

$$= \sin 2\pi n_1 t [a_1 + a_2 \cos 2\pi (n_1 - n_2)t] - \cos 2\pi n_1 t [a_2 \sin 2\pi (n_1 - n_2)t]$$

Or, putting $a_1 + a_2 \cos 2\pi (n_1 - n_2)t = a \cos \theta$ and $a_2 \sin 2\pi (n_1 - n_2)t = a \sin \theta$ we have

$$y = a \sin 2\pi n_1 t \cos \theta - a \cos 2\pi n_1 t \sin \theta.$$

Or, $$y = a \sin (2\pi n_1 t - \theta).$$

*showing the displacement at the point due to the resultant wave too varies cyclically with time, with an amplitude a but lags an angle θ behind the displacement due to the first wave, where*

$$a = \sqrt{(a \cos\theta)^2 + (a \sin\theta)^2}$$

$$= \sqrt{\left(a_1^2 + a_2^2 \cos^2 2\pi(n_1 - n_2)t + 2a_1 a_2 \cos 2\pi(n_1 - n_2)t + a_2^2 \sin^2 2\pi(n_1 - n_2)t\right)}$$

Or, $$a = \sqrt{a_1^2 + a_2^2 + 2a_1 a_2 \cos 2\pi(n_1 - n_2)t}$$

and $$\tan \theta = \frac{a \sin 2\pi(n_1 - n_2)t}{a_1 + a_2 \cos 2\pi(n_1 - n_2)t},$$

indicating that *the values of both a and θ keep on changing with time.*

Clearly, the amplitude at the point in question will have its maximum value $(a_1 + a_2)$ when $\cos 2\pi (n_1 - n_2 t) = + 1$,

*i.e.*, when $2\pi (n_1 - n_2)t = 2rp$, where is an integer, 0, 1, 2, 3 etc. *i.e.*, at

$$t = \frac{r}{n_1 - n_2} = 0, \frac{1}{(n_1 - n_2)}, \frac{2}{(n_1 - n_2)}, \frac{3}{(n_1 - n_2)}$$

*The time-interval between two successive maxima is thus*

$$\frac{1}{(n_1 - n_2)}.$$

And, the amplitude will have its *minimum* value $(a_1 - a_2)$ when $\cos 2\pi(n_1 - n_2)t = - 1$, or $2\pi(n_1 - n_2)t = (2r + 1)\pi$, where, again, r = 0, 1, 2, 3, etc. *i.e.*, at

$$t = \frac{2r+1}{2(n_1 - n_2)} = \frac{r}{(n_1 - n_2)} + \frac{1}{2(n_1 - n_2)} = \frac{1}{2(n_1 - n_2)},$$

$$\frac{3}{2(n_1 - n_2)}, \frac{5}{2(n_1 - n_2)} \text{ etc.,}$$

the *time-interval between two successive minima again being* $\frac{1}{(n_1 - n_2)}$, *although they occur in between the maxima.*

Thus, maxima and minima of amplitude alternate at the point in question as the two waves travel onwards, producing alternate loudness and faintness of sound there. T*hese alternations of loudness and faintness constitute beats, one loudness and one succeeding faintness constituting one beat.*

*And, clearly, the frequency of beats, i.e., the number of beats produced per second =* $(n_1 - n_2)$ *= the difference between the frequencies of the two sound waves.*

## EXERCISES

1. Discuss the changes (i) with respect to position, and (ii) with respect to time in the case of standing waves in air.

2. Discuss the solution of the differential wave equation $\frac{d^2y}{dx^2} = \left(\frac{1}{c^2}\right)\frac{d^2y}{dt^2}$ for a bounded system with boundary conditions y = 0 at x = 0 and x = L.

   N.B. Here c represents the wave velocity.

   **Ans.** $y = a \sin\left(\frac{r\pi x}{L}\right)\cos\left(\frac{r\pi vt}{L}\right)$

3. Show that there 'is no transference of energy across any section in the case of a wave in a linear bounded medium.

4. A sound wave in air is represented by the equation 5 sin 0.3142 (500 t – x), where t is in sec and x in cm. Write the equation of the wave which, on superposition with it, would produce a standing wave.

   **Ans.** – 5 sin 0.3142 (500t + x) or (ii) 5 sin 0.3142 (500t – x).

5. A progressive wave train in air, of amplitude 0.005 cm, frequency, 1000 cps and wavelength, 80 cm, and travelling along the

positive direction of x, is reflected normally at a rigid boundary. Obtain the values of (i) displacement amplitude, (ii) particle velocity amplitude, and (iii) amplitude of pressure variation at x = 50 cm. **Ans.** (i) 0.01cm (ii) 62.84 cm/sec, (iii) 0

6. Show that in the case of a sonometer (i.e. the medium with both boundaries rigid), the frequencies of the notes emitted form a harmonic series.

7. One end of a string, 100 cm long and of mass 5 gm, is attached to a prong or an electrically driven tuning fork, with weights suspended from the other so that it is quite horizontal and under tension. The fork vibrates in a direction perpendicular to the length of the string which is thrown into vibration in 4 loops emitting a note of frequency 60. Calculate the tension in the string.

   What would happen in the fork be turned so as to vibrate parallel to the length of the string.

   **Ans.** $4.5 \times 10^5$ *dynes; frequency of the note emitted* = 30

8. A sonometer wire is stretched by means of a piece of metal hanging from its other end. With the effective length of the wire (i.e., the distance between the two bridges) equal 100 cm, it emits its fundamental note, in unison with that of a given tuning fork. When the suspended piece of metal is immersed in water, the length of the wire has to be shortened by 7 cm in order that the fundamental note emitted by the wire may again be in unison with that of the fork. Obtain the value of the density of the metal piece. **Ans.** 7.4 gm/cc.

   [**Hint:** If m be the mass of the metal piece, tension T in the wire, in the first case (i.e., when the metal piece is in air) = m gm wt = mg dynes.

   When immersed in water, the weight of the metal piece, and therefore the tension in the wire = (m – m/ρ) = m(1 – 1/ρ) gm wt = m (1 – 1/ρ)g dynes, where ρ is the density of the metal piece.

   If n be the frequency of the tuning fork and hence that of the fundamental note emitted by the wire in either case, we have

$$n = \frac{1}{2l_1}\sqrt{\frac{mg}{\sigma}} = \frac{1}{2l_2}\sqrt{\frac{m(1-1/\rho)g}{\sigma}},$$

where $l_1$ and $l_2$ are the lengths of the wire in the two cases and, obviously, therefore, equal to 100 cm and 100 – 7 = 93 cm respectively.]

9. Show that when a given rod is clumped (i) at one end or (ii) in the middle, only odd harmonics can be excited but the frequencies of the various notes in case (i) are half those of the corresponding notes in case (ii).

10. Show that whether a rod be free at both end or clamped at both ends, both odd and even harmonics can be excited and that the frequencies of the corresponding notes are the same in either case.

11. A metal rod, 200 cm long, is clamped at one end and emits a fundamental note of frequency 800 when set into longitudinal vibration. If the value of Young's modulus for the material of the rod be $8 \times 10^{11}$ dynes/cm$^2$, obtain the value of its density.

    What would be the frequency of the fundamental note emitted by the rod if it were clamped (i) in the middle, (ii) at both ends or (iii) nowhere?

    **Ans.** 7.8 gm/cc; 1600 *cps in all three cases.*

12. (a) Show that the fundamental note given by an open pipe is the octave of the fundamental note given by the pipe if it were closed at one end.

    (b) Also show that the frequencies of the second harmonics of a closed and open pipe of the same length are in the ratio 3 : 4.

13. What is meant by the end correction of a pips? How may it be obtained for

    (i) a closed, and

    (ii) an open pipe?

14. (a) Tuning forks are usually mounted or their resonance boxes which are just rectangular boxes who one side open, so that the air column inside the box has the same fundamental frequency as that of the fork and is thus set into resonance with it. If the frequency of the fork be 320 and the velocity of sound in air, 1120 ft/sec, what should be the length of the box? (Ignore end correction).

(b) A tuning fork of frequency 256 is held above a column of air in a closed tube. The positions of the first and the second resonance are 33.0 cm and 100.5 cm respectively. Calculate the velocity of sound in air. **Ans.** (a) 10.5 in; (b) 345.6 m/sec.

15. State *Fourier's theorem* and mention the conditions under which it holds good. Also evaluate the various Fourier coefficients.

16. Obtain the Fourier series for a complex periodic function given by y = a from t = 0 to t = T/2 and y = 0 from T/2 to T.

**Ans.** $y = \frac{a}{2} + \frac{2a}{\pi}\left(\sin \omega t + \frac{1}{3}\sin 3\omega t + \frac{1}{5}\sin 5\omega t + \ldots\right).$

17. State the conditions tinder which Fourier's theorem may be applicable. Apply it to deduce the harmonic components of a saw-tooth curve of frequency $\frac{n}{2\pi}$ and form given by $y = -\frac{a}{2} + a\frac{n}{2\pi}t$ for $t = 0$ to $\frac{2\pi}{n}$.

**Ans.** $y = -\frac{a}{\pi}\left(\sin nt + \frac{1}{2}\sin 2nt + \frac{1}{2}\sin 3nt + \ldots\right).$

18. A quantity y has a value a from x = 0 to x = π and –a from x = π to x = 2π. Express y as a Fourier expansion in x.

**Ans.** $y = \frac{4a}{\pi}[\sin x + (\sin 3x)/3 + (\sin 5x)/5 + \ldots]$

19. In a displacement curve $y = \frac{4at}{T}$ from t = 0 to $T = \frac{T}{4}$ $y = -2a - 4\,at/T$ from T/4 to 3T/4 and y = 4 at/T – 4a from 3T/4 to T. Find the amplitude of the fundamental and the first two harmonics.

**Ans.** $\frac{8a}{\pi^2}, \frac{8a}{9\pi^2}, \frac{8a}{25\pi^2}.$

20. A function is displaced by half the range and the new function thus obtained is added to the first. Show that the result gives twice the sum of the independent term and the even harmonics. And, if the new function obtained is subtracted from the first, the result is twice the sum of the odd harmonics.

21. A plane progressive wave travelling along the +x direction has the following characteristics: a = 0.2 cm, v = 360 cm/sec and λ = 60 cm.

(a) Write down the equation for it (i) when displacement is zero at x = 0 and t = 0 and (ii) when displacement is maximum at x = 0 and t = 0.

(b) Obtain the displacement in either case at x = 120 cm and t = 2 sec.

**Ans.** (a) (i) $y = 0.2 \sin 2\pi (6t - x/60)$,

(ii) $y = 0.2 \cos 2\pi (6t - x/60)$

(b) (i) y = 0, (ii) y = 0.2 cm.

22. A plane progressive wave train of frequency 400 cps has a phase velocity of 480 m/sec. (a) How far apart are two points 30° out of phase? (b) What is the phase difference between two displacements at a given point at times $10^{-3}$ sec apart?

**Ans.** (a) 0.1 m, (b) 144°.

23. Which of the following are solutions of the one-dimensional wave equation? (i) $y = x^2 - v^2t^2$, (ii) $y = 7x - 10t$, (iii) $y = 2 \sin x \cos vt$ and (iv) $y = \sin 2x \cos vt$.

**Ans.** (a) 0.1 m, (b) 144°.

24. (a) Obtain the phase difference between two progressive waves represented by the following equations: (i) $y_1 = a \cos (\omega t - kx)$ and $y_2 = a \sin (\omega t - kx)$, (ii) $y_1 = f(Wt - kx)$ and $y_2 = F(Wt - kx - \phi)$.

(b) *Find the frequency, angular frequency, wave number and the propagation constant for light waves of wavelength* $6 \times 10^{-5}$ *cm.*

**Ans.** (a) (i) $\pi/2$ [∵ *phase difference* = $\cos (\omega t - kx) - \sin (\omega t - kx) = (\cos \omega t - kx) - \cos (\omega t - kx - \pi/2) = \pi/2$)

(ii) *The two being different types of functions of* $(t - x/v)$, *their phase difference cannot be determined.*

(b) $5 \times 10^{14}$ cps; $3.142 \times 10^{15}$ rad/sec;

$1.7 \times 10^{1}$ cm$^{-1}$; $1.05 \times 10^{5}$.

25. A plane progressive harmonic wave is travelling with a velocity of 340 m/sec in a fluid medium of density 0.0015 gm/c.c. If the amplitude of the wave be $10^{-4}$ cm and its frequency 300 cps, obtain the values of (i) *energy density* and (ii) *energy current* for it.

26. Determine (i) the velocity of sound in a gas in which two waves of lengths 50 cm and 50.5cm produce 6 beats per second and (ii) the velocity of sound in water in which waves of lengths 500 cm and 512 cm produce 6 beats per second.

**Ans.** (i) 303 m/sec; (ii) 1280 m/sec.

[**Hint:** Number of beats per second = $n_1 - n_2$, where $n_1 = v/\lambda_1$ and $n_2 = v/\lambda_2$, v being the velocity of sound in air in case (i) and in water in case (ii). So that, *number of beats per sec*

$$= (n_1 - n_2) = \frac{v}{\lambda_1} - \frac{v}{\lambda_2} = v\left(\frac{1}{\lambda_1} - \frac{1}{\lambda_2}\right),$$

whence, $(n_1 - a_2)$, $\lambda_1$ and $\lambda_2$ being known, v can be evaluated.]

27. Sound waves from a vibrating body reach a point by two paths. When the paths differ by 12 cm or by 36 cm, there is silence at the point. Calculate the frequency of the vibrating body if the velocity of sound in air is 330 m/sec. **Ans.** 1375 cps.

[**Hint:** It is clearly a case of *destructive interference*. For, when the path difference is 12 cm, the two waves arrive at the point in opposite phases and thus annul each other, producing silence. This path difference must be equal to an odd number of half wavelengths, *i.e.*, equal to $(2n + 1)\ \lambda/2$ or equal to $\lambda/2$, $3\lambda/2$ $5\lambda/2$ etc., the successive values of path difference differing from each other by $\lambda$. So that, again, when the path difference is 36 cm and there is silence at the point, it is clear that the difference between these two successive values of path difference must be equal to $\lambda$, the wavelength of each wave, i.e., $\lambda = (36 - 12) = 24$ cm. If therefore, n be the frequency of the vibrating body, or of the waves produced by it, we have $v = n\lambda$, whence, $n = v/\lambda = 33000/24 = 1375$.]

28. A source S and a detector D of high frequency waves area distanced apart on the ground (Fig. 3.26). The direct wave from S is found to be in phase at D with the wave from S that is reflected from a horizontal layer at an altitude H. The incident and reflected rays make the same angle with the reflecting layer. When the layer rises a distance h, no signal is detected at D. Neglect absorption in the atmosphere and find the relation between d, h, H and the wavelength $\lambda$ of the waves.

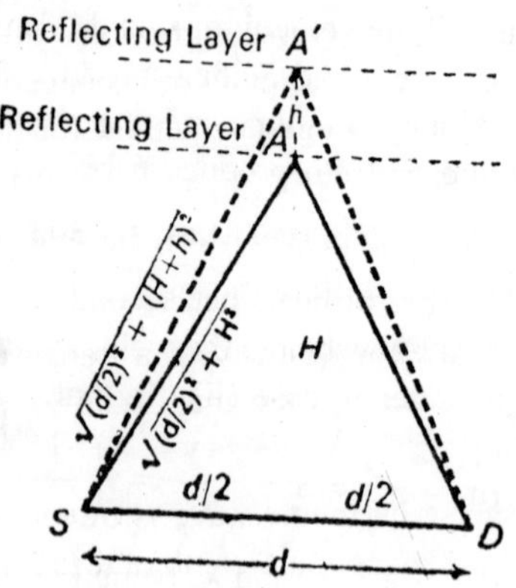


Fig. 3.26

**Ans.** $\lambda = 2\left[\sqrt{d^2 + 4(H+h)^2} - \sqrt{d^2 + 4H^3}\right]$

[**Hint:** Obviously, first the waves arrive at D, directly along SD and by reflection along SAD. The path difference between the two = (SA + AD) – SD. Since they arrive *in phase*, this path differenced = nλ.

*i.e.,* $2\sqrt{\left(\frac{d}{2}\right)^2 + H^2} - d = n\lambda.$

Next, the waves arrive at D, directly along SD and by reflection along SA'D. The path difference now is (SA' + A'D) – SD. Since they arrive *in opposite phases* and thus annual each other, this path difference

$$= (2n+1)\frac{\lambda}{2},$$

*i.e.,* $2\sqrt{\left(\frac{d}{2}\right)^2 + (H+h)^2} - d = \frac{(2n+1)\lambda}{2}.$

Therefore, $\frac{(2n+1)\lambda}{2} - n\lambda = \frac{\lambda}{2}.$

$$\left(2\sqrt{\left(\frac{d}{2}\right)^2 + (H+h)^2} - d\right) - \left(2\sqrt{\left(\frac{d}{2}\right)^2 + H^2} - d\right),$$

whence λ may be evaluated].

29. Discuss the transverse vibrations of strings and derive an expression for the velocity of propagation of a transverse wave along a string. Suggest a method for determining this velocity experimentally.

30. (a) Derive an expression for the velocity of transverse waves in a stretched uniform string.

    (b) Calculate the velocity of a transverse wave along a string of length 2 metres and mass 0.06 kg under a tension of 500 newton. **Ans.** 130 m/sec.

31. A transverse harmonic wave is travelling along a string. Show that the energy per unit length of the string is given by $2\pi^2 a^2n^2\sigma$, where a is the amplitude and n, the frequency of the wave and $\sigma$, the mass per unit length of the string. Also show that the power or the average rate of transfer of energy is given by $2\pi^2 a^2n^2\sigma$ v/sec, where v is the wave velocity.

32. Outline a method for the determination of the frequency of an alternating current with the help of a sonometer when the sonometer wire is (i) of a *magnetic* material, (b) of a *non-magnetic material.*

33. (a) Show that the time taken for a disturbance to pass along a string of length *l* cm and of mass m gm per cm is constant when the tension is $ml^2$ gm wt and calculate this time.

    [**Hint:** Wave velocity $v = \sqrt{T/m} = \sqrt{ml^2 g/m} = l\sqrt{g}$ cm/sec, because T = $ml^2$ gm wt = $ml^2g$ dynes. Therefore, *time taken to cover the length of the string*, say

    $$t = \frac{l}{l\sqrt{g}} = \frac{1}{\sqrt{g}} = \frac{1}{\sqrt{981}}.]$$

    (b) Calculate the speed of a transverse wave in a wire of 1.0 $mm^2$ cross section under the tension produced by 0.1 kilogram weight. (Specific gravity of material of the wire = 9.81 and g = 981 cm/sec$^2$.)

    **Ans.** (a) 0.3192 sec; (b) $10^3$ cm/sec.

34. (a) A piece of wire 50 cm long is stretched by a load of 25 kg and has a mass of 1.44 gin. Find the frequency of the second harmonic.

(b) If an addition of 25 *l*b to the tension of a string raises its pitch (or frequency) by a fifth (i.e., to 3/2 of its initial value), what was the original tension? **Ans.** (a) 583.2; (b) 20 *l*b wt.

35. A sonometer wire vibrates in unison with a tuning fork of frequency 320 when the stretching weight is 4 kg. What should be the stretching weight so that double the length of the wire may vibrate in unison with a tuning fork of frequency 240?

**Ans.** 9.0 kg wt.

36. (a) Show that if a transverse wave be travelling along a string, the slope of the string at any given point is equal to the ratio of the particle velocity at that point to the wave velocity.

(b) A transverse wave is travelling along a stretched wire of density 7.5 gm/c.c. with a velocity of $10^5$ cm/see. Obtain the value of the tensile stress in the wire.

**Ans.** $7.5 \times 10^{10}$ dynes/cm$^2$.

[**Hint:** *Tensile stress* $= \dfrac{\text{tension in the wire}}{\text{area of cross section of the wire}} = \dfrac{T}{a}$.

Since $v = \sqrt{\dfrac{T}{\sigma}}$, clearly, $T = \sigma v^2$, where $\sigma$ = *mass per unit length of the wire* = $(1 \times a)\,\rho$. $\therefore$ $T = a\rho v^2$. Hence *tensile stress* = a $\rho v^2/a = \rho v^2$.]

37. (a) A long thin wire has a bob, weighing 2 kg, suspended from its lower free end and functions as a pendulum. When the wire is plucked at its mid-point, it emits a note of frequency 256. What should be the weight of the bob in order that the wire, when plucked in the middle, gives an octave of the first note (*i.e.*, a note of twice its frequency)?

(b) A string of mass 2 gm/metre carries progressive waves of amplitude 1.5 cm, frequency, 60 sec$^{-1}$ and speed, 200 m/sec. Calculate (i) the energy per unit length per metre length of the wave, (ii) the rate of energy propagation in the wave.

**Ans.** (a) 8 kg wt.; (b), (i) $3.2 \times 10^{-2}$ joule, (ii) 6.4 joules/sec.

38. Obtain Newton's expression for the velocity of sound in a gaseous medium and explain La'place's correction of the same.

39. Show that the excess pressure p and particle velocity S in plane longitudinal waves in a medium of volume elasticity E are

related by p = E/ξ/v, where v is the speed of the waves.

[**Hint:** We have $p = -E.\frac{d\xi}{dx}$ and $\xi = -v\frac{d\xi}{dx}$. So that, $\frac{d\xi}{dx} = -\frac{\xi}{v}$.]

40. Show that in the case of a soundwave in a gas, the pressure amplitude ($p_m$) is $2\pi\, v^2\rho/\lambda$ times the displacement amplitude (a). Hence, or otherwise, show that the intensity of a sound wave in terms of displacement amplitude is given by $I = 2\pi^2 a^2 n^2 \rho v$ and in terms of pressure amplitude it is given by $I = p_m^{\,2}/2\rho v$, where the symbols have their usual meanings.

[**Hint:** $p = -\frac{K\,dy}{dx}$ and since $y = a \sin \frac{2\pi}{\lambda}(vt - x)$,

$$\frac{dy}{dx} = -\frac{2\pi}{\lambda}\, a \cos \frac{2\pi}{\lambda}(vt - x).$$

So that, $p = K.\frac{2\pi}{\lambda} a \cos\frac{2\pi}{\lambda}(vt - x) = v^2\rho\frac{2\pi}{\lambda} a \cos\frac{2\pi}{\lambda}\,(vt - x)$,

because $v = \sqrt{\frac{K}{\rho}}$ and $\therefore$ $K = v^2\rho$. Thus, *pressure amplitude,*

$$p_m = \frac{2\pi}{\lambda} v^2 \rho a\,]$$

41. Particle displacement in a plane longitudinal wave in a medium of density ρ is given by $\xi = A \cos 2\pi\,(t - x/v)$. Deduce expressions for (i) pressure amplitude, and (ii) energy flux.

42. One sound wave is travelling in water and another in air. (a) If their intensities be equal, obtain the ratio between their pressure amplitudes and (b) if their pressure amplitudes be equal, obtain the ratio between their intensities.

**Ans.** (a) $\sqrt{\rho_w\, v_w} : \sqrt{\rho_a\, w_a}$, (b) $\rho_a\, v_a : \rho_w\, v_w$,

where $\rho_w$ and $\rho_a$ are the densities of wafer and air and $v_w$, and $v_a$, the velocities of sound in water and air respectively.

43. (a) Discuss the effect of temperature and pressure on the velocity of sound in air.

(b) The velocity of sound in air at normal temperature and pressure is 330 m/sec. Find the velocity at a temperature of 27°C and a pressure of 74 cm of mercury. **Ans.** 345.9 m/sec.

44. The planet Jupiter has an atmosphere of a mixture of ammonia and methane at a temperature of – 130°C. Calculate the velocity of sound on this planet, assuming $\gamma$ for the mixture to be 1.3 and the molecular weight of the mixture to be 16.5. (Gas constant R = 8.3 *joules per deg. per gm. mol.*) **Ans.** 305.8 m/sec.

45. Show that in the case of a plane acoustic wave (i.e., sound wave), (i) the displacement, (ii) the particle velocity, (iii) the strain and (iv) the excess pressure each satisfy the differential wave equation

$$\frac{d^2\phi}{dt^2} = \frac{v^2 d^2\phi}{dx^2}.$$

(ii) Putting the wave equation in the form y = a sin k(vt – x), where $k = \frac{2\pi}{\lambda}$, we have *particle velocity*

$$U = \frac{dy}{dt} = kva \cos k(vt - x) = \phi, \text{ say.}$$

Then, $\frac{d^2\phi}{dt^2} = -k^3v^3 a \cos k(vt - x)$

and $\frac{d^2\phi}{dx^2} = -k^3 va \cos k(vt - x).$

Or, $\frac{d^2\phi}{dt^2} = \frac{v^2 d^2\phi}{dx^2}.$

(iii) Putting $\frac{dy}{dx} = -k a \cos (vt - x) = \phi$, we have

$\frac{d^2\phi}{dt^2} = k^3v^2 a \cos k (vt - x)$ and $\frac{d^2\phi}{dx^2} = k^3 a \cos (vt - x).$

Again, therefore, $\frac{d^2\phi}{dt^2} = \frac{v^2 d^2\phi}{dx^2}.$

(iv) Since $p = -\frac{K dy}{dx}$, proceeding as in (iii) above, we again have $\frac{d^2\phi}{dt^2} = \frac{v^2 d^2\phi}{dx^2}$.]

46. Show that in a plane harmonic sound wave, the excess pressure (p) is ahead of the displacement in phase by $\pi/2$.

[**Hint:** Displacement $y = a \sin \frac{2\pi}{\lambda}(vt - x)$ and, therefore, *excess pressure*

$$p = -K\frac{dy}{dx} = -\rho v^2 \frac{dy}{dx} = \rho v^2 \frac{2\pi}{\lambda} a \cos\frac{2\pi}{\lambda}(vt - x) = p_m \cos\frac{2\pi}{\lambda}(vt - x)$$

$= p_m \sin\frac{2\pi}{\lambda}\left[(vt - x) + \frac{\pi}{2}\right]$, where $K = \rho v^2 \left(\because v = \sqrt{\frac{K}{\rho}}\right)$ and

$\rho v^2 \frac{2\pi}{\lambda} a = p_m$, the amplitude of pressure variation.]

47. Obtain an expression for the velocity of a longitudinal wave in a solid rod. If the frequency of the waves produced be 1000/sec, the density of the material of the rod, 9 gm/c.c., the value of Young's modulus for it, $9 \times 10^{12}$ dynes/cm$^2$, calculate the wavelength of the waves. **Ans.** 1000 cm.

[**Hint:** $v = n\lambda = \sqrt{\frac{Y}{\rho}}$ and $\therefore\ \lambda = \frac{1}{n}\sqrt{\frac{Y}{\rho}}$.]

48. In a plane progressive wave in a fluid, the angular frequency is $\omega$ sec$^{-1}$ and the displacement amplitude is A cm. Deduce the values of (i) pressure amplitude, (ii) energy density in terms of $\omega$, A and $\rho$.

**Ans.** (i) $\frac{\omega^2 A\rho}{k}$; (ii) $\frac{\omega^2 A^2 \rho}{2}$.

49. (a) The velocity of a longitudinal wave in a wire of area of cross section 0.004 sq.cm. is twice the velocity of a transverse wave along it. Determine the ratio between the Young's modulus for the material of the wire and the tension to which the wire is subjected.

(b) The velocities of the longitudinal and transverse waves along a wire are the same. Show that the stress in the wire is equal to the value of Y for its material. **Ans.** (a) Y : T :: 1000 : 1

[**Hint:** (a) $v = \sqrt{\frac{T}{m}} = \sqrt{\frac{T}{a} \times 1 \times \rho} = \sqrt{\frac{T}{0.004\rho}}$ and $2v = \sqrt{\frac{Y}{\rho}}$.

(b) $v = \sqrt{\frac{T}{a} \times 1 \times \rho} = \sqrt{\frac{T}{a} \times \rho} = \sqrt{\frac{Y}{\rho}}$. So that, *stress,* $\frac{T}{a} = Y$.]

50. Explain analytically the formation of waves in a linear bounded medium. Why are they referred to as stationary or standing waves?

    How do they differ from ordinary progressive waves?

51. Distinguish between phase velocity and *group velocity* of train of waves and establish a relationship between the two. Show that in a non-dispersive medium they are the same.

52. Deduce the relation $v = \frac{\omega}{k}$, where v is the phase (or wave) velocity and ω and k have their usual meanings, and show that its first derivative with respect to k gives the group velocity.

53. Show that when $\frac{dv}{d\lambda} = \frac{v}{\lambda}$, the group velocity is zero and when $v = A + B\lambda$ (where A and B are constants), the group velocity is equal to A.

54. For what wavelength will the group velocity of surface waves in a liquid, of surface tension T and density ρ, be a minimum?

    **Ans.** $\lambda - 16\sqrt{\frac{T}{g\rho}}$.

55. What is meant by a plane progressive harmonic wave? Distinguish between transverse and a longitudinal wave and obtain an expression for a plane progressive wave, in general.

56. Show that the particle velocity at a point affected by a plane progressive wave is equal to wave velocity x slope of the displacement curve at that point.

57. Show that the particle velocity dy/dt in the case of a plane progressive wave is given by dy/dt = – v dy/dx. Hence derive the differential equation of a wave motion.

58. Show that the most general differential equation of a one-dimensional wave is $d^2\xi/dt^2 = v^2\, d^2\xi/dx^2$. Discuss the solutions for this in the case of a medium with rigid boundaries separated by length L.

59. What do you understand by *energy density and energy current* (or *intensity*) of a plane progressive wave? Obtain the usual expressions for them.

60. Obtain the expression for the energy of a plane progressive wave and show that, at any given instant, it is, on an average, half kinetic and half potential in form.

61. What is meant by the *principle of superposition* of waves? Explain in brief how it gives rise to the phenomena of '*beats*' and '*stationary waves*'. Is any loss of energy involved in either of these cases?

62. Show that the wave equation $y = a \sin \frac{2\pi}{\lambda}(vt - x)$ for a wave travelling iong the +x direction may be written in any one of the following forms:

    (i) $y = a \sin (\omega t - kx)$,

    (ii) $y = a \sin 2\pi (vt - x/\lambda)$,

    (iii) $y = a \sin \omega(t - x/v)$,

    (iv) $y = a \sin 2\pi (t/T - x/\lambda)$, and

    (v) $y = a \sin 2\pi (nt - x/\lambda)$.

63. A longitudinal harmonic wave train is travelling along a spring in the positive direction of the x-axis. It the frequency of the wave be 50 cps and the distance between a condensation and an adjacent rarefaction in the spring, 10 cm, obtain the velocity of the wave.

    Assuming the amplitude of the wave to be 2.0 cm, write down the wave equation, taking y = 0 at x = 0 and t = 0.

    **Ans.** (i) 0.0166/cm, (ii) 0, (iii) 3142 cm/sec, (iv) 0.

64. A plane progressive harmonic wave, travelling along the +x direction has an amplitude of 5.0 cm and frequency, 100 cps. If its velocity be 6000 cm/sec, obtain the values of (i) the wave number (in the sense of waves per unit length), (ii) displacement, (iii) particular velocity and (iv) particle Wave Motion at x = 150 cm and t = 1 sec.

    **Ans.** (i) 0.0166/cm, (ii) 0, (iii) 3142 cm/sec, (iv) 0.